The Physics of Semiconductor Microcavities

Edited by
Benoit Deveaud

1807–2007 Knowledge for Generations

Each generation has its unique needs and aspirations. When Charles Wiley first opened his small printing shop in lower Manhattan in 1807, it was a generation of boundless potential searching for an identity. And we were there, helping to define a new American literary tradition. Over half a century later, in the midst of the Second Industrial Revolution, it was a generation focused on building the future. Once again, we were there, supplying the critical scientific, technical, and engineering knowledge that helped frame the world. Throughout the 20th Century, and into the new millennium, nations began to reach out beyond their own borders and a new international community was born. Wiley was there, expanding its operations around the world to enable a global exchange of ideas, opinions, and know-how.

For 200 years, Wiley has been an integral part of each generation's journey, enabling the flow of information and understanding necessary to meet their needs and fulfill their aspirations. Today, bold new technologies are changing the way we live and learn. Wiley will be there, providing you the must-have knowledge you need to imagine new worlds, new possibilities, and new opportunities.

Generations come and go, but you can always count on Wiley to provide you the knowledge you need, when and where you need it!

William J. Pesce
President and Chief Executive Officer

Peter Booth Wiley
Chairman of the Board

The Physics of Semiconductor Microcavities

From Fundamentals to Nanoscale Devices

Edited by
Benoit Deveaud

WILEY-VCH Verlag GmbH & Co. KGaA

The Editor

Prof. Benoit Deveaud
Institute of Quantum Electronics
and Photonics
École Polytechnique Fédérale de
Lausanne (EPFL)
1015 Lausanne
Switzerland

Cover illustration
The cover picture depicts a computed
Rayleigh figure showing the letters M.I., the
initials of Marc Ilegems.

All books published by Wiley-VCH are carefully produced. Nevertheless, authors, editors, and publisher do not warrant the information contained in these books, including this book, to be free of errors. Readers are advised to keep in mind that statements, data, illustrations, procedural details or other items may inadvertently be inaccurate.

Library of Congress Card No.: applied for
British Library Cataloguing-in-Publication Data
A catalogue record for this book is available from the British Library.

Bibliographic information published by the Deutsche Nationalbibliothek
The Deutsche Nationalbibliothek lists this publication in the Deutsche Nationalbibliografie; detailed bibliographic data are available in the Internet at http://dnb.d-nb.de.

© 2007 WILEY-VCH Verlag GmbH & Co. KGaA, Weinheim

All rights reserved (including those of translation into other languages). No part of this book may be reproduced in any form – by photoprinting, microfilm, or any other means – nor transmitted or translated into a machine language without written permission from the publishers. Registered names, trademarks, etc. used in this book, even when not specifically marked as such, are not to be considered unprotected by law.

Typesetting	Druckhaus Thomas Müntzer, Bad Langensalza
Printing	betz-druck, Darmstadt
Bookbinding	Litges & Dopf Buchbinderei GmbH, Heppenheim

Printed in the Federal Republic of Germany
Printed on acid-free paper

ISBN 978-3-527-40561-9

Contents

Preface *XI*

List of Contributors *XVII*

1   Fifteen Years of Microcavity Polaritons *1*
     Vincenzo Savona 1

1.1   Introduction *1*
1.2   The Past *4*
1.2.1 The Beginning of the Microcavity Polariton Era *4*
1.2.2 Energy Relaxation and Polariton Photoluminescence *6*
1.2.3 The Problem of the Polariton Spectral Linewidth *7*
1.2.4 Influence of Structural Disorder *8*
1.3   The Present *10*
1.3.1 Parametric Amplification and Photoluminescence *10*
1.3.2 Quantum Correlation and Non-Classical Properties *13*
1.4   The Future *16*
1.4.1 Polariton Quantum Collective Phenomena *16*
1.4.2 Engineering Quantum Confined Polariton Nanodevices *19*
1.5   Conclusions *21*
       References *21*

2   MBE Growth of High Finesse Microcavities *31*
     Ursula Oesterle, Ross P. Stanley, and Romuald Houdré 31

2.1   Introduction *31*
2.2   Principles of MBE Growth *31*
2.2.1 Growth of $Al_xGa_{1-x}As/AlAs$ DBRs *32*
2.2.2 Growth of (In,Ga)As Quantum Wells *33*
2.2.3 Growth of Vertical Cavity Structures *33*
2.3   Characterization and Properties of Vertical Cavity Structures *35*
2.3.1 Error Tolerance *35*
2.3.2 Structural Properties *36*
2.3.3 Optical Measurements of High Finesse Microcavity Structures *39*
2.4   Conclusion *42*
       References *43*

Physics of Semiconductor Microcavities: From Fundamentals to Nanoscale Devices
Edited by Benoit Deveaud
Copyright © 2007 WILEY-VCH Verlag GmbH & Co. KGaA, Weinheim
ISBN: 978-3-527-40561-9

3	**Early Stages of Continuous Wave Experiments on Cavity-Polaritons** 45
	Romuald Houdré 45
3.1	Introduction (1992) 45
3.2	First Liquid Nitrogen and Room Temperature Observation (1993) 48
3.3	Cavity-Polariton Dispersion Curve (1994) 50
3.4	Bleaching of the Oscillator Strength (1995) 56
3.5	Continuous Wave Photoluminescence Experiments (1995–1996) 58
3.5.1	Nonresonant Excitation 58
3.5.2	Resonant Excitation 62
3.6	Linewidth, Disorder Effects and Linear Dispersion Modelling (1995–1997) 63
3.6.1	More on Linear Dispersion Modelling 63
3.6.2	Disorder Effects and Inhomogeneous Broadening (1995) 65
3.6.3	The Second Generation of Samples (1996) 68
3.6.4	Inhibition of Acoustic Phonon Broadening (1996) 70
3.6.5	Test of Linear Dispersion Theory (1997) 73
3.7	Rayleigh Scattering (2000) 76
3.8	Nonlinear Continuous Wave Effects (1999–2000) 79
3.9	Conclusion 82
	References 83
4	**Exciton-Polaritons and Nanoscale Cavities in Photonic Crystal Slabs** 87
	Lucio Claudio Andreani, Dario Gerace, and Mario Agio 87
4.1	Introduction 87
4.2	Mode Dispersion and Linewidths in Photonic Crystal Slabs 88
4.3	Exciton-Polaritons in Photonic Crystal Slabs 91
4.4	Nanoscale Cavities in Photonic Crystal Slabs 96
4.5	Strong Exciton-Light Coupling in Nanocavities 98
4.6	Conclusions 101
	References 102
5	**Parametric Amplification and Polariton Liquids in Semiconductor Microcavities** 105
	Jeremy J. Baumberg and Pavlos G. Lagoudakis 105
5.1	Introduction 105
5.2	Parametric Scattering at the Magic Angle 106
5.2.1	Ultrafast Experiments on Semiconductor Microcavities 106
5.2.2	Simple Pair Scattering 106
5.2.3	Quasimode Theory of Parametric Amplification 109
5.2.4	Multiple Scattering at the Magic Angle 111
5.2.5	Double Resonant On-Branch Multiple Scattering 112

5.3	Local Deformations of the Dispersion: Beyond Pair Scattering *114*
5.3.1	Polariton Liquids at the Bottom of the Polariton Trap *114*
5.3.2	Local Oscillator Strength Model *116*
5.3.3	Direct Time-Resolved Emission During Parametric Amplification *117*
5.4	Historical Perspective (JJB) *118*
	References 121

6	**Quantum Fluid Effects and Parametric Instabilities in Microcavities** *123*
	Cristiano Ciuti and Iacopo Carusotto 123
6.1	Preface *123*
6.2	Introduction *123*
6.3	Hamiltonian and Polariton Mean-Field Equations *124*
6.4	Stationary Solutions in the Homogeneous Case *126*
6.5	Linearized Bogoliubov-Like Theory *127*
6.5.1	Stability of the Stationary Solutions *128*
6.5.2	Complex Energy of the Collective Excitations *130*
6.5.2.1	Excitation Near the Inflection Point of the LP Dispersion *132*
6.5.2.2	Excitation Near the Bottom of the LP Dispersion *134*
6.5.2.3	Simplified Analytical Model for Excitation Close to the Bottom of the LP Dispersion *137*
6.6	Response to a Static Potential: Resonant Rayleigh Scattering *138*
6.6.1	Weak Excitation Regime and Elastic RRS Ring *139*
6.6.2	Superfluid Regime *142*
6.6.3	Precursors of Parametric Instabilities and Branch Sticking *144*
6.6.4	Cherenkov Regime *146*
6.7	Conclusions *149*
	References 149

7	**Non-Linear Dynamical Effects in Semiconductor Microcavities** *151*
	Jean-Louis Staehli, Stefan Kundermann, Michele Saba, Christiano Ciuti, Augustin Baas, Thierry Guillet, and Benoit Deveaud 151
7.1	Introduction *151*
7.2	Experimental *154*
7.2.1	The Microcavity *154*
7.2.2	Pump–Probe Experiments and Parametric Amplification *156*
7.3	A Simple Theoretical Model *159*
7.4	Coherent Control *161*
7.5	Measurements Resolved in Real Time *165*
7.6	Conclusions *168*
	References 168

8	**Polariton Correlation in Microcavities Produced by Parametric Scattering** *171*	
	Wolfgang Langbein 171	
8.1	Introduction *171*	
8.2	Investigated Sample and Experimental Details *172*	
8.3	Parametric Scattering for a Single Pump Direction *173*	
8.4	Parametric Scattering for Two Pump Directions *176*	
8.5	Polariton Quantum Complementarity by Parametric Scattering *180*	
8.6	Conclusions *184*	
	References *185*	
9	**Spin Dynamics of Exciton Polaritons in Microcavities** *187*	
	Ivan A. Shelykh, Alexei V. Kavokin, and Guillaume Malpuech 187	
9.1	Introduction *187*	
9.2	Experimental Results *189*	
9.3	Pseudospin Formalism and Pseudospin Rotation *193*	
9.4	Interplay Between Spin and Energy Relaxation *198*	
9.5	Spin-Dynamics of Polariton–Polariton Scattering *205*	
9.6	Perspective: Toward "Spin-Optronic" Devices *209*	
	References *210*	
10	**Bose–Einstein Condensation of Microcavity Polaritons** *211*	
	Vincenzo Savona and Davide Sarchi 211	
10.1	Introduction *211*	
10.2	Bose–Einstein Condensation: Basic Facts *212*	
10.3	Review of Exciton and Polariton BEC *216*	
10.4	Some Considerations on Microcavity Polariton BEC *222*	
10.5	Afterword *224*	
	References *224*	
11	**Polariton Squeezing in Microcavities** *227*	
	Antonio Quattropani and Paolo Schwendimann 227	
11.1	Introduction *227*	
11.2	Squeezed States *228*	
11.3	Intrinsic Squeezing of Polaritons *230*	
11.3.1	Intrinsic Squeezing of Bulk Polaritons *230*	
11.3.2	Intrinsic Squeezing of Polaritons in Confined Systems *234*	
11.4	Squeezing for Interacting Microcavity Polaritons *236*	
	References *242*	

12	**High Efficiency Planar MCLEDs** *245*
	Reto Joray, Ross P. Stanley, and Marc Ilegems 245

12.1	Introduction *245*
12.2	Microcavities *246*
12.2.1	Fundamentals *246*
12.2.2	State of the Art Planar Semiconductor MCLEDs *248*
12.3	Novel Concepts *252*
12.3.1	Phase-Shift Cavity *252*
12.3.2	Oxide DBR *254*
12.4	Conclusions *256*
	References 257

13	**Progresses in III-Nitride Distributed Bragg Reflectors and Microcavities Using AlInN/GaN Materials** *261*
	Jean-François Carlin, Cristof Zellweger, Julien Dorsaz, Sylvain Nicolay, Gabriel Christmann, Eric Feltin, Raphael Butté, and Nicolas Grandjean 261

13.1	Introduction *261*
13.2	AlInN Alloy: Growth and Characterization *262*
13.2.1	AlInN: Motivation and Difficulty *262*
13.2.2	Growth of AlInN and AlInN/GaN DBRs *263*
13.2.3	Structural Characteristics: X-ray Evaluations and TEM Images *264*
13.2.4	Optical Index Contrast to GaN *266*
13.2.5	Bandgap and Dispersion of Refractive Index *269*
13.2.6	Photoluminescence and Stokes Shift *272*
13.3	Microcavity Light Emitting Diode *273*
13.4	High Reflectivity DBR and Residual Absorption *277*
13.5	Epitaxial Microcavities *281*
13.6	Conclusion *284*
	References 285

14	**Microcavities in Ecole Polytechnique Fédérale de Lausanne, Ecole Polytechnique (France) and Elsewhere: Past, Present and Future** *287*
	Claude Weisbuch and Henri Benisty 287

14.1	Introduction *287*
14.1.1	The Light–matter Interaction in Semiconductors *287*
14.1.2	The Impact of Electronic Motion Quantization *288*
14.1.3	The Impact of Photon Mode Quantization *290*
14.2	The Interplay of Photon and Electron Dimensionalities *290*
14.3	Looking Backwards: a Short History of Microcavities in Solids *293*
14.4	The Birth of the Microcavity Effort in Lausanne *296*
14.5	Why We Like Microcavities! *298*
14.6	The Future: What Are We Looking For? *300*
	References 301

Subject Index   *303*

Preface

This book has been initially prepared as a special edition of Physica Status Solidi, published as a festschrift in honor of Marc Ilegems, on the occasion of his retirement. The editors however considered that the topic was important enough to deserve a publication per se. Let me therefore state in a few words why should this volume be of interest for a large audience, and then rephrase briefly what were the achievements of Marc Ilegems that deserved to see his name associated to such a book. I will then summarize the different chapters so the the reader may have a proper overview of the content of this book.

The idea of going from a standard semiconductor laser to a vertical surface emitting system appeared at the end of the 80th (the so-called VCSEL – Vertical Cavity Surface Emitting Laser). Interesting preliminary work was accomplished by the group of Franz Karl Reinhart at Ecole Polytechnique Fédérale, in Lausanne. The idea is in fact quite simple and resides in the very high gain achievable in semiconductor active layers together with the possibility to grow monolithic distributed Bragg reflectors exhibiting reflectivities very close to 100%. It was then possible to imagine, and later indeed realize, semiconductor lasers with a very small thickness for the active layer.

What was not planned at the beginning of such a process was that the coupling between light and matter would be strong enough to bring the whole system in the strong coupling regime. It is only with the seminal work of Claude Weisbuch, during a stay in Japan, that he realized a very simple reflectivity experiment allowing him to evidence, for the first time, the existence of a strong coupling regime in a VCSEL structure, that we now call microcavities because the first aim of the system is not to make a laser. The strong coupling between excitons in a quantum well and the confined photon modes in the optical cavity, gives rise to a new quasiparticle, called polariton, that is built from half an exciton plus half a photon.

The interest on the physics of polaritons as constantly been rising since then because of the numerous and very specific properties that these new quasiparticles show. In fact, such properties come from the double nature of polaritons, bearing some resemblance to photons, and therefore behaving as bosons, as well as remembering that they are made up with electrons and holes, and therefore being able to interact. The result is that we have true bosons in the low-density limit, with an effective mass of the order of 10^{-5} times the mass of an electron. At the same

Physics of Semiconductor Microcavities: From Fundamentals to Nanoscale Devices
Edited by Benoit Deveaud
Copyright © 2007 WILEY-VCH Verlag GmbH & Co. KGaA, Weinheim
ISBN: 978-3-527-40561-9

time, these quasi bosons may interact through their electronic constituents, possibly opening the route towards real life devices. They also have a very strangely shaped dispersion relation, which allows non-linear parametric effects to occur.

In the process between the discovery of microcavity polaritons, the understanding of their complex physical properties, and the possible applications, the collaboration between the group of Claude Weisbuch, and the group of Marc Ilegems has been essential. The two teams have worked together towards first the realization of very high quality microcavities, and second the realization of clear-cut experiments allowing a very precise understanding of the properties of microcavity polaritons.

The present book aims at describing the latest advances in the physics of microcavity polaritons. It will be, I hope, useful for students and researchers in the field, allowing them to understand the basic properties of polaritons that make them so attractive and interesting.

This book is then organized in the following way.

In Chapter 1 "*Introduction*" Vincenzo Savona provides a tutorial presentation of the basics of the field and overviews what has already been published on the subject, but needs to be recalled here for sake of completeness of the volume. Indeed, some of the students interested in microcavities might not have studied these early developments.

In Chapter 2, entitled "*MBE Growth of High Finesse Microcavities*", Ursula Oesterle, Ross Stanley, and Romuald Houdré describe the growth process that has allowed them to obtain very high finesse microcavities, Such microcavities can consist of hundreds of different layers and can be several microns thick, with the absolute thickness of each layer and the smoothness of each interface having an impact on the optical quality of the structure.

In Chapter 3, entitled "*Early Stages of Continuous Wave Experiments on Cavity Polaritons*", Romuald Houdré presents a description of the basics of cavity-polariton physics. He describes several key experiments that he has been realizing and highlights the historical context of such experiments performed on cavity-polariton during the period 1992–2000. In particular, the angle resolved emission properties are described in great detail.

In Chapter 4, entitled "*Exciton-Polaritons and Nanoscale Cavities in Photonic Crystals*", Lucio Claudio Andreani, Dario Gerace, and Mario Agio give an overview of recent theoretical work on exciton-light coupling in waveguide-embedded photonic crystals is reviewed. The following issues are discussed: (1) a quantum-mechanical formulation of the interaction between photonic modes and quantum-well excitons, leading to a description of photonic crystal polaritons; (2) calculations of variable-angle reflectance spectra, which show that radiative polaritons can be excited by an optical beam incident on the slab surface; (3) a description of nanoscale cavities with extremely high Q-factors and low mode volumes in photonic crystal slabs; (4) a quantum-mechanical model of the interaction between confined nanocavity modes and single quantum-dot transitions, leading again to a strong-coupling regime of light-matter interaction.

In Chapter 5 entitled "*Parametric Amplification and Polariton Liquids in Semiconductor Microcavities*", Jeremy J. Baumberg and Pavlos G. Lagoudakis provide a description of parametric amplification in semiconductor microcavities as an example in which nonlinear optical interactions produced by the exchange interaction of excitons become so large that multiple scattering of polaritons becomes important. They review time-resolved observations of the polariton interactions in a number of different geometries including pumping at either the magic angle, or the bottom of the polariton trap. Situations in which the polariton dispersion is multiply occupied by large populations give rise to k-dependent energy shifts, modifying the dispersion dynamically, a situation termed by them *"the strongly-interacting polariton liquid."*

In Chapter 6, called *"Quantum Fluid Effects and Parametric Instabilities in Microcavities"*, Cristiano Ciuti and Iacopo Carusotto present a description of the nonequilibrium properties of a microcavity polariton fluid, injected by a nearly-resonant continuous wave pump laser. In the first part, they point out the interplay between the peculiar dispersion of the Bogoliubov-like polariton excitations and the onset of polariton parametric instabilities. They show how collective excitation spectra having no counterpart in equilibrium systems can be observed by tuning the excitation angle and frequency. In the second part, the impact of these collective excitations on the in-plane propagation of the polariton fluid is explained. The authors show that the resonant Rayleigh scattering induced by artificial or natural defects is a very sensitive tool allowing to evidence fascinating effects such as polariton superfluidity or polariton Cherenkov effect.

In Chapter 7, named *"Non-Linear Dynamical Effects in Semiconductor Microcavities"*, Jean-Louis Staehli, Stefan Kundermann, Michele Saba, Cristiano Ciuti, Augustin Baas, Thierry Guillet, and Benoit Deveaud describe an investigation of the parametric amplification and its coherent control in a semiconductor microcavity. The time and angle resolved pump and probe experiments show that several picoseconds after pumping the polaritons are still coherent and parametric scattering is still going on. The experimental data concerning the time integrated measurements are in qualitative agreement with the numerical data obtained from a relatively simple theoretical model based on three polarization components, pump, probe, and idler. The dynamics of parametric amplification is also studied in real time, the measurements reveal that stimulation may be considerably delayed with respect to the arrival of pump and probe.

In Chapter 8, entitled *"Polariton Correlation in Microcavities Produced by Parametric Scattering"*, Wolfgang Langbein describes the measurements of the spontaneous and self-stimulated parametric emission from a semiconductor microcavity after resonant pulsed excitation. The emission of the lower polariton branch is resolved in two-dimensional momentum space, using either time-resolved or spectrally resolved detection. The polariton-polariton scattering dynamics is generally in good agreement with the theory using the nonlinearity due to the excitonic part and the dispersion due to the photonic part of the polariton. The peculiar figure-8 shaped distribution in momentum space of the final states of the parametric scattering is observed. Renormalization of the dispersion due to the bound biexciton state is found to influence the final state distribution. Using two pump directions the shape of the final state distribution can be changed to a peanut or oval for the mixed parametric processes. In this dual pump configuration, he finds that polaritons in two distinct idler-modes interfere if, and only if, they share the same signal-mode, showing the existence of polariton pair correlations that store the "which-way" information.

In Chapter 9, named *"Spin Dynamics of Exciton-Polaritons in Microcavities"*, Ivan A. Shelykh, Alexei V. Kavokin, and Guillaume Malpuech address a complex set of optical phenomena linked to the spin dynamics of exciton-polaritons in semiconductor microcavities. Their state can be fully characterized by a so-called "pseudospin" accounting for both spin and dipole moment orientation. The pseudospin dynamics of exciton-polaritons is quite rich and complex, giving rise to non-trivial changes in polarization of light emitted by the cavity versus time, pumping energy, pumping intensity and polarization. The authors overview the essential experimental results in this field before presenting a formalism allowing to interpret the key experimental findings. The strong coupling regime leaving aside all polarization effects in VCSELs is also discussed.

In Chapter 10, entitled *"Bose–Einstein Condensation of Microcavity Polaritons"*, Vincenzo Savona and Davide Sarchi first summarize the basic facts about Bose–Einstein condensation of weakly interacting systems. Then, they review the main experimental and theoretical works aimed at the realization of this intriguing phenomenon in the system of microcavity polaritons. All the experimental results until present suggest that Bose– Einstein condensation is still eluding the experimentalists' efforts. On the theoretical side, only recently polaritons have been described within a quantum field theory of an interacting Bose gas, suggesting that polariton condensation might occur at least as a consequence of polariton parametric scattering under resonant excitation.

In Chapter 11, named *"Polariton Squeezing in Microcavities"*, Antonio Quattropani and Paolo Schwendimann describe the squeezed states that may be found in solids when considering solid state excitations like polaritons. Squeezing may be observed through the polariton radiation field component, whose statistics reproduces the statistics of the polaritons. The squeezing of polaritons in different semiconductor systems and excitations regimes is discussed. For a sufficiently low density of excitation, an intrinsic but very small squeezing is found for bulk as well as for quantum well and microcavity polaritons. When the density of excitation becomes larger, a large amount of squeezing of microcavity polaritons induced by polariton–polariton scattering is demonstrated both in theory and in experiment.

In Chapter 12, *"High Efficiency Planar MCLEDs"*, Reto Joray, Ross P. Stanley, and Marc Ilegems extend the discussion of microcavities towards experimental and theoretical work on the optimization of the light extraction properties of light-emitting diodes. They demonstrate that such works have led to the current high efficiency microcavity LEDs but also to the high brightness LEDs based on other approaches, which are presently available on the market. An overview of the state of the art of planar semiconductor microcavity LEDs is presented.

In Chapter 13, entitled *"Progresses in III-Nitride Distributed Bragg Reflectors and Microcavities Using AlInN/GaN Materials"*, Jean-François Carlin, Cristof Zellweger, Julien Dorsaz, Sylvain Nicolay, Gabriel Christmann, Eric Feltin, Raphael Butté, and

Nicolas Grandjean propose to use lattice-matched AlInN/GaN to replace the Al(Ga)N/GaN material system for III-nitride Bragg reflectors, despite the poor material quality of AlInN reported until very recently. They report an improvement of AlInN material that allowed for successful fabrication of a microcavity light emitting diode, a distributed Bragg reflector with 99.4% reflectivity and microcavities with a quality factor over 800. These results establish state-of-the-art values for III-nitrides, and announce the future importance of AlInN in GaN-based optoelectronics.

In Chapter 14, that closes this book, entitled *"Microcavities in Ecole Polytechnique Fédérale de Lausanne, Ecole Polytechnique (France) and Elsewhere: Past, Present and Future"*, Claude Weisbuch and Henri Benisty present an overview of semiconductor microcavity research, with emphasis on the activities carried out at Ecole Polytechnique Fédérale de Lausanne and Ecole Polytechnique (Palaiseau, France). They give clues for the understanding of the past as well as indications for future possible developments.

List of Contributors

Claudio Adreani
Dipartimento di Fisica Alessandro
Volta
Università di Pavia
via Bassi 6
27100 Pavia
Italy

Mario Agio
Dipartimento di Fisica Alessandro
Volta
Università di Pavia
via Bassi 6
27100 Pavia
Italy

Augustin Baas
Laboratoire d'Optoélectronique
Quantique
Institut de Photonique
et d'Optoélectronique Quantiques
Ecole Polytechnique Fédérale de
Lausanne
EPFL-PH Bâtiment de Physique
1015 Lausanne
Switzerland

Jeremy Baumberg
School of Physics and Astronomy
University of Southampton
Southampton SO17 1BJ
United Kingdom

Henri Benisty
Laboratoire Charles Fabry de l'Institut
d'Optique
91403 Orsay
France

Raphael Butté
Institute of Quantum Electronics
and Photonics
École Polytechnique Fédérale de
Lausanne (EPFL)
1015 Lausanne
Switzerland

Jean-François Carlin
Institute of Quantum Electronics
and Photonics
École Polytechnique Fédérale de
Lausanne (EPFL)
1015 Lausanne
Switzerland

Iacopo Carusotto
CRS BEC-INFM and Dipartimento
di Fisica
Università di Trento
38050 Povo
Italy

Physics of Semiconductor Microcavities: From Fundamentals to Nanoscale Devices
Edited by Benoit Deveaud
Copyright © 2007 WILEY-VCH Verlag GmbH & Co. KGaA, Weinheim
ISBN: 978-3-527-40561-9

Gabriel Christmann
Institute of Quantum Electronics
and Photonics
École Polytechnique Fédérale de
Lausanne (EPFL)
1015 Lausanne
Switzerland

Cristiano Ciuti
Laboratoire Pierre Aigrain
École Normale Supérieure
24 rue Lhomond
75005 Paris
France

Benoit Deveaud
Institute of Quantum Electronics
and Photonics
École Polytechnique Fédérale de
Lausanne (EPFL)
1015 Lausanne
Switzerland

Julien Dorsaz
Institute of Quantum Electronics
and Photonics
École Polytechnique Fédérale de
Lausanne (EPFL)
1015 Lausanne
Switzerland

Eric Feltin
Institute of Quantum Electronics
and Photonics
École Polytechnique Fédérale de
Lausanne (EPFL)
1015 Lausanne
Switzerland

Dario Gerace
Dipartimento di Fisica Alessandro
Volta
Università di Pavia
via Bassi 6
27100 Pavia
Italy

Nicolas Grandjean
Institute of Quantum Electronics
and Photonics
École Polytechnique Fédérale de
Lausanne (EPFL)
1015 Lausanne
Switzerland

Thierry Guillet
Groupe d'Etude des Semiconducteurs
Case courrier 074
Université de Montpellier II
34095 Montpellier cedex 5
France

Romuald Houdré
Institut de Micro et Optoélectronique
Ecole Polytechnique Fédérale de
Lausanne
1015 Lausanne
Switzerland

Marc Ilegems
Institute of Quantum Electronics
and Photonics
Ecole Polytechnique Fédérale de
Lausanne (EPFL)
Station 3
1015 Lausanne
Switzerland

Reto Joray
Institute of Quantum Electronics
and Photonics
Ecole Polytechnique Fédérale de
Lausanne (EPFL)
Station 3
1015 Lausanne
Switzerland

Alexei Kavokin
Chair of Nanophysics and Photonics
School of Physics and Astronomy
University of Southampton
Highfield, Southampton SO17 1BJ
United Kingdom

Stefan Kundermann
Laboratoire d'Optoélectronique
Quantique
Institut de Photonique et
d'Optoélectronique Quantiques
Ecole Polytechnique Fédérale de
Lausanne
EPFL-PH Bâtiment de Physique
1015 Lausanne
Switzerland

Pavlos G. Lagoudakis
School of Physics and Astronomy
University of Southampton
Southampton SO17 1BJ
United Kingdom

Wolfgang Langbein
Department of Physics
and Astronomy
Cardiff University
5 The Parade
Cardiff CF24 3YB
United Kingdom

Guillaume Malpuech
LASMEA, UMR 6602 CNRS
Université Blaise-Pascal
24, av des Landais
63177 Aubiere
France
and
St. Petersburg State Polytechnical
University
Polytekhnicheskaya ul. 29
195251 St. Petersburg
Russia

Sylvain Nicolay
Institute of Quantum Electronics
and Photonics
École Polytechnique Fédérale de
Lausanne (EPFL)
1015 Lausanne
Switzerland

Ursula Oesterle
Swisscom
730 Montgomery Street
San Francisco, CA 94303
USA

Antonio Quattropani
Institute of Theoretical Physics
Ecole Polytechnique Fédérale de
Lausanne (EPFL)
1015 Lausanne
Switzerland

Michele Saba
Dipartimento di Fisica
Universita degli Studi di Cagliari
09042 Monserrato
Italy

Davide Sarchi
Institute of Theoretical Physics
Ecole Polytechnique Fédérale de
Lausanne (EPFL)
1015 Lausanne
Switzerland

Vincenzo Savona
Institute of Theoretical Physics
Ecole Polytechnique Fédérale de
Lausanne (EPFL)
1015 Lausanne
Switzerland

Paolo Schwendimann
Institute of Theoretical Physics
Ecole Polytechnique Fédérale de
Lausanne (EPFL)
1015 Lausanne
Switzerland

Ivan A. Shelykh
St. Petersburg State Polytechnical
University
Polytekhnicheskaya ul. 29
195251 St. Petersburg
Russia

Jean-Louis Staehli
Laboratoire d'Optoélectronique
Quantique
Institut de Photonique
et d'Optoélectronique Quantiques
Ecole Polytechnique Fédérale de
Lausanne
EPFL-PH Bâtiment de Physique
1015 Lausanne
Switzerland

Ross P. Stanley
Swiss Center for Electronics and
Microtechnology (CSEM)
Rue Jaquet-Droz 1
2007 Neuchatel
Switzerland

Claude Weisbuch
Laboratoire de Physique de la Matière
Condensée
Ecole Polytechnique
91128 Palaiseau
France

Christoph Zellweger
Institute of Quantum Electronics
and Photonics
École Polytechnique Fédérale de
Lausanne (EPFL)
1015 Lausanne
Switzerland

Chapter 1
Fifteen Years of Microcavity Polaritons

Vincenzo Savona

1.1
Introduction

In semiconductors with a direct interband optical transition, the fundamental electronic excitations above the ground state are excitons, namely hydrogen-like bound electron-hole pairs. In a perfect bulk semiconductor material, however, a correct description of the excited states must include the linear coupling of excitons to the electromagnetic field. This coupling gives rise to normal modes that are linear superpositions of one exciton and one photon mode, called *exciton-polaritons*. Exciton-polaritons are the actual excited states of a bulk semiconductor.

Bulk polaritons were suggested in 1958 by J. Hopfield [58] and measured a few years later by means of non-linear optical spectroscopy [2, 52, 57, 152]. The normal-mode coupling for polaritons is a result of momentum conservation in the exciton-photon interaction. This selection rule imposes a one-to-one coupling between exciton and photon modes having the same momentum. As a consequence, within the linear coupling regime, the situation is analogous to that of two linearly-coupled harmonic oscillators. Two normal modes at different energies, the upper and the lower polariton, are formed. Due to the very different exciton and photon energy dispersion, as a function of momentum, polariton modes display an anticrossing with a minimum energy separation that can be as large as 16 meV in bulk GaAs [2]. Within the anticrossing region, polaritons are full admixtures of exciton and photon modes, while they have pure exciton or photon character far from it.

The concept of polaritons remained linked to the physics of bulk semiconductors for more than two decades. The progress in fabrication of epitaxial semiconductor heterostructures, particularly quantum wells (QWs) led naturally to a description of their electronic excitations in terms of the polariton concept. However, because of the mismatch in the dimensionality of excitons (2D) and photons (3D), momentum conservation applies only to the in-plane component, and one exciton mode couples to a continuum of photon modes, resulting in an irreversible radiative decay instead of a normal-mode coupling [6, 44]. Only at larger in-plane momenta, photon modes that are evanescent in the direction orthogonal to the QW can form surface polaritons, which however result in a vanishingly small deviation from the bare exciton and photon modes [144].

To make the leap from three to two-dimensions, polaritons had to wait until 1992, when C. Weisbuch published the first successful measurement of normal-mode coupling in a semiconductor microcavity (MC). Planar semiconductor MCs were developed in the 80's basically for producing vertical-cavity surface emitting lasers (VCSEL's). Below the lasing threshold, a VCSEL behaves as a light-emitting diode and the spontanous emission rate is determined by the details of the MC mirrors. These are stacks of layers of different semiconductor material, called distributed Bragg reflectors (DBRs), that are able to produce a high reflectivity thanks to multiple interference. One of the main problems that were being studied in connection with the MC-QW system was the change of the total spontaneous photon emission rate, starting from an initial exciton state [18], that should be enhanced according to the Purcell effect [114]. The main effect of DBRs, however, is to concentrate the angle-dependent emission rate within a narrow cone around the direction normal to the sample. This led to develop mirrors of increasingly high reflectivity, resulting in a longer photon lifetime at small in-plane momenta, inside the planar structure (see e.g. the contributions by Oesterle et al. [105] and by Joray et al. [70] to the present volume). It was thanks to these advances that the exciton-photon coupling rate became faster than the damping, and the first strong-coupling sample was produced. The story behind this discovery is told by C. Weisbuch in his contribution to this volume [150].

However, normal-mode coupling was being widely investigated in another system, that of Rydberg atoms in metal resonators [16, 72]. The differences between this system and that of polaritons are very substantial and led, in the past, to a some misunderstandings in the context of microcavity polaritons. First, an atom is a strongly non-linear object, well described in terms of a two-level system rather than a harmonic oscillator. The Rydberg transition is saturated after a single quantum of excitation has been absorbed (not accounting for spin). This intrinsic non-linearity makes the Rabi frequency, namely the rate at which the excitation is exchanged between the electromagnetic field and the atomic system, depend on the number of atoms present in the cavity. The well known Jaynes-Cummings quantum-mechanical model [69] describes this behaviour as proportional to the square root of the number of atomic two-level systems. In this respect, the basic Rabi frequency corresponding to a single atom exchanging its energy with one cavity photon, is named vacuum-field Rabi frequency, to highlight the fact that the initial condition (and one through which the system ideally cycles during its time evolution) is that of an excited atomic transition in presence of the photon vacuum. The so called self-induced Rabi oscillations, corresponding to the regime of several atoms, had been measured already in 1983 by Kaluzny et al. The first observation of vacuum-field Rabi oscillations for a single atom in a cavity was instead made by Thompson et al. in 1992 [148].[1] The intrinsic non-linearity of the atom-cavity system makes the Rabi

---

1) Curiously enough, in the same year the semiconductor-microcavity polaritons have been also observed for the first time by Weisbuch et al.

oscillations and their dependence on the number of atoms a purely quantum phenomenon, as described by the Jaynes-Cummings model. Excitons, on the other hand, are an almost non-saturable system at low density, due to the infinite spatial extension of their total wave-function in a semiconductor QW. Ideally, nonlinear effects in a QW become important when the excitons approach the saturation density [133]. Recently, however, more subtle nonlinear effects such as the polariton parametric scattering [33, 35, 131, 141] have been observed in the limit of the lowest densities accessible in an optical experiment (see also the contributions by Staehli et al. [139], by Baumberg et al. [13], and by W. Langbein [85] to the present volume). The second important difference between atom-cavity normal-mode coupling and bulk polaritons is the way the single-mode selection is performed. In the polariton case, the translational invariance provides the momentum selection rule that ensures single-mode coupling between photons and excitons. In the atom-cavity case, the atom is a point-like system and momentum conservation does not hold. The single-mode selection must then be engineered by photon confinement, with the requirement of a high quality factor in order for the photon escape rate to be slower than the Rabi frequency. Very recently, the same kind of vacuum-field Rabi splitting as in the atom-cavity case, was achieved by embedding one semiconductor quantum dot into a semiconductor microresonator where photons were confined in three dimensions [79, 110, 116, 157] (see also the contribution by Andreani et al. [3] to the present volume, where strong cavity-quantu-dot coupling is discussed in the case of photonic-crystal microresonators). This result is the true semiconductor analogous to theatom-cavity vacuum-field Rabi splitting.

Strong light-matter interaction and normal-mode coupling in solid-state devices are objects of increasingly intense research. The reason lies in the perspective of engineering new kinds of electronic excitations with unique quantum coherence properties, long correlation both in space and time, and robustness to environment-induced decoherence, thanks to their hybrid nature sharing the properties of light and electrons. Polaritons are the oldest manifestation of normal-mode coupling, and also one of the more promising in view of these developments. This volume collects papers by the most prominent actors in the field, providing a broad overview of the state of the art and of the directions in which polariton research is moving. The readers interested in the past evolution of the research on MC-polaritons, might look at several review articles, each focusing on different aspects of the problem [75, 78, 122, 126]. The present introduction aims at providing the reader witha general overview of the MC-polariton physics as it progressed during the last fifteen years. This is not, however, an exhaustive review, as certainly many important topics in polariton research are partially or not at all covered (for example the role of spin and light polarization, for which a starting point might be the contribution by Shelykh et al. [136] to the present volume, or the recent progress in developing room-temperature polariton systems, for which the reader can refer to the contribution by Carlin et al. [24]), while many others are repeated in the individual contributions to this volume (see e.g. the contributions by R. Houdré [59] and C. Weisbuch [150] to the present volume). It will nevertheless help the reader go through the following chapters, that are individual research reports by the most active researchers in the field.

1.2
The Past

1.2.1
The Beginning of the Microcavity Polariton Era

Back in 1992, Claude Weisbuch, at the time visiting Tokyo University, proposed and realized the first semiconductor microcavity device displaying vacuum-field Rabi splitting [151]. The early work by Weisbuch et al. already contained all the essential features of the MC-polariton system, including the anticrossing as a function of the exciton-cavity detuning, the dependence of the Rabi splitting on the number of QWs embedded in the cavity and on the exciton and photon linewidths, and finally an analysis in terms of linear response theory. An exciting account of this discovery is given by Claude Weisbuch in this volume [150], together with a short review of the early MC-polariton physics.

The first observation of polariton vacuum-field Rabi splitting generated great excitement within a restricted community of researchers who were in quest of cavity quantum electrodynamics (CQED) effects [93] in semiconductor planar systems. Within a few months, further evidence of the 2-D polariton physics was provided by Houdré et al. with the observation of the vacuum-field Rabi splitting up to room temperature [62], and with the measurement of the energy-momentum dispersion curve by means of angle-resolved emission spectroscopy. Finally, the vacuum-field Rabi oscillations were directly time-resolved in an experiment of ultrafast photoluminescence upconversion spectroscopy by Norris et al. [102] and later by Jacobson et al. [67].

Closely related to the observation of polaritons in MC-embedded QWs was the measurement, by Tredicucci et al., of bulk polariton modes in a semiconductor microcavity in which the whole cavity layer provided the excitonic transition [26, 27, 149]. In this system, the cavity layer displays a series of closely spaced exciton resonances originating from the energy quantization of the exciton center-of-mass motion, confined along the growth direction within the λ-cavity slab.

Very soon after the first measurement of MC-polaritons, the very basic theoretical framework for the description of the polariton modes was developed. The appearance of a vacuum-field Rabi splitting in the polariton spectrum was understood in terms of a simple linear disperision model [61], in full analogy to the case of atoms in an optical cavity [159]. In this model, the exciton linear response function is described in terms of a Lorentz resonance, and the overall response of the cavity–exciton system is evaluated within linear response theory of the classical electromagnetic field. In the case of atoms in a cavity, the analysis by Zhu et al. [159] suggested that the appearance of a spectral doublet in a classical linear dispersion model would somewhat ruled out the idea of vacuum-field Rabi splitting, which should be instead related to the quantum fluctuations of the field vacuum. Today we know that the linear response theory is fully equivalent to the quantum model of a Bose (non saturable) excitation coupled to the quantum field of radiation [5, 123]. In this case, the vacuum field Rabi splitting is equally well described in terms of a semiclassical linear-response model. Purely quantum effects are only present if the system Hamiltonian contains many-body interaction terms (beyond the terms quadratic in the quantum fields). For the atomic case, in which the two-level system is the most appropriate description of a single atom transition, CQED effects appear as soon as the number N of atoms is larger than one. This implies, in particular, an increase of the Rabi splitting – the so called self-induced Rabi splitting – proportional to $\sqrt{N}$ [72]. For polaritons, on the other hand, nonlinear effects on the Rabi splitting appear only when the exciton density approaches the saturation density, namely when $na_B^2 \sim 1$ [133], where n is the exciton areal density in the QW and a_B the exciton Bohr radius which, in GaAs QWs, is of the order of 10 nm. In this limit, the exciton oscillator strength and consequently the Rabi splitting vanish [68, 78], contrarily to the atom-cavity case [72].

A microscopic model of the microcavity polariton modes and their energy-momentum dispersion relation was derived from the diagonalization of the linear exciton-photon coupling Hamiltonian [107, 124], within the assumption of ideal cavity mirrors, and later by including the detailed frequency response of DBRs [71, 130]. These latter works presented a full three-dimensional treatment of the electromagnetic field, thus including leakage through the cavity mirrors. Within this description, polaritons arise from the linear coupling of an exciton, having a given in-plane momentum $k_\parallel$, to the continuum of photon modes having the same in-plane momentum component – as required by the in-plane translational invariance – and all the possible values of the remaining component k_z. In the case of a bare QW, this photon continuum has a smooth density of states, resulting in the intrinsic exciton radiative lifetime [6, 44, 144]. In presence of a planar MC, the photon continuum in the z-direction displays a sharp peak in the density of states, corresponding to the resonant cavity mode, and the normal-mode coupling arises. The work by Savona et al. points out to the presence of *leaky modes* in a DBR-MC. Leaky modes arise due to resonances within the multi-layered structure formed by the cavity and the DBR layers, in which the electromagnetic field can penetrate. A particular role is played by leaky modes at frequency lower than the main cavity mode. These, due to their energy-momentum dispersion, become resonant with the exciton energy at some finite value of the in-plane momentum, typically outside the external emission cone. The leaky modes appear as sharp peaks in the emission rate of the lower polariton at large momenta [130], and in case of small exciton and photon linewidths they can even produce strong coupling with an anticrossing of normal modes. Given the two-dimensional density of states, the theory predictsthat typically more than 80% of the luminescence is emitted into the leaky modes and absorbed in the sample substrate [130]. Their presence is thus very relevant in determining the overall dynamics of the polariton photoluminescence. Strong coupling between the exciton and leaky modes was measured only very recently in a II–VI sample by Richard et al. [119].

In the same years, the equivalence between semiclassical and full quantum descriptions of MC-polaritons was finally established for the linear coupling limit [5, 107, 123]. The work by Pau et al. [107] was the first to highlight the effect of an inhomogeneous distribution of exciton resonances, aimed at modeling the effect of excitons localized by interface disorder in the QW. Within the linear dispersion theory, the polariton spectrum including an inhomogeneous exciton distribution was thoroughly studied by Andreani et al. [4]. This was however only a phenomenological way of including disorder and exciton localization into the polariton problem, as the in-plane momentum conservation is lifted and a full 3-D calculation of the coupled exciton-photon modes is in principle required. This problem is briefly reviewed in the Section on structural disorder.

1.2.2
Energy Relaxation and Polariton Photoluminescence

Most of the spectral properties of MC-polaritons are detected by angle-resolved photoluminescence, which gives access to the energy- and momentum-resolved population of the polariton states within the external emission cone. In this process, the semiconductor is optically excited at high energy, producing a population of free electron-hole pairs that eventually relax and populate the polariton states. A polariton can relax energy basically in two ways. First, through the exciton-phonon interaction, by emitting optical or acoustic phonons. Second, via many-body Coulomb interactions with other polaritons or with free carriers present in the system.

The photoluminescence dynamics is governed by the balance between energy relaxation and spontaneous emission rates. If the relaxation rates are much slower than the polariton radiative emission rates, then relaxation to the lowest states is suppressed andthe spontaneous emission occurs most likely from the higher energy states. This phenomenon, called *relaxation bottleneck*, is expected already in the case of QW excitons [111]. For polaritons, the bottleneck effect should in principle be even more effective, due to the faster radiative rates and to the steep energy-momentum dispersion in the strong-coupling region which typically suppresses inelastic scattering rates.

The bottleneck effect was observed for the first time in MCs based on II–VI materials [100, 101], where the Rabi splitting is larger, and soon afterwards in III–V systems [142]. In these works it was clear how the bottleneck effect was less pronounced than expected from theoretical calculations based on polariton-phonon scattering [145], and rather strongly dependent on the excitation density. This observation suggested that many-body scattering effects take part very effectively in the energy relaxation process. Indeed, a model of the relaxation dynamics in terms of a Boltzmann equation predicts stimulated scattering to the lowest energy states in presence of polariton-polariton Coulomb scattering [146] and also if only polariton-phonon processes are taken into account [47]. Stimulated polariton scattering was actually observed in experiments [38, 135]. It was however clear that while stimulated scattering predicted the population buildup in the ground polariton state at high density, it could not explain the rather efficient thermalization of the polariton distribution that was typically observed at lower density.

The present understanding of this behaviour is that polariton relaxation at low to medium density is governed by at least three mechanisms. The first was proposed by Porras et al. [113] and consists in an interplay between Coulomb scattering and phonon relaxation. According to this mechanism, polaritons in the strong coupling region can undergo Coulomb scattering, ending up in one polariton in a lower-energy state and the other in the exciton-like region of the dispersion, at higher energy. Within this region, which acts as a thermal reservoir, the excess energy is relaxed by acoustic phonon emission. The Porras mechanism is not driven by final-state stimulation and can therefore explain thermalization into a smooth distribution of populations. The second mechanism involved in polariton relaxation is very likely the scattering on free carriers originating from charged defects. This mechanism has been modeled within a Boltzmann formalism [94] and evidence of it was found in experiments [12, 109, 143]. This mechanism is however strongly sample dependent, as it relies on the quality of the sample fabrication, and an accurate quantitative characterization of it is presently unavailable. As a third possible mechanism, we mention structural disorder that should give rise to polariton localization and to a lifting of the momentum conservation in the relaxation processes. Disorder in MC-polaritons is discussed in more detail below. Again, however, an accurate characterization of its influence on the relaxation process, either theoretical or experimental, is still unavailable.

1.2.3
The Problem of the Polariton Spectral Linewidth

The spectral linewidth of polaritons is determined by many factors, among which the most effective are the photon escape rate, the polariton-phonon scattering, the polariton-polariton Coulomb scattering, and the inhomogeneous spectral distribution due to structural disorder of the interfaces. The problem of polariton spectral linewidths, in the early years of MC-polaritons, has always been object of intense debate (see e.g. the account given by R. Houdré in his contribution to this volume [59]).

Within the simplest possible description, consisting in a two-coupled-oscillator model, linewidths can be included as damping rates for the exciton and photon oscillators [123]. Then, the frequency eigenvalues of the problem are expressed as

$$\omega = \frac{\omega_x + \omega_c - i(\gamma_x + \gamma_c)}{2} \pm \frac{1}{2}\sqrt{\Omega_R^2 + (\omega_x - \omega_c - i(\gamma_x - \gamma_c))^2} \tag{1}$$

where ω_x and ω_c are the bare exciton and cavity mode frequencies, Ω_R is the vacuum-field Rabi splitting, and γ_x and γ_c are, respectively, the exciton and photon damping rates. Within this purely phenomenological description, the linewidth arises as a result of a damping phenomenon and the corresponding lineshape is a Lorentzian. Equation (1) leads, in particular, to equal polariton linewidths for zero exciton-cavity detuning $\omega_x = \omega_c$, given by the arithmetic average of the two initial damping rates.

Although this idea was widely used in the literature, it was also strongly debated, leading to many studies of the actual polariton linewidth. In samples with bad interface quality, the actual linewidth was dominated by inhomogeneous broadening due to disorder. For high quality samples, on the other hand, experiments showed that upper and lower polariton linewidths were different even at zero detuning, with a general tendency of the lower polariton mode to have a smaller linewidth with unusually small dependency on external parameters like temperature or density [8, 63, 64].

This observation was understood as a true effect of the steep polariton dispersion. The polariton damping rate of a given state is the result of outscattering processes from that state. Scattering can occur because of phonon emission or absorption, or because of collision with another polariton. In both cases, if we assume an ideally planar system, the total energy and the total momentum are conserved in the process. Then, polaritons in the strong-coupling region of the energy momentum dispersion are characterized by a very steep dispersion curve and therefore by a very small density of final states available for an outscattering process. This mechanism of course is not included in the simple two-oscillator picture and can lead to a lower polariton mode that is extremely robust to line broadening. This was predicted by theoretical analyses in terms of the Fermi golden rule, in both cases of phonon [125] and collisional [32] broadening.

In most modern samples, because of this suppression of collisional broadening, the lower polariton linewidth is fully determined by the cavity mode linewidth.

1.2.4
Influence of Structural Disorder

Structural disorder in heterostructures can dramatically influence their optical response. This is the case of excitons in QWs [160], where interface roughness and alloy disorder always produce localization of the center-of-mass exciton wave function over tens of nanometers. Correspondingly, an inhomogeneous distribution of the energy spectrum of the localized eigenstates arises. Exciton localization is responsible of two phenomena. The first, as already mentioned, is an inhomogeneous broadening of the optical spectrum. The second is the resonant Rayleigh scattering, namely the resonant scattering of a plane electromagnetic wave in all directions due to the breaking of the in-plane translational invariance.

In the case of MC-polaritons, the inhomogeneous broadening problem was the object of an intense and, in the light of our present understanding, overenphasized debate. All the existing works have focused on the effect of QW disorder on the exciton, while assuming an ideally planar MC. In a first work, Whittaker [156] suggested a mechanism called *motional narrowing* in which the small polariton effective mass results in a long-range polariton center-of-mass wave function that averages out the exciton disorder potential, resulting in a suppression of the inhomogeneous polariton linewidth. The first theory by Whittaker was however fawled by an erroneous application of the Born approximation in perturbation theory, and Whittaker published soon a correct analysis of the problem within similar assumptions [153]. In the meantime, Savona et al. presented a numerical result of the full diagonalization of the coupled exciton-photon Hamiltonian in presence of QW disorder [127]. Though technically correct and successful in reproducing the asymmetry in the polariton linewidth, this work included an incorrect qualitative interpretation of the numerical result in terms of interbranch scattering on disorder, that affected exclusively the upper polariton branch. In the same work, however, it was correctly pointed out that the effect of motional narrowing on the lower polariton was of the same extent as for a bare QW exciton. A more sound and simple explanation of the asymmetry funally came with the following work by Whittaker [153]. The polariton linewidth in absorption or reflectivity experiments can be modeled within linear dispersion theory, provided one adopts the microscopically computed exciton response function in presence of disorder. This latter, because of excitonic motional narrowing, typically shows an asymmetry with a sharp low-energy tail and a more slowly decaying high-energy one [160]. This can easily explain the observed asymmetry and was later shown to be the correct approximation [154] to apply when starting from the general problem [127]. This was also confirmed by an accurate simultaneous measurement of the exciton and polariton linear response [50].

Similar considerations apply to the resonant Rayleigh scattering. In this case, the MC scattering spectrum can be modeled as the bare QW exciton scattering spectrum filtered by the MC optical response [55, 154].

As a general consideration, we conclude that the exciton-photon coupling does not change significantly the wave functions of localized exciton states in presence of disorder. The polariton spectral features can therefore be thought of as the result of strong coupling between the photon mode and the ensemble of all localized exciton states.

We conclude this section by pointing out that very few works have been devoted to study the effect of disorder at the interfaces of the MC. Given the strong resonant character of a planar resonator, it should be expected that thickness fluctuations of thecavity layer should give rise to an inhomogeneous photon spectrum and, to some extent, to lateral photon localization. As a result, polaritons should also be localized and inhomogeneously broadened. Evidence of the influence of photon disorder is found when comparing the measured cavity mode linewidth with the one calculated from the nominal value of the DBR reflectivity. The former is always significantly larger than the latter, suggesting an additional inhomogeneous broadening. More direct evidence of polariton localization was provided by Langbein et al. [86], who measured the intrinsic momentum broadening of resonant Rayleigh scattering of polaritons, deducing a polariton localization length of the order of 30 µm in a typical GaAs/AlGaAs MC. Further evidence comes from the ubiquitous cross-shaped pattern present in resonant Rayleigh scattering in momentum space [54, 83]. This pattern is originated from the crosshatch pattern in real space caused by misfit dislocations in the epitaxial growth [15]. Given the GaAs/AlAs lattice misfit, the typical length scale of planar regions bounded by cross hatches should be of the order of a few tens of µm, in agreement with the estimated polariton localization length in these systems. Finally, recent measurements of spatially resolved emission in presence of a large occupation of the lowest-energy polariton states in a II–VI MC [117], gave direct visual access to the localized polariton states, whose typical size in this material is of the order of a few microns.

1.3
The Present

Most of the present fundamental research on MC-polaritons is focused on nonlinear optical phenomena, particularly those involving parametric polariton scattering and quantum polariton optics. The early experiments at high excitation density had prompted the experimentalists with a quite straightforward phenomenology, including an increase of the polariton linewidth and a corresponding bleaching of the Rabi splitting as a function of density [60, 78]. It was soon understood that the basic excitonic nonlinearities can reasonably explain these observation. In particular, the saturation of the exciton oscillator strength [133], well described by a Hartree–Fock model of the many-body exciton system, is responsible for the disappearance of the Rabi splitting. To the next order of scattering theory, many-body interactions also give rise to an excitation-induced dephasing, explaining the increase in linewidth. This scenario and the relevant research works are discussed in great detail in the review article by G. Khitrova et al. [78]. In this work, the reader will also find a thorough account of the so called "Boser controversy", that we will briefly discuss in the following section.

1.3.1
Parametric Amplification and Photoluminescence

Great excitement was brought by the discovery in 2000, by the groups by J. J. Baumberg and M. S. Skolnick, of two phenomena that would mark a new era of MC-polariton physics: the polariton parametric amplification [131] and its spontaneous counterpart [141], the parametric photoluminescence.

Polariton parametric processes bear a very close analogy with parametric downconversion of photons in nonlinear crystals, as known from the quantum optics domain [97]. In this latter case, by pumping the crystal with photons at energy $\hbar\omega$, the $\chi^{(2)}$ nonlinearity can produce two new photons of energy $\hbar\omega_1$ and $\hbar\omega_2$, whose sum is equal to the initial energy $\hbar\omega$. Depending on the geometry of the crystal, phase-matching conditions imposed by total momentum conservation have to be additionally fulfilled. In the case of polaritons, parametric processes originate from the third-order nonlinearity characterizing the polariton system. It stems from the exciton-exciton Coulomb scattering and from the density-dependent saturation of the oscillator strength, both resulting in a contribution to the Hamiltonian of fourth order in the polariton field [34]. In the traditional setup, one laser beam is used to resonantly pump polaritons in one given mode along the lower-polariton energy-momentum dispersion curve. Given the peculiar shape of this curve – which displays an inflection point at half energy distance from the band bottom and the bare exciton energy – the scattering of two pump polaritons to two new polaritons, respectively at smaller and larger energy, is made possible by energy-momentum conservation. The two final states are called *signal* and *idler* polariton respectively, as in photon parametric downconversion.

The polariton parametric amplification is the process by which, in presence of a pump, a weak probe probe beam resonant with the signal mode is amplified by stimulating the parametric process. Correspondingly, the stimulated scattering generates a polariton field in the idler mode. It is a process involving intense fields that can be fully described in terms of classical polariton field amplitudes [34], as in classical nonlinear optics. The first observations were obtained at the so called "magic angle", namely by choosing a pump momentum that results in the signal mode being at zero momentum, so that the probe beam enters the sample at normal incidence. As we will see below, however, this is not the only allowed configuration. Polariton parametric amplification was observed in samples made of different materials and can result in extremely large gain on the probe beam, reaching a few thousands at low temperature [120]. It has a sharp resonant character and hence is extremely fast, allowing switching times below 1 ps even at liquid nitrogen temperature in II–VI samples. In samples with small inhomogenous broadening, it persists up to high temperature, thanks to the polariton robustness to loss mechanisms, and it can be coherently controlled [81] (see also the contributions by Baumberg et al. [13] and by Staehli et al. [139] to the present volume). When GaN-based polariton samples will finally be available [22, 147] (see also the contribution by Carlin et al. [24] to the present volume), the amplification should persist at room temperature, thus opening the way to possible applications as fast optical switches or amplifiers.

Polariton parametric photoluminescence is a spontaneous process based on the same scattering amplitude and on the same selection rules as parametric amplification. When a polariton mode is resonantly pumped, emission is measured at the signal and idler modes even in the absence of a probe beam. The most intuitive explanation of this phenomenon is that the parametric scattering is stimulated by the vacuum field fluctuations at the signal and idler modes in the low density limit. The polariton parametric photoluminescence is an extremely complex process, one for which the theoretical modeling was essential for understanding the rich phenomenology it displays. The most considerable contribution to the theory of parametric processes and photoluminescence in particular was given by C. Ciuti and is reviewed in Ref. [34]. In the basic theory, the pump polariton field is assumed to be a purely classical field, driven by the resonant laser beam. By further neglecting polariton scattering terms that do not involve the pump mode, the resulting Hamiltonian is quadratic in the signal and idler polariton operators $\hat{p}_k$ and $\hat{p}_{k_i}$, however containing anomalous terms that do not conserve the particle number. Here k is the signal momentum, $k_i = 2k_p - k$ is the idler momentum and k_p is the pump momentum fixed by the external pump angle. Then, the dynamical equations for the signal and idler polariton populations $N_k = \langle \hat{p}_k^\dagger \hat{p}_k \rangle$ and $N_{k_i} = \langle \hat{p}_{k_i}^\dagger \hat{p}_{k_i} \rangle$ are linearly coupled to an analogous equation for the anomalous quantum correlation $A_k = \langle \hat{p}_k^\dagger \hat{p}_{k_i}^\dagger \rangle$. Here we report these equations for clarity (restricted to the lower polariton branch),

$$\frac{dN_k(t)}{dt} = -\frac{2\gamma_k}{\hbar} N_k + \frac{2}{\hbar} \operatorname{Im}\left[g P_{k_p}^2 \, e^{-2i\omega_p t} A_k + \mathcal{L}_{k,\mathcal{N}} \right] \qquad (2)$$

$$i\hbar \frac{dA_{\boldsymbol{k}}(t)}{dt} = \left[-E_{\boldsymbol{k}} - E_{\boldsymbol{k}_i} - i(\gamma_{\boldsymbol{k}} + \gamma_{\boldsymbol{k}_i}) \right] A_{\boldsymbol{k}} - g P_{\boldsymbol{k}_p}^{*2} e^{2i\omega_p t} (1 + N_{\boldsymbol{k}} + N_{\boldsymbol{k}_i}) + \mathcal{L}_{\boldsymbol{k},A} \tag{3}$$

where $E_{\boldsymbol{k}}$ is the polariton dispersion, $\gamma_{\boldsymbol{k}}$ the parametrized polariton damping rate (accounting for radiative and nonradiative mechanisms), ω_p is the pump frequency, $P_{\boldsymbol{k}_p}$ is the classical pump polariton field, and g is the nonlinear coupling strength assumed k-independent for simplicity. The quantities $\mathcal{L}_{\boldsymbol{k},(A,\mathcal{N})}$ derive from quantum Langevin random forces that are needed in the formalism, if the damping terms are included. From these equations, it appears clearly that the polariton populations are driven by the pump only through the anomalous correlation $A_{\boldsymbol{k}}$. This proves that parametric photoluminescence is a purely quantum process and cannot be interpreted in terms of a Boltzmann equation involving solely population variables, as was done by some authors [51]. A very detailed experimental characterization by W. Langbein [84] on a high-quality sample, in particular, showed that the increase of signal and idler populations in a pulsed experiment is not instantaneous, following the increase of the anomalous term (see also the contribution by W. Langbein to this volume [85]). Many features can be inferred from Eqs. (2) and (3). For a fixed pump momentum $\boldsymbol{k}_p$, the parametric process is allowed for a set of values of the signal and idler momentum, forming an "eight"-shaped curve in momentum coordinates. This pattern was measured for the first time very recently by W. Langbein [84]. This "eight"-shaped pattern was never observed before, as a sample of extremely good quality was required. In particular, the polariton energy-momentum dispersion must be well defined not only in the strong coupling but also in the exciton-like region. For QWs with some amount of disorder, the inhomogeneous distribution both in energy and momentum, in the exciton-like region of the dispersion, will completely wash out the selection rule for the parametric process. The sample adopted by W. Langbein was designed with state-of-the-art exciton inhomogeneous broadening and is, up to now, the only sample on which the parametric momentum pattern has been measured. If two pump beams are used to pump different polariton modes, than mixed parametric processes arise in which two polaritons, each from a different pump, scatter to a signal and a idler mode. The energy-momentum conservation in this case can give rise to a rich variety of shapes of the momentum pattern, that were measured on the same sample [121] (see also W. Langbein's contribution to this volume [85]).

If Fourier-transformed to the frequency domain, Eqs. (2) and (3) predict the resonant frequencies of the parametric photoluminescence. It turns out [34] that for large pump intensity these resonances are shifted with respect to the bare polariton dispersion, and can display bifurcations or a change in the sign of the first derivative of the dispersion curve. Evidence of this behaviour has been found in experiments [132].

In Eq. (3), the term $(1 + N_{\boldsymbol{k}} + N_{\boldsymbol{k}_i})$ shows that for $N > 1$ the parametric photoluminescence displays a stimulated behaviour, which was experimentally characterized with high accuracy [84]. Far above the stimulation threshold, Eqs. (2) and (3) no longer hold and will diverge for a finite value of the pump polariton amplitude. In more realistic terms, the stimulated behaviour will deplete the pump polariton mode at a rate faster than the input rate provided by the external pump beam. In this case the equations can be generalized by introducing a third equation for the classical pump field, with the external laser field amplitude as the pump parameter. If this is done, then the pump mode inthe model is depleted before reaching the critical value for which the equations diverge. A very interesting issue however arises in connection with the stimulated behaviour. It is expected that above threshold a spontaneous symmetry breaking occurs and the polariton field at the signal and idler modes acquires a classical complex amplitude $\langle \hat{p}_k \rangle \neq 0$. This is called *polariton parametric oscillation*. It can be interpreted as a poor men's Bose–Einstein condensation (BEC), with the classical field amplitude at the signal mode being the order parameter for the phase transition [14]. Indeed, below threshold the phases of the signal and idler quantum fields are purely fluctuating, while only their *relative* phase is fixed by the link to the classical pump amplitude. C. Ciuti was the first to suggest the possible occurrence of parametric oscillations by noticing that a singular solution with a non-zero classical field is admitted by the parametric photoluminescence equations, if the pump polariton amplitude takes exactly the critical value. Since however the parametric photoluminescence Eqs. (2) and (3) assume purely quantum fluctuating fields and diverge for the pump amplitude approaching this critical value, they cannot describe the way this spontaneous symmetry breaking occurs for increasing pump intensity. Later, Savona et al. [129] extended the parametric photoluminescence theory in the spirit of the Hartree–Fock–Bogoliubov approximation [137], commonly adopted to model BEC of an interacting gas. This model was limited by the restriction to three polariton modes, but allowed for the first time a description of the transition to the parametric oscillation regime. A quantum montecarlo study closely followed [25], generalizing this result to all polariton modes. The polariton field amplitudes in the parametric oscillation are expected to behave in all respects as a quantum fluid, in particular featuring superfluidity [11, 29].

If parametric oscillations at the signal and idler modes occur, the corresponding classical field amplitudes can in turn act as pump fields for parametric photoluminescence involving other polariton modes. This gives rise to off-branch resonances and multiple scattering [34, 155]. This behaviour was observed in experiments [132], and represents a proof – though indirect – that parametric oscillations with spontaneous symmetry breaking do take place. Further indirect evidence was found in the observation of optical bistability in parametric oscillations [9, 10].

Finally, we point out to a recent experiment [45] where parametric oscillation was observed in a very special setup of three vertically stacked semiconductor MCs in the weak coupling regime.

1.3.2
Quantum Correlation and Non-Classical Properties

Probably the most appealing feature of polariton parametric photoluminescence is that signal-idler polariton pairs are produced in non-classical states with quantum correlations. These have their origin in the expression of the parametric Hamiltonian, that reads

$$\hat{H} = \sum_{\mathbf{k}} E_{\mathbf{k}} \hat{p}_{\mathbf{k}}^{\dagger} \hat{p}_{\mathbf{k}} + \sum_{\mathbf{k},\mathbf{k}',\mathbf{k}_s,\mathbf{k}_i} \left[G_{\mathbf{k},\mathbf{k}'} \hat{p}_{\mathbf{k}_s}^{\dagger} \hat{p}_{\mathbf{k}_i}^{\dagger} + \text{h.c.} \right] \delta_{\mathbf{k}_s + \mathbf{k}_i, \mathbf{k} + \mathbf{k}'} \qquad (4)$$

where

$$G_{\mathbf{k},\mathbf{k}'} = g P_{\mathbf{k}} P_{\mathbf{k}'} (\delta_{\mathbf{k},\mathbf{k}_{p1}} + \delta_{\mathbf{k},\mathbf{k}_{p2}})(\delta_{\mathbf{k}',\mathbf{k}_{p1}} + \delta_{\mathbf{k}',\mathbf{k}_{p2}}) \qquad (5)$$

is a parametric coupling term generalized to the case of two pump fields at momenta $\mathbf{k}_{p1}$ and $\mathbf{k}_{p2}$. The signal and idler momenta are now denoted $\mathbf{k}_s$ and $\mathbf{k}_i$. If one evaluates the time-evolution of the system, starting from the vacuum polariton state as the initial state, then in the limit of low pump amplitude (or equivalently in the limit of short time) the quantum state of the system will be

$$|\psi(t)\rangle = \alpha(t)|0\rangle + \sum_{\mathbf{k}_s} \beta_{\mathbf{k}_s}(t) |1_{\mathbf{k}_s} 1_{\mathbf{k}_i}\rangle + \ldots \qquad (6)$$

where $\beta \ll \alpha$. In this state, signal and idler polaritons are pair-correlated, namely the joint probability of having a signal polariton *and* and idler polariton is equal to the probability of just having one of them [73, 121]. Again, this scenario is closely related to what happens in quantum optics, where the analogous property of parametric photon downconversion has been used to produce pair correlated and entangled photons, even in the limit of intense laser beams [82].

Very recently there were several proposals of taking advantage of this feature of polariton parametric photoluminescence for observing non-classical states of polaritons. One of the first observations of this kind was the amplitude squeezing of light emitted from parametric oscillation, by Karr et al. [74]. Indeed the parametric Hamiltonian, with the assumption of a classical pump field, is identical to that of a nonlinear Kerr medium [97] and is therefore able to produce a quantum state for which the Heisenberg uncertainty of one quadrature operator is smaller (by only 5% in this experiment) than that of its canonical conjugate, always fulfilling the uncertainty principle. The experiment by Karr et al. was performed in the special setup in which pump, signal and idler polariton modes coincide at zero momentum – a singular point in momentum space for which parametric scattering is always allowed. This result was confirmed by a theoretical analysis [134]. It should be pointed out that polariton squeezing was already predicted, although in the limit of negligible many-body interactions, long before by Hradil et al. [65]. Intrinsic polariton squeezing can in principle occur because of the anti-resonant terms in the minimal coupling Hamiltonian between radiation and matter in the Coulomb gauge. However, the theory predicts an extremely small amount of squeezing in this case, that cannot be detected in realistic situations on GaAs-based materials. A full account of polariton squeezing, both instrinsic and in parametric oscillations, can be found in the contribution by A. Quattropani and P. Schwendimann to the present volume [115].

Another expression of non-classical physics that can arise from parametric photoluminescence is polariton entanglement. By inspection of the quantum state (6), we see that it is a linear combination of the vacuum field and of an entangled state all possible signal-idler pair-states allowed by the conservation rules. Hence, momentum entanglement is a natural outcome of polariton parametric photoluminescence. In a similar way, spin entanglement is also possible. In fact, the polariton-polariton scattering is spin conserving, and the exciton-radiation selection rules are such that spin-up (-down) polaritons couple to clockwise (counterclockwise) circularly polarized light [31]. By using a linearly polarized pump beam then, simple linear superposition principle shows that the signal-idler polariton pair will be generated in a linear combination of the two possible spin states. This kind of polarization entanglement would require photon-coincidence detection to be measured and is still awaiting experimental confirmation. Ciuti recently proposed a special kind of momentum entanglement called branch entanglement [28]. In this case the entanglement arises with respect to the polariton branch index, and can be obtained by setting the pump resonant with the upper polariton branch. However, as Ciuti pointed out, a pump beam resonant with the upper polariton will generate a large amount of outscattering to the exciton-like part of the lower polariton branch at large momenta, implying a fast loss rate that might make an experimental verification challenging. The only evidence of polariton momentum entanglement was obtained by W. Langbein in an experiment where a two-pump setup was used [121], following a theoretical prediction by Savasta et al. (an account of this experimental result can also be found in the contribution by W. Langbein to this volume [85]). The scope of this experiment was to prove Bohr's quantum complementarity principle on polaritons. To this purpose, the two idler polaritons produced by the two pump beams were made interfere on a detector. In the low intensity limit, below the parametric oscillation threshold, at most one signal-idler polariton pair at a given time is present. Then, the same situation as in Young's two-slit experiment exists. The two idler-polariton paths will produce quantum interference if and only if the two paths are indistinguishable. This happens only if the two corresponding signal modes coincide. Differently, it would be possible – by independently detecting one of the two signal polaritons – to gather "which-way" information on the two idler channels and, quantum interference could not arise. A similar experiment was successfully carried out in the case of parametrically downcoverted photons by Mandel et al. [96]. The experimental proof of this mutual exclusion between "which-way" information and quantum interference was an indirect evidence that, in the case of two distinguishable signal polaritons, the polariton pair is produced in a momentum-entangled state. The possibility of fabricating polariton quantum boxes (see next section) could make it possible, by the same principle, to produce spatially entangled pairs of confined polaritons over distances of several microns.

It should be pointed out that a polariton is a many-body electronic excitation of a semiconductor device, existing in a densely packed medium and therefore in principle subject to strong decoherence induced by coupling to the environment. The ability to generate nonclassical states of electronic excitations holds great promise for future implementations of quantum information technology in semiconductor devices. Polaritons, due to their hybrid nature and robustness to decoherence, might be the best candidate for achieving this purpose.

To conclude this section, we point out to a general difficulty in measuring the idler-polariton photoluminescence in parametric experiments in the standard single-pump configuration. This is due basically to three factors, all related to the fact that the idler mode is located in the exciton-like region of the polariton dispersion. First, the idler is almost fully exciton-like, with a small photon component, thus suppressing the coupling to the external electromagnetic field and the photoluminescence intensity. Second, outscattering mechanisms are more favoured by the flat dispersion curve, and the damping rate of the idler polariton mode is consequently faster than that of the signal mode. Third, disorder induced inhomogeneous broadening, both in energy and momentum, is only effective in the exciton-like part of the polariton dispersion, thus lifting the momentum selection rule and partially washing out the sharp resonance of the idler photoluminescence. This feature is very general, as already remarked. More emphasis should therefore be devoted to experiments in the double pump geometry or using quantum-confined polariton levels in quantum boxes (see below), where these limitations are not present.

1.4
The Future

1.4.1
Polariton Quantum Collective Phenomena

The perspective of observing polariton quantum collective phenomena is the driving force of present and future research on MC-polaritons. By quantum collective phenomenon we mean a phase transition resulting in a spontaneous symmetry breaking with formation of a quantum mechanical order parameter. Such a situation is very appealing because it would imply the existence of a macroscopic quantum state of many polaritons, displaying properties such as superfluidity, Josephson oscillations in confined geometry, robustness to decoherence. In addition to the fundamental interest of this novel quantum collective phenomenon in solids, it would also bear a great potential for applications in devices exploiting the quantum phase, like quantum information processing.

In the previous section we have already mentioned the occurrence of spontaneous symmetry breaking in the transition from the regime of polariton parametric photoluminescence to parametric oscillation. If, on one side, this is at present the only polariton quantum collective phenomenon backed up by convincing experimental evidence, its behaviour is not governed by the physics of phase transitions in thermal equilibrium. In particular, temperature and chemical potential cannot be defined for this system, although in the limit of the generalized Gross–Pitaevskii, analogous to the zero-temperature Bogoliubov model of BEC, the pump energy can be interpreted as a pinned chemical potential, in the sense of the ground-state energy of the quantum fluid. Further, the signal and idler polariton gases below – threshold, though being each an incoherent gas of quasiparticles, still have quantum pair-correlations. Nevertheless, the parametric oscillator is expected to display quantum fluid behaviour and superfluidity, as was described by Ciuti and Carusotto [25] (see also their contribution to this volume [30]).

An intense research activity is being instead devoted to the quest for polariton BEC [42, 43, 46, 48, 49, 76, 77, 87, 88, 95, 98, 99, 113, 117, 118, 128, 138, 158] (see also the contributions by Baumberg et al. [13], Ciuti et al. [30], Savona et al. [128], and Shelykh et al. [136] to the present volume). As in the case of excitons, for which BEC has been sought for about fourty years (see e.g. [53] and the special issue of Solid State Communications Vol. 134, 2005), the idea was stimulated by the quasi-bosonic behaviour of polaritons in the limit of low density, and by their extremely small effective mass. This topic is widely reviewed later in this volume by Savona and Sarchi [128], and we would not indulge in a repetition of the same concepts. Let us just highlight a few important points concerning the two-dimensional character, the non-equilibrium nature and what would be conclusive experimental evidence in a spectroscopy experiment.

The Hohenberg–Mermin–Wagner theorem [56] states that conventional BEC with the formation of off-diagonal long-range order cannot take place in two dimensions. The reason is that thermal fluctuations of the quantum phase would dominate at any finite temperature, fully destroying the condensate. No long-range order can then arise. However, it can be shown that for an interacting gas, still a superfluid behaviour can be achieved below a critical temperature T_c. In this case, a different kind of phase transition is instead expected: the Berezinskii–Kosterlitz–Thouless transition [112]. This is a topological phase transition in which, above a critical temperature T_{BKT}, vortices can spontaneously form (thanks to the fact that a vortex in two dimensions costs a finite amount of energy), destroying superfluidity. The critical temperature T_{BKT} is always lower than the superfluid critical temperature T_c of the spatially homogeneous gas, and is therefore the phase transition that should be observed in two-dimensions. The Berezinskii–Kosterlitz–Thouless transition has been discussed in the context of MC-polaritons [76]. This interpretation should however be revisited in the light of disorder and polaritonlocalization, always occurring in MCs. The Berezinskii–Kosterlitz–Thouless is a topological transition and makes sense only in a perfetly homogeneous medium. Polariton localization, as mentioned above, occurs within the range of a few to a few tens of microns [86, 117]. A rigorous proof holds [89] that whenever in a two-dimensional system a finite energy gap separates the ground single-particle state from the excited states, then thermal fluctuations no longer dominate and conventional BEC with a macroscopic occupation of the ground state is recovered. Such a gap can typically arise in confined systems (see discussion below), or even in presence of disorder because locally, within the area interested by optical excitation, polariton localization will give rise to a discrete energy spectrum for the lowest-lying states. Even in the limit of a system of infinite extension, it can be rigorously proved [91] that an ideal Bose gas can undergo conventional BEC, thanks to the Lifshitz tail in the lowest part of the density of states, causing condensation to occur in some localized state at finite values of temperature and density. Both confinement and polariton localization implie a finite spatial extension of the condensate.

The deviations from thermal distribution of polaritons, in connection to the relaxation bottleneck effect, have already been discussed. A debate is currently ongoing, on whether polariton BEC can be understood in terms of standard equilibrium thermodynamics or if deviations from equilibrium occur. A conclusive answer to this question is not yet available. However, if on one hand it is clear that at sufficiently high excitation density the relaxation mechansims taking place through many-body interactions will favour thermalization within the polariton branches [113, 143], this does not exclude that a large non-equilibrium population of hot excitons or electron-hole pairs at large energy and momentum exists. This hot population, while not affecting the condensate thermalization, could nevertheless modify the polariton spectrum via many-body scattering and by saturating the exciton oscillator strength [133]. Another deviation from thermal equilibrium are quantum fluctuations of the condensate, that are expected to occur even at zero temperature. These typically tend to deplete the condensate by occupying single-particle excited states [37, 90, 112]. In the case of strong many-body interactions, this condensate depletion can be very effective and produce deviations from the standard Bose–Einstein distribution of polariton populations.

Finally, let us briefly discuss what would be the "smoking gun" of polariton BEC in an optical experiment. In the case of diluted alkali atoms, the striking observation was a peaked distribution in momentum space of the atomic population [1]. This was very remarkable as, for a common atomic gas, classical statistical mechanics and the Gibbs principle make this particular configuration in the system phase space extremely unlikely. There are however other unique features characterizing a Bose condensate. First, the energy-momentum dispersion of collective excitations, displaying the typical Bogoliubov linear behaviour at small momenta [140]. Second, the off-diagonal long-range order, expressed in the spatial quantum correlations, that are nonzero up to infinite distance [20]. These correlations are today considered as the true distinctive feature of an interacting condensate, according to the Penrose–Onsager criterion [108]. Third, related to the spatial coherence, there is the ability of two condensates to produce quantum interference when overlapping [7]. All these features were clearly observed in the case of diluted atoms, but were simply considered as further confirmations of theoccurrence of BEC. For polaritons, observing a macroscopic occupation of the ground state might be not enough. The reason is twofold. First, a coherent pump beam is present in the system at high energy, and might induce a coherent excitation of the ground state by some unexpected many-body scattering effect (we have just seen how this can happen in parametric oscillations). Second, a semiconductor at high enough excitation density will start lasing, with the result of emitting light in a single mode of the momentum space. This ambiguity, in particular, was at the basis of the "boser" controversy, that followed an observation of strong single-mode emission under high density excitation [106] (for a detailed account, see Khitrova et al. [78]). The observation was later explained in terms of a bleaching of the exciton-photon coupling and a subsequent lasing effect [80], and the same authors withdrew their initial interpretation [23]. Credit must be given, however, to the work by Imamoglu and Yamamoto [66], who have for the first time suggested the possibility of MC-polariton BEC. Later, a similar strong nonlinear emission at zero momentum under high-energy nonresonant excitation was reported by several authors, in more controlled experiments where simultaneous evidence of polaritons in the strong coupling regime was provided [19, 21, 38, 135]. Although very promising, these observations are not sufficient to claim BEC, as explained. More recently, polariton BEC was claimed by Deng et al. [42, 43], again on the basis of a momentum redistribution of polaritons, but with simultaneous measurement of the spatial intensity profile and of the second-order time coherence of the emitted light. Yet, these observations are not conclusive, as none of the distinguishing features mentioned above has been evidenced, while the measured second-order time coherence does not display a sharp transition atthe supposed condensation threshold, casting therefore doubts on the validity of this claim. The most recent experiments are focusing on coherence properties in real and momentum space [117, 118], and leave hope that a conclusive observation of polariton BEC is not very far ahead.

1.4.2
Engineering Quantum Confined Polariton Nanodevices

The history of semiconductor physics has been marked by a continuous progress towards reducing the dimensionality of semiconductor structures, from bulk semiconductors, to quantum wells, quantum wires, eventually producing a rich variety of zero-dimensional nanostructures. These advances were driven by the idea of confining electrons in order to obtain a quantized spectrum with discrete energy levels that would display novel optical and transport properties to be exploited in nanodevice technology.

Polaritons in bulk materials were first described by Hopfield in 1958 [58], but it took until 1992 [151] before the idea of two-dimensional polaritons in MCs was conceived and realised! The reason is probably to be attributed to the need for high-quality semiconductor optical resonators that were made available only with the advances in epitaxial growth. The two further steps, towards one-dimensional and zero-dimensional polariton systems, are the object of very recent research.

Zero-dimensional confinement of polaritons would be extremely attractive both for fundamental studies and in view of applications. On the fundamental side, we have already remarked that a discrete polariton spectrum could be favorable to BEC and to parametric polariton photoluminescence or oscillations with an even balance between signal and idler. Pairs of polariton quantum boxes with tunnel coupling might display Josephson oscillations as soon as a quantum fluid – either a Bose condensate or a parametric oscillation – is formed. They would also allow the design of several configurations in which polariton spatial entanglement can be achieved. It should be also pointed out that polaritons, due to their very steep dispersion, would display quantum confinement and a discrete energy spectrum already when confined over a few microns. This would make fabrication, optical and electronic addressing of such a microdevice an extremely simple task, if compared to semiconductor quantum dots [17]. Among the possible applications, one could imagine devices for single photon emission, as well as for the generation of entangled pairs of electronic excitations or the storage of quantum information, that could be employed in quantum information technology.

The studies on confined polaritons are very recent and mostly based on the idea of etching a *micropillar* out of a planar semiconductor MC. This approach is justified by the fact that the micropillar fabrication technology can benefit of the progress in vertical-cavity surface-emitting lasers (VCSELs). The experimental efforts were also motivated by the quest for strong coupling and the formation of normal modes between a zero-dimensional cavity mode and an interband transition in a quantum dot [116]. The transition from two-dimensional polaritons to one and zero dimension was studied for the first time by Dasbach et al. [39, 40] and by Obert et al. [103, 104]. In the experiments by Dasbach et al., both one-dimensional wires and pillars were studied by photoluminescence spectroscopy. The one-dimensional result clearly displayed polariton dispersion features, with the simultaneous presence of upper and lower polariton branches [39, 41]. The single-micropillar spectrum clearly showed discrete energy modes from which it was possible to observe parametric photoluminescence [39, 40]. In this case, however, a clear signature ofsimultaneous upper and lower polariton modes was not reported. Evidence of upper and lower polariton is essential to claim the formation of strong coupling and exciton-photon normal modes, as parametric scattering can be equally obtained in the weak coupling regime [45]. The reason why polariton features were not observed in these early studies is perhaps to be attributed to the low quality factor of the resonator, introduced by etching the whole body of the microcavity. This technology has however progressed very quickly in the last years and presently micropillars can be fabricated with quality factors approaching 10^5 [92]. Although these new microresonators would probably display striking polariton features, no samples with embedded quantum wells have been produced so far.

It is interesting to notice, however, that the common scheme used in today's microcavity samples, consisting in producing two-dimensional confinement of both excitons and photons, is redundant, as polaritons already exist in the bulk material and their energy-momentum dispersion is dominated by the steep photon mode dispersion. Ideally, by reducing the dimensionality of a bulk polariton system, confined polaritons could be obtained. Indeed, this idea is the starting point of the realisation of bulk MC-polaritons, by Tredicucci et al. [149], where only the photon mode is confined to two dimensions, the excitonic transition being provided by the bulk cavity layer. An analogous idea to produce polariton quantum boxes has recently been introduced by El. Daïf et al. [36]. Starting from a planar MC, photon modes can be confined by engraving a very shallow pattern on top of the cavity layer and growing the top mirror afterwards. Thicker regions on the MC plane will have aphoton-mode resonance at a frequency lower than in thinner regions. Thus, by patterning e.g. a circular region thicker than the surrounding, this will act as a lateral potential well for the photon modes, eventually producing confinement. If the planar MC has one or more embedded quantum wells and displays polariton modes, then confined polaritons will arise by virtue of this mechanism. A very shallow pattern is enough to produce confinement of a few polariton modes. As an example, the sample in Ref. [36] has circular mesa patterns of 6 nm height and three different diameters: 3.6, 9 and 19 µm. The 6 nm height corresponds in a GaAs MC to a variation of the photon mode energy of about 9 meV. As a result, the smallest diameter mesa clearly shows three confined modes of the lower polariton, and three corresponding confined modes of the upper polariton. The energy quantization amounts to about 1 meV in this sample, and it is smaller (with correspondingly more extended modes) for the two larger mesas. The interest in this kind of samples lies in the presence of both confined and extended modes, due to the finite height of the energy barrier in the confining potential. The photoluminescence measurements clearly show both lower and upper confined polaritons, displaying an anticrossing behaviour as the exciton-cavity detuning is varied. The spectral lines are very narrow, suggesting that the interfaces are very flat within the mesa region. These polariton quantum boxes can thus reduce the polariton linewidth to the minimum value determined by the cavity mode lifetime. Thanks to the ease in fabrication and optical addressing, this new technique is very promising for the realisation of systems of coupled polariton quantum boxes that are an ideal system for the observation of polariton quantum fluids and nonclassical states.

1.5
Conclusions

We have reviewed the most important developments in the field of MC-polaritons during the last fifteen years. Started as a straightforward extension of three-dimensional polariton physics, this area of research has now become extremely active after the recent developments in polariton parametric effects, the possibility of Bose-Einstein condensation, and the evolution towards polariton nanodevices. It involves an increasingly large number of research groups in the world, who are currently exploring new frontiers along the directions of nanostructure engineering and the use of new materials. The feeling within the community – beautifully expressed by the various contributions to the present volume – is that the time is not far when polariton devices will become a perfect workbench for studying fundamental quantum physics, and a prominent technology for modern quantum optoelectronics.

References

[1] M. H. Anderson, J. R. Ensher, M. R. Matthews, C. E. Wieman, and E. A. Cornell, Observation of Bose–Einstein condensation in a dilute atomic vapor. Science **269**(5221), 198–201 (1995).
[2] L. C. Andreani, Optical transitions, excitons, and polaritons in bulk and low-dimensional semiconductor structures. In E. Burstein and C. Weisbuch, editors, Confined Electrons and Photons, volume 340 of NATO Advanced Study Institute, Series B: Physics, p. 57 (Plenum Press, New York 1995).
[3] L. C. Andreani, D. Gerace, and M. Agio, Exciton-polaritons and nanoscale cavities in photonic crystal slabs. phys. stat. sol. (b) **242**(11), 2197–2209 (2005).

[4] L. C. Andreani, G. Panzarini, A. V. Kavokin, and M. R. Vladimirova, Effect of inhomogeneous broadening on optical properties ofexcitons in quantum wells. Phys. Rev. B **57**(8), 4670–4680 (1998).

[5] L. C. Andreani, V. Savona, P. Schwendimann, and A. Quattropani, Polaritons in high reflectivity microcavities – semiclassical and full quantum treatment of opticalproperties. Superlattices Microstruct. **15**(4), 453–458 (1994).

[6] C. Andreani, F. Tassone, and F. Bassani, Radiative lifetime of free-excitons in quantumwells. Solid State Commun. **77**(9), 641–645 (1991).

[7] R. Andrews, C. G. Townsend, H. J. Miesner, D. S. Durfee, D. M. Kurn, and W. Ketterle, Observation of interference between two bose condensates. Science **275**(5300), 637–641 (1997).

[8] T. Baars, M. Bayer, A. Forchel, F. Schafer, and J. P. Reithmaier, Polariton-polariton scattering in semiconductor microcavities: Experimental observation of thresholdlike density dependence. Phys. Rev. B **61**(4), R2409– R2412 (2000).

[9] A. Baas, J. P. Karr, H. Eleuch, and E. Giacobino, Optical bistability in semiconductor microcavities. Phys. Rev. A **69**(2), 023809 (2004).

[10] A. Baas, J. P. Karr, M. Romanelli, A. Bramati, and E. Giacobino, Optical bistability in semiconductor microcavities in the nondegenerate parametric oscillation regime: Analogy with the optical parametric oscillator. Phys. Rev. B **70**(16), 161307 (2004).

[11] A. Baas, J.-Ph. Karr, M. Romanelli, A. Bramati, and E. Giacobino, Quantum degeneracy of microcavity polaritons. Phys. Rev. Lett. **96**(17), 176401 (2006).

[12] D. Bajoni, M. Perrin, P. Senellart, A. Lemaitre, B. Sermage, and J. Bloch, Dynamics of microcavity polaritons in the presence of an electron gas. Phys. Rev. B **73**(20), 205344 (2006).

[13] J. J. Baumberg and P. G. Lagoudakis, Parametric amplification and polariton liquids in semiconductor microcavities. phys. stat. sol. (b) **242**(1 1), 2210–2223 (2005).

[14] J. J. Baumberg, P. G. Savvidis, R. M. Stevenson, A. I. Tartakovskii, M. S. Skolnick, D. M. Whittaker, and J. S. Roberts, Parametric oscillation in a vertical microcavity: A polariton condensate or micro-optical parametric oscillation. Phys. Rev. B **62**(24), R16247–R16250 (2000).

[15] R. Beanland, M. Aindow, T. B. Joyce, P. Kidd, M. Lourenco, and P. J. Goodhew, A study of surface cross-hatch and misfit dislocation-structure in in0.15ga0.85as/gaas grown by chemical beam epitaxy. J. Crystal Growth **149**(1–2), 1–11 (1995).

[16] P. R. Berman, Cavity quantum electrodynamics. Academic Press, Boston etc. (1994).

[17] D. Bimberg, M. Grundmann, and N. N. c. Ledencov. Quantum dot heterostructures (Wiley, Chichester 1999).

[18] Gunnar Bjrk, Henrich Heitmann, and Yoshihisa Yamamoto, Spontaneous-emission coupling factor and mode characteristics of planar dielectric microcavity lasers. Phys. Rev. A **47**(5), 4451–4463 (1993).

[19] J. Bleuse, F. Kany, A. P. de Boer, P. C. M. Christianen, R. Andre, and H. Ulmer-Tuffigo, Laser emission on a cavity-polariton line in a ii-vi microcavity. J. Cryst. Growth **185**, 750–753 (1998).

[20] I. Bloch, T. W. Hansch, and T. Esslinger, Measurement of the spatial coherence of a trapped bose gas at the phase transition. Nature **403**(6766), 166–170 (2000).

[21] F. Boeuf, R. Andre, R. Romestain, L. S. Dang, E. Peronne, J. F. Lampin, D. Hulin, and A. Alexandrou, Evidence of polariton stimulation in semiconductor microcavities. Phys. Rev. B **62**(4), R2279–R2282 (2000).

[22] R. Butte, G. Christmann, E. Feltin, J. F. Carlin, M. Mosca, M. Ilegems, and N. Grandjean, Room-temperature polariton luminescence from a bulk gan microcavity. Phys. Rev. B **73**(3), 033315 (2006).

[23] H. Cao, S. Pau, J. M. Jacobson, G. Bjork, Y. Yamamoto, and A. Imamoglu, Transition from a microcavity exciton polariton to a photon laser. Phys. Rev. A **55**(6), 4632–4635 (1997).

[24] J.-F. Carlin, C. Zellweger, J. Dorsaz, S. Nicolay, G. Christmann, E. Feltin, R. Butt, and N. Grandjean, Progresses in iii-nitride distributed bragg reflectors and microcavities using alinn/gan materials. phys. stat. sol. (b) **242**(11), 2326–2344 (2005).

[25] I. Carusotto and C. Ciuti, Spontaneous microcavity-polariton coherence across the parametric threshold: Quantum monte carlo studies. Phys. Rev. B **72**(12), 125335 (2005).

[26] Y. Chen, A. Tredicucci, and F. Bassani, Polaritons in semiconductor microcavities – effect of bragg confinement. Journal De Physique Iv **3**(C5), 453–456 (1993).

[27] Y. Chen, A. Tredicucci, and F. Bassani, Bulk exciton polaritons in gaas microcavities. Phys. Rev. B **52**(3), 1800 (1995).

[28] C. Ciuti, Branch-entangled polariton pairs in planar microcavities and photonic wires. Phys. Rev. B **69**(24), 245304 (2004).

[29] C. Ciuti and I. Carusotto, Quantum fluid effects and parametric instabilities in microcavities. J. Phys.: Condens. Matter **242**(11), 2224–2245 (2005).

[30] C. Ciuti and I Carusotto, Quantum fluid effects and parametric instabilities in microcavities. phys. stat. sol. (b) **242**(11), 2224–2245 (2005).

[31] C. Ciuti, V. Savona, C. Piermarocchi, A. Quattropani, and P. Schwendimann, Role of the exchange of carriers in elastic exciton-exciton scattering in quantum wells. Phys. Rev. B **58**(12), 7926–7933 (1998).

[32] C. Ciuti, V. Savona, C. Piermarocchi, A. Quattropani, and P. Schwendimann, Threshold behavior in the collision broadening of microcavity polaritons. Phys. Rev. B **58**(16), 10123–10126 (1998).

[33] C. Ciuti, P. Schwendimann, B. Deveaud, and A. Quattropani, Theory of the angle-resonant polariton amplifier. Phys. Rev. B **62**(8), R4825–R4828 (2000).

[34] C. Ciuti, P. Schwendimann, and A. Quattropani, Theory of polariton parametric interactions in semiconductor microcavities. Semicond. Sci. Technol. **18**(10), S279–S293 (2003).

[35] Cristiano Ciuti, Paolo Schwendimann, and Antonio Quattropani, Parametric luminescence of microcavity polaritons. Phys. Rev. B **63**(4), 041303, Jan 2001.

[36] O. El Daif, A. Baas, T. Guillet, J. P. Brantut, R. I. Kaitouni, J. L. Staehli, F. Morier-Genoud, and B. Deveaud, Polariton quantum boxes in semiconductor microcavities. Appl. Phys. Lett. **88**(6), 061105 (2006).

[37] F. Dalfovo, S. Giorgini, L. P. Pitaevskii, and S. Stringari, Theory of Bose–Einstein condensation in trapped gases. Rev. Mod. Phys. **71**(3), 463–512 (1999).

[38] L. S. Dang, D. Heger, R. Andre, F. Boeuf, and R. Romestain, Stimulation of polariton photoluminescence in semiconductor microcavity. Phys. Rev. Lett. **81**(18), 3920–3923 (1998).

[39] G. Dasbach, M. Bayer, M. Schwab, and A. Forchel, Spatial photon trapping: tailoring the optical properties of semiconductor microcavities. Semicond. Sci. Technol. **18**(10), S339–S350 (2003).

[40] G. Dasbach, M. Schwab, M. Bayer, and A. Forchel, Parametric polariton scattering in microresonators with three-dimensional optical confinement. Phys. Rev. B **6420**(20), 201309 (2001).

[41] G. Dasbach, M. Schwab, M. Bayer, D. N. Krizhanovskii, and A. Forchel, Tailoring the polariton dispersion by optical confinement: Access to a manifold of elastic polariton pair scattering channels. Phys. Rev. B **66**(20), 201201 (2002).

[42] H. Deng, G. Weihs, C. Santori, J. Bloch, and Y. Yamamoto, Condensation of semiconductor microcavity exciton polaritons. Science **298**(5591), 199–202 (2002).

[43] H. Deng, G. Weihs, D. Snoke, J. Bloch, and Y. Yamamoto, Polariton lasing vs. photon lasing in a semiconductor microcavity. Proc. Natl. Acad. Sci. U. S. A. **100**(26), 15318–15323 (2003).

[44] B. Deveaud, F. Clerot, N. Roy, K. Satzke, B. Sermage, and D. S. Katzer, Enhanced radiative recombination of free-excitons in gaas quantum-wells. Phys. Rev. Lett. **67**(17), 2355–2358 (1991).

[45] C. Diederichs, J. Tignon, G. Dasbach, C. Ciuti, A. Lematre, J. Bloch, Ph. Roussignol, and C. Delalande, Parametric oscillation in vertical triple microcavities. Nature **440** (7086), 904–907, April 2006.

[46] T. D. Doan, H. T. Cao, D. B. Thoai, and H. Haug, Condensation kinetics of microcavity polaritons with scattering by phonons and polaritons. Phys. Rev. B **72**(8), 085301 (2005).
[47] T. D. Doan and D. B. T. Thoai, Suppression of the bottleneck in semiconductor microcavities. Solid State Commun. **123**(10), 427–430 (2002).
[48] P. R. Eastham and P. B. Littlewood, Bose condensation of cavity polaritons beyond the linear regime: The thermal equilibrium of a model microcavity. Phys. Rev. B **64**(23), 235 101 (2001).
[49] P. R. Eastham and P. B. Littlewood, Finite-size fluctuations and photon statistics near the polariton condensation transition in a single-mode microcavity. Phys. Rev. B **73**(8), 085306 (2006).
[50] C. Ell, J. Prineas, T. R. Nelson, S. Park, H. M. Gibbs, G. Khitrova, and S. W. Koch, Influence of structural disorder and light coupling on the excitonic response of semiconductor microcavities. Phys. Rev. Lett. **80**(21), 4795–4798 (1998).
[51] J. Erland, V. Mizeikis, W. Langbein, J. R. Jensen, and J. M. Hvam, Stimulated secondary emission from semiconductor microcavities. Phys. Rev. Lett. **86**(25), 5791–5794 (2001).
[52] D. Fröhlich, E. Mohler, and P. Wiesner, Observation of exciton polariton dispersion in cucl. Phys. Rev. Lett. **26**(10), 554–556, Mar 1971.
[53] A. Griffin, D. W. Snoke, and S. Stringari, editors. Bose–Einstein condensation (Cambridge University Press 1995).
[54] M. Gurioli, F. Bogani, D. S. Wiersma, P. Roussignol, G. Cassabois, G. Khitrova, and H. Gibbs, Experimental study of disorder in a semiconductor microcavity. Phys. Rev. B **64**(16), 165309 (2001).
[55] G. R. Hayes, S. Haacke, M. Kauer, R. P. Stanley, R. Houdre, U. Oesterle, and B. Deveaud, Resonant rayleigh scattering versus incoherent luminescence in semiconductor microcavities. Phys. Rev. B **58**(16), R10 175– R10178 (1998).
[56] P. C. Hohenber, Existence of long-range order in 1 and 2 dimensions. Phys. Rev. **158**(2), 383 (1967).
[57] B. Honerlage, R. Levy, J. B. Grun, C. Klingshirn, and K. Bohnert, The dispersion of excitons, polaritons and biexcitons in direct-gap semiconductors. Phys. Rep. **124**(3), 161–253 (1985).
[58] J. J. Hopfield, Theory of the contribution of excitons to the complex dielectic constant of crystals. Phys. Rev. **112**(5), 1555–1567 (1958).
[59] R. Houdre, Early stages of continuous wave experiments on cavity-polaritons. phys. stat. sol. (b) **242**(1 1), 2167– 2196 (2005).
[60] R. Houdre, J. L. Gibernon, P. Pellandini, R. P. Stanley, U. Oesterle, C. Weisbuch, J. Ogorman, B. Roycroft, and M. Ilegems, Saturation of the strong-coupling regime in a semiconductor microcavity – free-carrier bleaching of cavity polaritons. Phys. Rev. B **52**(11), 7810–7813 (1995).
[61] R. Houdre, R. P. Stanley, U. Oesterle, M. Ilegems, and C. Weisbuch, Room-temperature exciton-photon rabi splitting in a semiconductor microcavity. Journal De Physique Iv **3**(C5), 51–58 (1993).
[62] R. Houdre, R. P. Stanley, U. Oesterle, M. Ilegems, and C. Weisbuch, Room-temperature cavity polaritons in a semiconductor microcavity. Phys. Rev. B **49**(23), 16761–16764 (1994).
[63] R. Houdre, R. P. Stanley, U. Oesterle, and C. Weisbuch, Linear response and rayleigh scattering of cavitypolaritons. Physica E **11**(2–3), 198–204 (2001).
[64] R. Houdre, R. P. Stanley, U. Oesterle, and C. Weisbuch, Strong coupling regime in semiconductor microcavities. C. R. Phys. **3**(1), 15–27 (2002).
[65] H. Hradil, A. Quattropani, V. Savona, and P. Schwendimann, Squeezed polaritons in confined systems. Journal De Physique IV **3**(C5), 393–396, October 1993.
[66] A. Imamoglu, R. J. Ram, S. Pau, and Y. Yamamoto, Nonequilibrium condensates and lasers without inversion: Exciton-polariton lasers. Phys. Rev. A **53**(6), 4250–4253 (1996).
[67] J. Jacobson, S. Pau, H. Cao, G. Bjork, and Y. Yamamoto, Observation of exciton-polariton oscillating emission in a single-quantum-well semiconductor microcavity. Phys. Rev. A **51**(3), 2542–2544 (1995).

[68] F. Jahnke, M. Kira, S. W. Koch, G. Khitrova, E. K. Lindmark, T. R. Nelson, D. V. Wick, J. D. Berger, O. Lyngnes, and H. M. Gibbs, Excitonic nonlinearities of semiconductor microcavities in the nonperturbative regime. Phys. Rev. Lett. **77**(26), 5257–5260 (1996).

[69] E. T. Jaynes and F. W. Cummings, Comparison of quantum and semiclassical radiation theories with application to beam maser. Proceedings of the Ieee **51**(1), 89 (1963).

[70] R. Joray, R. P. Stanley, and M. Ilegems, High efficiency planar mcleds. phys. stat. sol. (b) **242**(11), 2315–2325 (2005).

[71] S. Jorda, Dispersion of exciton polaritons in cavity-embedded quantum-wells. Phys. Rev. B **51**(15), 10185–10188 (1995).

[72] Y. Kaluzny, P. Goy, M. Gross, J. M. Raimond, and S. Haroche, Observation of self-induced rabi oscillations in 2-level atoms excited inside a resonant cavity – the ringing regime of super-radiance. Phys. Rev. Lett. **51**(13), 1175–1 178 (1983).

[73] J. P. Karr, A. Baas, and E. Giacobino, Twin polaritons in semiconductor microcavities. Phys. Rev. A **69**(6), 063807 (2004).

[74] J. P. Karr, A. Baas, R. Houdre, and E. Giacobino, Squeezing in semiconductor microcavities in the strongcoupling regime. Phys. Rev. A **69**(3), 03 1802 (2004).

[75] A. Kavokin and G. Malpuech, Cavity polaritons. Elsevier, Amsterdam (2003).

[76] A. Kavokin, G. Malpuech, and F. P. Laussy, Polariton laser and polariton superfluidity in microcavities. Phys. Lett. A **306**(4), 187–199 (2003).

[77] J. Keeling, P. R. Eastham, M. H. Szymanska, and P. B. Littlewood, Polariton condensation with localized excitons and propagating photons. Phys. Rev. Lett. **93**(22), 226403 (2004).

[78] G. Khitrova, H. M. Gibbs, F. Jahnke, M. Kira, and S. W. Koch, Nonlinear optics of normal-mode-coupling semiconductor microcavities. Rev. Mod. Phys. **71**(5), 1591–1639 (1999).

[79] G. Khitrova, H. M. Gibbs, M. Kira, S. W. Koch, and A. Scherer, Vacuum rabi splitting in semiconductors. Nature Physics **2**(2), 81–90 (2006).

[80] M. Kira, F. Jahnke, S. W. Koch, J. D. Berger, D. V. Wick, T. R. Nelson, G. Khitrova, and H. M. Gibbs, Quantum theory of nonlinear semiconductor microcavity luminescence explaining "boser" experiments. Phys. Rev. Lett. **79**(25), 5170–5173 (1997).

[81] S. Kundermann, M. Saba, C. Ciuti, T. Guillet, U. Oesterle, J. L. Staehli, and B. Deveaud, Coherent control of polariton parametric scattering in semiconductor microcavities. Phys. Rev. Lett. **91**(10), 107402 (2003).

[82] A. Lamas-Linares, J. C. Howell, and D. Bouwmeester, Stimulated emission of polarization-entangled photons. Nature **412**(6850), 887–890 (2001).

[83] W. Langbein, Energy and momentum broadening of planar microcavity polaritons measured by resonant light scattering. J. Phys.: Condens. Matter **16**, S3645 (2004).

[84] W. Langbein, Spontaneous parametric scattering of microcavity polaritons in momentum space. Phys. Rev. B **70**(20), 205301 (2004).

[85] W. Langbein, Polariton correlation in microcavities produced by parametric scattering. phys. stat. sol. (b) **242**(1 1), 2260–2270 (2005).

[86] W. Langbein and J. M. Hvam, Elastic scattering dynamics of cavity polaritons: Evidence for time-energy uncertainty and polariton localization. Phys. Rev. Lett. **88**(4), 047401 (2002).

[87] F. P. Laussy, G. Malpuech, A. Kavokin, and P. Bigenwald, Spontaneous coherence buildup in a polariton laser. Phys. Rev. Lett. **93**(1), 016402 (2004).

[88] F. P. Laussy, I. A. Shelykh, G. Malpuech, and A. Kavokin, Effects of Bose–Einstein condensation of exciton polaritons in microcavities on the polarization of emitted light. Phys. Rev. B **73**(3), 035315 (2006).

[89] J. Lauwers, A. Verbeure, and V. A. Zagrebnov, Bose–Einstein condensation for homogeneous interacting systems with a one-particle spectral gap. Journal of Statistical Physics, **112**(1–2), 397–420 (2003).

[90] A. J. Leggett, Bose–Einstein condensation in the alkali gases: Some fundamental concepts. Rev. Mod. Phys. **73**(2), 307–356 (2001).

[91] O. Lenoble, L. A. Pastur, and V. A. Zagrebnov, Bose–Einstein condensation in random potentials. C. R. Phys. **5**(1), 129–142 (2004).

[92] A. Loffler, J. P. Reithmaier, G. Seak, C. Hofmann, S. Reitzenstein, M. Kamp, and A. Forchel, Semiconductor quantum dot microcavity pillars with high-quality factors and enlarged dot dimensions. Appl. Phys. Lett. **86**(11), 111105 (2005).

[93] H. Mabuchi and A. C. Doherty, Cavity quantum electrodynamics: Coherence in context. Science **298**(5597), 1372–1377 (2002).

[94] G. Malpuech, A. Kavokin, A. Di Carlo, and J. J. Baumberg, Polariton lasing by exciton-electron scattering in semiconductor microcavities. Phys. Rev. B **65**(15), 153310 (2002).

[95] G. Malpuech, Y. G. Rubo, F. P. Laussy, P. Bigenwald, and A. V. Kavokin, Polariton laser: thermodynamics and quantum kinetic theory. Semicond. Sci. Technol. **18**(10), S395–S404 (2003).

[96] L. Mandel, Quantum effects in one-photon and two-photon interference. Rev. Mod. Phys. **71**(2), S274–S282 (1999).

[97] L. Mandel and E. Wolf, Optical coherence and quantum optics (Cambridge University Press, Cambridge 1995).

[98] F. M. Marchetti, J. Keeling, M. H. Szymanska, and P. B. Littlewood, Thermodynamics and excitations of condensed polaritons in disordered microcavities. Phys. Rev. Lett. **96**(6), 066405 (2006).

[99] F. M. Marchetti, B. D. Simons, and P. B. Littlewood, Condensation of cavity polaritons in a disordered environment. Phys. Rev. B **70**(15), 155327 (2004).

[100] M. Muller, J. Bleuse, and R. Andre, Dynamics of the cavity polariton in cdte-based semiconductor microcavities: Evidence for a relaxation edge. Phys. Rev. B **62**(24), 16886–16892 (2000).

[101] M. Muller, J. Bleuse, R. Andre, and H. Ulmer-Tuffigo, Observation of bottleneck effects on the photoluminescence from polaritons in ii-vi microcavities. Physica B **272**(1–4), 476–479 (1999).

[102] T. B. Norris, J. K. Rhee, C. Y. Sung, Y. Arakawa, M. Nishioka, and C. Weisbuch, Time-resolved vacuum rabi oscillations in a semiconductor quantum microcavity. Phys. Rev. B **50**(19), 14663–14666 (1994).

[103] M. Obert, J. Renner, G. Bacher, A. Forchel, R. Andre, and L. S. Dang, Nonlinear emission from ii–vi photonic dots in the strong coupling regime. Physica E **21**(2–4), 835–839 (2004).

[104] M. Obert, J. Renner, A. Forchel, G. Bacher, R. Andre, and D. L. S. Dang, Nonlinear emission in ii-vi pillar microcavities: Strong versus weak coupling. Appl. Phys. Lett. **84**(9), 1435–1437 (2004).

[105] U. Oesterle, R. P. Stanley, and R. Houdré, Mbe growth of high finesse microcavities. phys. stat. sol. (b) **242**(11), 2157–2166 (2005).

[106] S. Pau, H. Cao, J. Jacobson, G. Bjork, Y. Yamamoto, and A. Imamoglu, Observation of a laserlike transition in a microcavity exciton polariton system. Phys. Rev. A **54**(3), R1789–R1792 (1996).

[107] Stanley Pau, Gunnar Bjrk, Joseph Jacobson, Hui Cao, and Yoshihisa Yamamoto, Microcavity exciton-polariton splitting in the linear regime. Phys. Rev. B **51**(20), 14437 (1995).

[108] O. Penrose and L. Onsager, Bose–Einstein condensation and liquid helium. Phys. Rev. **104**(3), 576–584 (1956).

[109] M. Perrin, P. Senellart, A. Lemaitre, and J. Bloch, Polariton relaxation in semiconductor microcavities: Efficiency of electron-polariton scattering. Phys. Rev. B **72**(7), 075340 (2005).

[110] E. Peter, P. Senellart, D. Martrou, A. Lemaitre, J. Hours, J. M. Gerard, and J. Bloch, Exciton-photon strong-coupling regime for a single quantum dot embedded in a microcavity. Phys. Rev. Lett. **95**(6), 067401 (2005).

[111] C. Piermarocchi, F. Tassone, V. Savona, A. Quattropani, and P. Schwendimann, Nonequilibrium dynamics of free quantum-well excitons in time-resolved photoluminescence. Phys. Rev. B **53**(23), 15834–15841 (1996).

[112] L. P. Pitaevskii and S. Stringari, Bose–Einstein Condensation (Clarendon Press, Oxford 2003).
[113] D. Porras, C. Ciuti, J. J. Baumberg, and C. Tejedor, Polariton dynamics and Bose–Einstein condensation in semiconductor microcavities. Phys. Rev. B **66**(8), 085304 (2002).
[114] E. M. Purcell, Spontaneous emission probabilities at radio frequencies. Phys. Rev. **69**(11-1), 681–681 (1946).
[115] A. Quattropani and P. Schwendimann, Polariton squeezing in microcavities. phys. stat. sol. (b) **242**(1 1), 2302– 23 14 (2005).
[116] J. P. Reithmaier, G. Sek, A. Loffler, C. Hofmann, S. Kuhn, S. Reitzenstein, L. V. Keldysh, V. D. Kulakovskii, T. L. Reinecke, and A. Forchel, Strong coupling in a single quantum dot-semiconductor microcavity system. Nature **432**(7014), 197–200 (2004).
[117] M. Richard, J. Kasprzak, R. Andre, R. Romestain, L. S. Dang, G. Malpuech, and A. Kavokin, Experimental evidence for nonequilibrium bose condensation of exciton polaritons. Phys. Rev. B **72**(20), 201301 (2005).
[118] M. Richard, J. Kasprzak, R. Romestain, R. Andre, and L. S. Dang, Spontaneous coherent phase transition of polaritons in cdte microcavities. Phys. Rev. Lett. **94**(18), 187401 (2005).
[119] M. Richard, R. Romestain, R. Andre, and L. S. Dang, Consequences of strong coupling between excitons and microcavity leaky modes. Appl. Phys. Lett. **86**(7), 071916 (2005).
[120] M. Saba, C. Ciuti, J. Bloch, V. Thierry-Mieg, R. Andre, L. S. Dang, S. Kundermann, A. Mura, G. Bongiovanni, J. L. Staehli, and B. Deveaud, High-temperature ultrafast polariton parametric amplification in semiconductor microcavities. Nature **414**(6865), 731–735 (2001).
[121] S. Savasta, O. D. Stefano, V. Savona, and W. Langbein; Quantum complementarity of microcavity polaritons. Phys. Rev. Lett. **94**(24), 246401 (2005).
[122] V. Savona, Linear optical properties of semiconductor microcavities with embedded quantum wells. In H. Benisty, editor, Confined photon systems: Fundamentals and applications, pages 173–242. Springer Verlag, Berlin, New York (1999).
[123] V. Savona, L. C. Andreani, P. Schwendimann, and A. Quattropani, Quantum-well excitons in semiconductor microcavities – unified treatment of weak and strong-coupling regimes. Solid State Commun. **93**(9), 733–739 (1995).
[124] V. Savona, Z. Hradil, A. Quattropani, and P. Schwendimann, Quantum-theory of quantum-well polaritons in semiconductor microcavities. Phys. Rev. B **49**(13), 8774–8779 (1994).
[125] V. Savona and C. Piermarocchi, Microcavity polaritons: Homogeneous and inhomogeneous broadening in the strong coupling regime. phys. stat. sol. (a) **164**(1), 45–51 (1997).
[126] V. Savona, C. Piermarocchi, A. Quattropani, P. Schwendimann, and F. Tassone, Optical properties of microcavity polaritons. Phase Transitions **68**(1), 169–279 (1999).
[127] V. Savona, C. Piermarocchi, A. Quattropani, F. Tassone, and P. Schwendimann, Microscopic theory of motional narrowing of microcavity polaritons in a disordered potential. Phys. Rev. Lett. **78**(23), 4470–4473 (1997).
[128] V. Savona and D. Sarchi, Bose–Einstein condensation of microcavity polaritons. phys. stat. sol. (b) **242**(1 1), 2290–2301 (2005).
[129] V. Savona, P. Schwendimann, and A. Quattropani, Onset of coherent photoluminescence in semiconductor microcavities. Phys. Rev. B **71**(12), 125315 (2005).
[130] V. Savona, F. Tassone, C. Piermarocchi, A. Quattropani, and P. Schwendimann, Theory of polariton photoluminescence in arbitrary semiconductor microcavity structures. Phys. Rev. B **53**(19), 13051–13062 (1996).
[131] P. G. Savvidis, J. J. Baumberg, R. M. Stevenson, M. S. Skolnick, D. M. Whittaker, and J. S. Roberts, Angleresonant stimulated polariton amplifier. Phys. Rev. Lett. **84**(7), 1547–1550 (2000).
[132] P. G. Savvidis, C. Ciuti, J. J. Baumberg, D. M. Whittaker, M. S. Skolnick, and J. S. Roberts, Off-branch polaritons and multiple scattering in semiconductor microcavities. Phys. Rev. B **64**(7), 07531 1 (2001).

[133] Schmitt-Rink, D. S. Chemla, and D. A. B. Miller, Theory of transient excitonic optical nonlinearities in semiconductor quantum-well structures. Phys. Rev. B **32**(10), 6601–6609, Nov 1985.

[134] P. Schwendimann, C. Ciuti, and A. Quattropani, Statistics of polaritons in the nonlinear regime. Phys. Rev. B **68**(16), 165324 (2003).

[135] P. Senellart and J. Bloch, Nonlinear emission of microcavity polaritons in the low density regime. Phys. Rev. Lett. **82**(6), 1233–1236 (1999).

[136] A. Shelykh, A. V. Kavokin, and G. Malpuech, Spin dynamics of exciton polaritons in microcavities. phys. stat. sol. (b) **242**(11), 2271–2289 (2005).

[137] H. Shi and A. Griffin, Finite-temperature excitations in a dilute bose-condensed gas. Phys. Rep. **304**(1–2), 2–87 (1998).

[138] D. Snoke, Spontaneous bose coherence of excitons and polaritons. Science **298**(5597), 1368–1372 (2002).

[139] L. Staehli, S. Kundermann, M. Saba, C. Ciuti, A. Baas, T. Guillet, and B. Deveaud, Non-linear dynamical effects in semiconductor microcavities. phys. stat. sol. (b) **242**(11), 2246–2259 (2005).

[140] J. Steinhauer, R. Ozeri, N. Katz, and N. Davidson, Excitation spectrum of a Bose–Einstein condensate. Phys. Rev. Lett. **88**(12), 120407 (2002).

[141] R. M. Stevenson, V. N. Astratov, M. S. Skolnick, D. M. Whittaker, M. Emam-Ismail, A. I. Tartakovskii, P. G. Savvidis, J. J. Baumberg, and J. S. Roberts, Continuous wave observation of massive polariton redistribution by stimulated scattering in semiconductor microcavities. Phys. Rev. Lett. **85**(17), 3680–3683 (2000).

[142] A. I. Tartakovskii, M. Emam-Ismail, R. M. Stevenson, M. S. Skolnick, V. N. Astratov, D. M. Whittaker, J. J. Baumberg, and J. S. Roberts, Relaxation bottleneck and its suppression in semiconductor microcavities. Phys. Rev. B **62**(4), R2283–R2286 (2000).

[143] A. I. Tartakovskii, D. N. Krizhanovskii, G. Malpuech, M. Emam-Ismail, A. V. Chernenko, A. V. Kavokin, V. D. Kulakovskii, M. S. Skolnick, and J. S. Roberts, Giant enhancement of polariton relaxation in semiconductor microcavities by polariton-free carrier interaction: Experimental evidence and theory. Phys. Rev. B **67**(16), 165302 (2003).

[144] F. Tassone, F. Bassani, and L. C. Andreani, Resonant and surface-polaritons in quantum-wells. Nuovo Cimento Della Societa Italiana Di Fisica D-Condensed Matter Atomic Molecular and Chemical Physics Fluids Plasmas Biophysics 12(12), 1673–1687 (1990).

[145] F. Tassone, C. Piermarocchi, V. Savona, A. Quattropani, and P. Schwendimann, Bottleneck effects in the relaxation and photoluminescence of microcavity polaritons. Phys. Rev. B **56**(12), 7554–7563 (1997).

[146] F. Tassone and Y. Yamamoto, Exciton-exciton scattering dynamics in a semiconductor microcavity and stimulated scattering into polaritons. Phys. Rev. B **59**(16), 10830–10842 (1999).

[147] Tawara, H. Gotoh, T. Akasaka, N. Kobayashi, and T. Saitoh, Cavity polaritons in ingan microcavities at room temperature. Phys. Rev. Lett. **92**(25), 256402 (2004).

[148] R. J. Thompson, G. Rempe, and H. J. Kimble, Observation of normal-mode splitting for an atom in an optical cavity. Phys. Rev. Lett. **68**(8), 1132–1135 (1992).

[149] A. Tredicucci, Y. Chen, V. Pellegrini, M. Borger, L. Sorba, F. Beltram, and F. Bassani, Controlled excitonphoton interaction in semiconductor bulk microcavities. Phys. Rev. Lett. **75**(21), 3906–3909 (1995).

[150] C. Weisbuch and H. Benisty, Microcavities in ecole polytechnique federale de Lausanne, ecole polytechnique (France) and elsewhere: past, present and future. phys. stat. sol. (b) **242**(11), 2345–2356 (2005).

[151] C. Weisbuch, M. Nishioka, A. Ishikawa, and Y. Arakawa, Observation of the coupled exciton-photon mode splitting in a semiconductor quantum microcavity. Phys. Rev. Lett. **69**(23), 3314–3317 (1992).

[152] C. Weisbuch and B. Vinter, Quantum semiconductor structures fundamentals and applications. Academic Press, Boston, 1991.

[153] D. M. Whittaker, What determines inhomogeneous linewidths in semiconductor microcavities? Phys. Rev. Lett. **80**(21), 4791–4794 (1998).

[154] D. M. Whittaker, Resonant rayleigh scattering from a disordered microcavity. Phys. Rev. B **61**(4), R2433– R2435 (2000).

[155] D. M. Whittaker, Effects of polariton-energy renormalization in the microcavity optical parametric oscillator. Phys. Rev. B **71**(11), 115301 (2005).

[156] D. M. Whittaker, P. Kinsler, T. A. Fisher, M. S. Skolnick, A. Armitage, A. M. Afshar, M. D. Sturge, and J. S. Roberts, Motional narrowing in semiconductor microcavities. Phys. Rev. Lett. **77**(23), 4792–4795 (1996).

[157] T. Yoshie, A. Scherer, J. Hendrickson, G. Khitrova, H. M. Gibbs, G. Rupper, C. Ell, O. B. Shchekin, and D. G. Deppe, Vacuum rabi splitting with a single quantum dot in a photonic crystal nanocavity. Nature **432**(7014), 200–203 (2004).

[158] M. Zamfirescu, A. Kavokin, B. Gil, G. Malpuech, and M. Kaliteevski, Zno as a material mostly adapted for the realization of room-temperature polariton lasers. Phys. Rev. B **65**(16), 161205 (2002).

[159] Yifu Zhu, Daniel J. Gauthier, S. E. Morin, Qilin Wu, H. J. Carmichael, and T. W. Mossberg, Vacuum rabi splitting as a feature of linear-dispersion theory: Analysis and experimental observations. Phys. Rev. Lett. **64**(21), 2499–2502, (1990).

[160] R. Zimmermann, E. Runge, and V. Savona, Theory of resonant secondary emission: Rayleigh scattering versus luminescence. In T. Takagahara, editor, Quantum coherence, correlation, and decoherence in semiconductor nanostructures, page 89 (Academic PressPress, Amsterdam, Boston, 2003).

Chapter 2
MBE Growth of High Finesse Microcavities

Ursula Oesterle, Ross P. Stanley, and Romuald Houdré

2.1
Introduction

Fabry–Perot resonators, consisting of two Distributed Bragg Reflectors (DBRs), separated by a cavity with active medium such as (In,Ga)As quantum wells, are the basis of a wide range of devices, i.e. lasers, enhanced photodiodes, and light emitting diodes [1, 2]. These vertical microcavity structures containing at times hundreds of different layers require highly stable growth conditions, with precise control over compositions and individual layer thickness, while maintaining smooth interfaces. Although advanced growth techniques such as molecular beam epitaxy (MBE) and metal-organic chemical vapour deposition (MOCVD) can provide the critical tolerances, they are being pushed to their limits. In this chapter the growth by MBE and characterization of such high finesse microcavity structures using a combination of different diagnostic tools will be discussed.

2.2
Principles of MBE Growth

MBE is a cold wall vacuum evaporation technique [3]. True to its name, molecular beams generated from heated effusion cells interact on a heated crystalline substrate to produce a single-crystal epitaxial layer. Because of the ultra high vacuum environment (starting at a base pressure of $<1 \times 10^{-10}$ mbar and going up to 5×10^{-7} mbar during growth), the flow of components from the sources to the substrate is in the molecular flow regime. Thus the beams can be considered unidirectional with negligible interaction between them. Shutters in front of the source effusion cells control the atom beam flow towards the substrate.

The atom arrival rate, which is related to the element vapour pressure, is controlled through the temperature of the effusion cells. Typical growth rates in MBE are in the order of one monolayer per second (1 µm per hour). This allows epitaxial layers with different compositions to be deposited a single monolayer at a time with

Physics of Semiconductor Microcavities: From Fundamentals to Nanoscale Devices
Edited by Benoit Deveaud
Copyright © 2007 WILEY-VCH Verlag GmbH & Co. KGaA, Weinheim
ISBN: 978-3-527-40561-9

atomically abrupt interfaces between layers. Since the typically growth temperature is low (550–650 °C for GaAs), interdiffusion between layers and surface accumulation effects can be considered negligible. Under normal growth conditions, all the incident group III atoms stick to the substrate and only enough group V atoms adhere to satisfy the group III atoms giving rise to stochiometric growth. The excess group V species are desorbed. In this case, the epitaxial growth rate and composition are proportional to the group III flux rates, which are controlled through the temperature of the source effusion cells.

At high enough substrate temperature, desorption of the group III element atoms can become significant and their sticking coefficients are no longer unity. The surface migration and desorption of the group III elements depends on the bond strength, which varies from element to element; aluminium has a higher bond strength than gallium, gallium in turn has a higher bond strength than indium. Consequently aluminium migrates less easily than gallium, and therefore requires a higher substrate temperature and less arsenic for growth. Indium, in InAs on the other hand, starts to desorb at rather low temperatures, around 560 °C, compared to 640 °C for gallium in GaAs and 750 °C for aluminium in AlAs [4, 5]. Desorption of the group III elements changes the growth rates but stochiometric growth persists as long as there is sufficient arsenic to incorporate.

2.2.1
Growth of $Al_xGa_{1-x}As$/AlAs DBRs

Distributed Bragg Reflectors (DBRs) are stacks of alternating high and low refractive index materials, where every layer has a $\lambda/4$ optical thickness. They can easily contain at least 25 pairs and be several microns thick. Smooth interfaces and precise accuracy in the individual layer thickness are key param-eters for very high reflecting DBRs. As the thickness of an epitaxial layer is controlled through the III-material source flux, very stable temperature of the source effusion cells and a continuous supply of source material are imperative throughout the whole growth time, which can easily take 12 hours. Smooth interfaces are only formed when two-dimensional growth, which is related to the substrate temperature and the arsenic flux, persists throughout the whole growth process.

DBR mirrors with the best optical properties are grown at relatively high substrate temperatures (620–630 °C) to overcome the limited surface migration of aluminium, yet still be below the temperature at which gallium starts to desorb. The optical pyrometer used to measure the temperature of the substrate works around the same wavelength (950 nm) as the centre wavelength of the DBR structures being grown. As soon as a few pairs of the DBR mirror are grown they interfere with the pyrometer reading, leading to erroneous temperature measurements. To overcome this problem, the power setting of the substrate heating filaments is calibrated against the temperature measured by the pyrometer before initiating growth. These values are then used to change the substrate temperature during the growth, independent of the pyrometer reading.

2.2 Principles of MBE Growth

The interference of the pyrometer reading with the reflection of the DBR structure being grown leads to an oscillation in the pyrometer reading correlating the period of the mirrors, which means the number of periods grown can be counted from the number of oscillations. A quarter wavelength oscillation in the pyrometer reading corresponds to an optical thickness of the growing layer of $\lambda/4$ for this wavelength. Therefore, oscillation in pyrometer temperature readings can be used for in-situ control of the thickness of DBRs centred on the same wavelength [6].

2.2.2
Growth of (In,Ga)As Quantum Wells

Two factors need to be taken into consideration when growing (In,Ga)As structures:

1. (In,Ga)As has a larger lattice parameter than GaAs and is therefore compressively strained in the plane. When growing (In,Ga)As on GaAs, a correction factor in the growth rate, which takes into account the lattice deformation due to the strain has to be introduced.
2. Indium starts to re-evaporate at relatively low substrate temperatures and a correction for the lost indium has to be taken into consideration when growing at higher substrate temperatures.

Although, higher substrate temperatures (around 600 °C) may give better (In,Ga)As quantum wells [7], precise control of indium incorporation is lost. When accurate composition and thickness of the quantum wells is needed (i.e. emission at a well defined wavelength), the substrate temperature should be below the indium desorption temperature. In this work, the $In_{0.2}Ga_{0.8}As$ quantum wells on GaAs were typically grown at low substrate temperatures, 520–530 °C using high growth rates under excess arsenic conditions.

2.2.3
Growth of Vertical Cavity Structures

A typical microcavity structure used in our work, consists of two sets of 20–27 pairs (Ga,Al)As/AlAs DBR mirrors separated by a 3/2 lambda GaAs cavity with two sets of 3 (In, Ga)As quantum wells, see Fig. 2.1. When growing a complete vertical cavity structure, the growth of the DBR mirrors, the GaAs cavity spacer and the (In,Ga)As quantum wells all have to be compatible. This means making compromises in the substrate temperatures and growth rates of the different materials. As the (In,Ga)As quantum wells in the cavity need to be grown around 520 °C, the DBR mirrors are grown at a substrate temperature as low as possible without losing optical quality, i.e. 590–600 °C.

When designing microcavity structures for optical pumping of the GaAs cavity, the DBR mirrors need to be transparent to pump beam. Thus the GaAs in the DBR mirror is replaced with $Al_{0.1}Ga_{0.9}As$. To grow such $Al_{0.1}Ga_{0.9}As$/AlAs DBRs at reasonable growth rates requires two Al sources: one for the AlAs layers and one with a

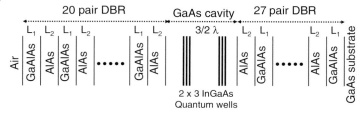

Fig. 2.1 Typical microcavity structure consisting of a bottom (Al,Ga)As/AlAs DBR, a $3\lambda/2$ GaAs cavity with 2 sets of 3 (In,Ga)As quantum wells and a second (Al,Ga)As/AlAs DBR. The total thickness is over 7 μm.

much lower flux for the small amount of aluminium in the $Al_{0.1}Ga_{0.9}As$ layers. However, when only one aluminium source is available, the $Al_{0.1}Ga_{0.9}As$ can be grown as a pseudo-alloy, i.e. a short period superlattice of binaries. If the individual layers in the superlattice are thin enough, the carriers will tunnel through them and the pseudo-alloy will have the same properties as the bulk material [8]. In our case, this solution was chosen and the $Al_{0.1}Ga_{0.9}As$ was grown using a superlattice with a 40 Å period, consisting of 4 Å AlAs and 36 Å GaAs.

The calibration of the absolute growth rates as well as the short term fluctuations and long term stability of the effusion cells are crucial. In order to achieve the necessary precision, the flux from the effusion cells (equivalently the growth rate) has to be measured just before the actual growth of the structure, once the effusion cells have been stabilized at the growth temperatures. Classical calibration techniques such as RHEED oscillation measurements or measuring the fluxes using the ion beam gauge on the substrate manipulator (which may give precise ratios between the fluxes), are not sufficiently accurate for the absolute growth rate values required here.

In our case, a calibration sample consisting of a 5-pair AlAs/GaAs DBR and an (In,Ga)As quantum well is grown, once the cells had been given enough time to stabilize at growth temperatures, and just before commencing the growth of the final vertical cavity structure. The current growth rates of the aluminium and gallium sources are determined from X-ray diffraction measurements of the 5-pair DBR, while photoluminescence of the quantum well, is used to determine the indium growth rate. In order to not perturb the equilibrium in the cells, the gallium and aluminium cell temperatures are not changed. Instead thickness adjustments are made by changing the time for the growth of each layer. Composition corrections of the (In,Ga)As quantum wells are made by changing the indium cell temperature. Such adjustments, if any, are minute. This calibration procedure of characterizing a test structure before the growth of high finesse microcavities and does not take longer than RHEED oscillation measurements and proved to be more successful for our needs.

2.3
Characterization and Properties of Vertical Cavity Structures

2.3.1
Error Tolerance

A simultaneous alignment of the peak emission wavelength of the active medium, the peak reflectivity of the DBR mirrors and the Fabry–Perot mode is essential for a resonant microcavity structure. The peak emission wavelength of the strained quantum wells depends both on the indium composition and the thickness of the quantum wells. For the DBR mirror, the peak reflectivity wavelength depends on the thickness of all layers in the mirror stack. The cavity resonance depends both on the thickness of the cavity as well as the layers in the DBR mirror. Thus, the rate at which each feature of a microcavity structure shifts with changes in the growth rates varies. Small errors in thickness and composition of the individual layers in the structure can already lead to deteriorations of the optical properties.

The stopband has a maximum width and a maximum reflectivity when both types of layers in the DBR have the same optical length, and the thickness of each type of layer corresponds to roughly to a quarter wavelength optical path. The effect of a systematic error in the thickness of the layers in a DBR stack is shown in Fig. 2.2. These simulations show that the effect of a systematic error in one type of layers in the DBR stack shifts the stopband, but does not significantly degrade the peak reflectivity. A decrease in thickness of the layers shifts the peak reflectivity to a shorter wavelength. In case (b) of Fig. 2.2, a 5% reduction in the thickness of the AlAs layers reduces the peak reflectivity from 99.95% to 99.91% and shifts the

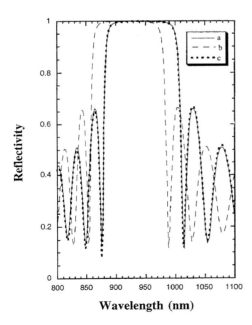

Fig. 2.2 Simulation of a 20-pair AlAs/GaAs DBR centred at 940 nm showing the effect of a systematic error. In case (a) all the layers have a $\lambda/4$ path, (b) all the AlAs layers are reduced in thickness by 5%, (c) all the AlAs layers are reduced in thickness by 5% while all the GaAs layers are increased by 5%, such that the period remains the same.

stopband by 30 nm to shorter wavelength. The width of the stopband depends on the thickness of the individual layer and reduces with any errors in the optical length. However, the position of the stopband depends on the optical thickness of a mirror pair and not on the thickness of the individual layers when there are many pairs in the DBR. Thus, the period thickness and not the individual layer thickness is the crucial parameter for the mirror to be centred at the right wavelength. The most likely error during growth is a drift of the source fluxes due to the emptying of the cells. Such a systematic error leads to inhomogeneous layer thickness and can destroy the properties of the Bragg mirror.

The wavelength at which the cavity resonance is positioned will ultimately determine the emission wavelength of a microcavity device. Assuming a non-dispersive medium, any error in the position of the Fabry–Perot mode is due to errors in the thickness of the layers. As the cavity consists of GaAs, an error in the gallium flux will shift the Fabry–Perot mode as well as the peak reflectivity of the DBR, which shifts at a slower rate than the Fabry–Perot mode. Due to the penetration of the optical wave into the DBR mirror an error in the aluminium flux will not only shift the peak reflectivity of the DBR accordingly, but also induce a slight shift of the Fabry–Perot mode. A systematic error of 5% in all the layers will shift the peak reflectivity of the DBR as well as the Fabry–Perot mode, such that it will remain roughly in the centre of the stopband.

Finally, the Fabry–Perot mode has to be aligned with the peak emission of the quantum well. For (In, Ga)As quantum wells, the band gap decreases with increasing indium composition. At the same time the confinement energy of the electrons and holes decreases with increasing well width. Taking a 75 Å $In_{0.2}Ga_{0.8}As$ quantum well as an example: A 10% increase in the gallium flux will increase the well thickness, shifting the emission wavelength to a longer wavelength by 3.8 nm. At the same time the extra gallium causes a decrease in the indium composition, which will shift emission wavelength to a shorter wavelength by 3.1 nm. Therefore a 10% error in the gallium flux will cause no significant net shift of the energy of the quantum well. Also a small error in the indium flux will have no significant effect on the emission wavelength, considering that the indium composition in the quantum well is <20%.

Independent of errors in the growth rates, the inhomogeneity of the fluxes from the effusion cells will automatically induce a thickness variation across a wafer. In some cases, this inhomogeneity can be turned into an advantage to voluntarily induce a thickness gradient in the cavity spacer layer by momentarily stopping the rotation of the substrate during growth. In these structures the DBR stopband and the gain spectrum of the quantum well stay roughly the same across the wafer, whereas the Fabry–Perot mode varies by about 30 nm across the wafer for a 2 minute rotation stop. This ensures that resonance conditions occur at some point on the wafer.

2.3.2
Structural Properties

The wedge transmission electron microscope (TEM) image of a microcavity structure, shown in Fig. 2.3, consists of three (In,Ga)As quantum wells in a GaAs cavity with an AlAs/Al$_{0.1}$Ga$_{0.9}$As DBR on both sides – 27 pairs on the substrate side and 20 pairs on top. When exposed to air, AlAs oxidizes strongly inducing the strong bending of the thickness fringes seen in the image. The images show how small the cavity (266 nm) is compared to the whole structure of over 7 μm. The insert shows an enlargement of the cavity with the three (In,Ga)As quantum wells, where the dispersion of the thickness fringes in the quantum wells is due to the built-in strain. While the AlAs on GaAs interfaces are flat, the GaAs on AlAs interfaces

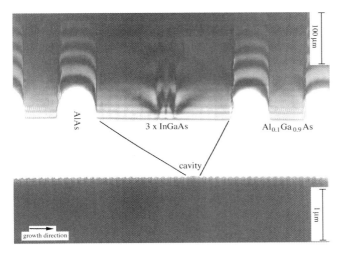

Fig. 2.3 Wedge TEM image showing a complete microcavity structure consisting of a 27-pair top DBR, a λ GaAs cavity and a 20-pair bottom DBR. The insert is an enlargement of the cavity region, showing 3 (In,Ga)As quantum wells and the superlattice structure of the Al$_{0.1}$Ga$_{0.9}$As layers in the DBR.

Fig. 2.4 X-section TEM of the cavity region, with a GaAs/AlAs DBR on one side and an Al$_{0.1}$Ga$_{0.9}$As/AlAs DBR on the other side of the cavity. The Al$_{0.1}$Ga$_{0.9}$As layers consist of a superlattice structure of 4 Å of AlAs and 36 Å of GaAs. The three thick dark lines in the centre are the (In,Ga)As quantum wells.

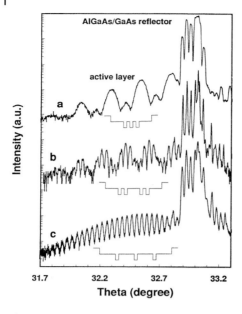

Fig. 2.5 X-ray diffraction images: (004) rocking curves of three different microcavity structures with different cavity spacing and quantum well profiles.

show roughness. Growth of GaAs tends to flatter the growing layer, while growth of AlAs induces some roughness due to the limited surface mobility of aluminium. The amplitude of the roughness is in the order of 20 Å and the undulation of the order of 50 Å. A closer look at the $Al_{0.1}Ga_{0.9}As$ layers shows their pseudo-alloy structure, where the dark lines are the 4 Å AlAs layers and the lighter layers between the dark lines correspond to 36 Å of GaAs, see Fig. 2.4.

X-ray diffraction measurements show two distinct groups of features; one set originating from the quantum well structure and the second set close to the substrate originating from the DBR structure, see Fig. 2.5. Although the overall shape of the features arising from the cavity and quantum wells can be easily understood, simulations are necessary to determine numeric values of the quantum thicknesses and composition. The periodicity and composition of the mirrors can be calculated from the angular position of the satellites around the substrate peak. To obtain more precise results, measurements are carried out close to the GaAs (002) as well as (004) directions and over a 180° spread in azimuth position to counteract the effects of residual substrate disorientation [9].

In the example in Fig. 2.6, consisting of a 20-pair $AlAs/Al_{0.1}Ga_{0.9}As$ DBR, the satellite peaks have a full width at half maximum (FWHM) of 15 arcsec, which is close to the theoretical values, and show no broadening up to the fourteenth order implying good crystal quality and little variation in thickness between individual layers. A fit to the X-ray satellite peak gives a mirror periodicity of 1438.6 Å for the satellites close to the GaAs(002) reflection and 1439.9 Å for those close to the GaAs(004) reflection, giving an average period of 1439 ± 1 Å. The Al concentration in the (Ga,Al)As pseudoalloy is in the range of 9.5–10% depending on the model used [10, 11].

2.3 Characterization and Properties of Vertical Cavity Structures

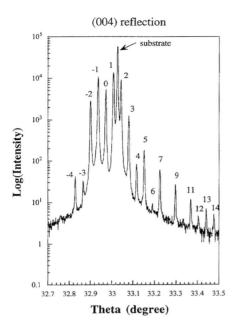

Fig. 2.6 X-ray diffraction image: (004) rocking curve of a DBR mirror showing the satellite peaks used to calculate the thickness and average composition of the DBR structure.

2.3.3
Optical Measurements of High Finesse Microcavity Structures

Reflectivity and photoluminescence [12] measurements are not only used to characterize the optical properties of the microcavity structure but can give a lot of insight to growth related parameters such as growth stability, uniformity of growth, material purity and upper limits of long-range interface roughness. At the same time they give a good estimate for internal absorption and for material composition when combined with X-ray diffraction measurements. An empty microcavity (with no active medium) is an excellent structure to extract growth related parameters through optical measurements. The width of the Fabry–Perot mode is directly related to the reflectivity of the DBR mirrors. However, if quantum wells are present the Fabry–Perot linewidth is limited by the quantum well absorption. The empty cavity structure used in the discussion below consists of a bottom DBR mirror consisting of 27 pairs of (675 Å $Al_{0.1}Ga_{0.9}As$)/(764 Å AlAs); a 2670 Å GaAs cavity and a 20-pair (675 Å $Al_{0.1}Ga_{0.9}As$) / (764 Å AlAs) top mirror. Due to the fact that only one aluminium source was available for the growth, the $Al_{0.1}Ga_{0.9}As$ was grown as a pseudo-alloy consisting of 4 Å AlAs and 36 Å GaAs.

Figure 2.7a shows a reflectivity spectrum taken at the centre of the sample. There is a wide stopband from 880 to 975 nm, with sideband minima to either side. In the middle a sharp Fabry–Perot mode can be seen. The Fabry–Perot mode is shown with greater resolution in Fig. 2.7b. The solid line is a Lorentzian with a FWHM of 0.95 Å. When including instrumental resolution this corresponds to a linewidth of

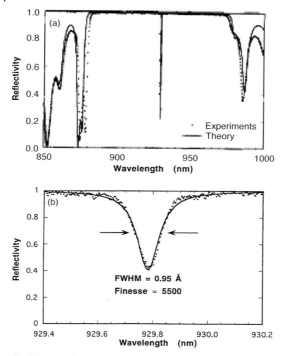

Fig. 2.7 (a) Reflectivity spectrum of an empty cavity structure. The solid line is a calculated curve. (b) Details of the Fabry–Perot mode. The solid line is a Lorentzian with a full width half maximum of 0.95 Å.

0.84 Å. The minimum reflectivity is 40% while the maximum transmission is 20% implying that 40% is absorbed or scattered by the sample. Taking into account the absorption of the n-doped substrate, measured to be ~12 cm^{-1}, the corrected reflection, transmission and absorption in the FP section are 37%, 48%, and 14% respectively. These values place a limit of 0.9 cm^{-1} on the internal losses due to absorption of scattering. Such a small number is indicative of material with very good optical quality.

The exact form of the reflectivity spectrum is highly sensitive to the periodicity and composition of the individual layers. When the structure deviates far from the ideal form, the spectrum can become complicated to analyse. In our case the structure is close to ideal, allowing a lot of information related to the growth to be extracted. However, to avoid any measurement errors a narrow and highly parallel beam has to be used.

The reflectivity spectrum was calculated using the known dispersion relations of GaAs, $Al_{0.1}Ga_{0.9}As$ and AlAs, along with the knowledge of the individual layer thicknesses measured by X-ray diffraction. Care was taken to measure the same spot on the sample with both X-ray and optical measurements. The standard method of transfer matrices was used to calculate reflectivity [13]. The results are shown as the solid line in Fig. 2.7a. The only unknown parameter was the position of the Fabry–Perot resonance, which could not be measured by X-ray diffraction. The most uncertain parameter is the concentration of Al in $Al_xGa_{1-x}As$. The effect of changing the concentration is to shift the position of the stopband and to change its width. The best agreement with measurement is obtained with a value of 9.7% Al which falls in the range 9.5–10% calculated from the X-ray data.

In a perfect DBR microcavity the pairs of side-lobes on each side of the Fabry–Perot resonance are symmetric. Any variation from the ideal structure leaves well-defined traces in the side-lobes and so the side-lobes can be used to determine any deviations from ideal. In other words, the side-lobes are characteristic for a sample. In Fig. 2.7a there is a strong asymmetry, with the short wavelength pair having equal heights while the long wavelength pair has very different heights. This can be simulated by assuming that the second mirror is thinner than the first mirror. The agreement with the measured spectra is excellent if the second mirror is assumed to be 0.25% thinner than the first mirror. This thickness variation is small considering that the growth for this sample lasted 20 h, i.e. 10 h from the middle of the first mirror to the middle of the second. The reflectivity spectrum of a DBR Fabry–Perot is very sensitive to any relative change between the two mirrors because it acts as a differential measurement, where the reflection minima are due to interplay of both mirrors.

More information can be extracted from the linewidth of the Fabry–Perot mode, when comparing cavities with high mirror reflectivities with theoretical calculations. The finesse can be calculated from the linewidth $\Delta\lambda$ of the Fabry–Perot mode at the wavelength λ by [13]

$$F = \frac{\text{free spectral range}}{\Delta\lambda} = \frac{\lambda}{m\,\Delta\lambda} \tag{1}$$

where m is the order of the fringe. A λ sized cavity is a second order cavity ($m = 2$) so with $\lambda = 9300$ Å and $\Delta\lambda = 0.84$ Å this gives a finesse of 5530. However, as the phase shift depends on the wavelength for dielectric mirrors the cavity is effectively longer than the spacer layer and this increases the finesse over a similar structure with ideal or metallic mirrors. This may be taken into account by using an effective order, m_{eff}. For our sample, $m = 2$ and $m_{eff} = 7.54$. The reason for using m_{eff} is that many authors quote the finesse in terms of a local free spectral range divided by the linewidth, which is equivalent to using m_{eff} in Eq. (1). Taking this definition the effective finesse, F_{eff}, of the cavity becomes $F_{eff} = 1470$.

For ideal mirrors the finesse can be related to the reflectivity of the front and back mirrors in the limit of high reflectivities by

$$F = \pi\sqrt{R}/(1-R) \tag{2}$$

Using the effective finesse determined above, the experimental mirror reflectivity will become $R = 99.82\%$. The theoretical linewidth of a Fabry–Perot resonance calculated both analytically and using the transfer matrix method for multilayer structures, is equal to 0.4 Å as opposed to the measured linewidth of 0.84 Å. Besides measurement errors due to using a convergent probe beam, the difference between these values can be attributed to several factors: (a) residual absorption or scattering at interfaces, (b) diffraction losses due to mirror roughness, and (c) cavity width fluctuations.

The possibility of internal loss, due to either residual absorption or scattering at interfaces, can be calculated by including a distributed loss in the theoretical calculations. The internal losses of <0.9 cm^{-1} assessed above set an upper limit of a linewidth of 0.5 Å. This factor is not as important in the observed linewidth, as expected from high purity undoped material.

Secondly it is known that MBE growth leads to interface roughness in the order of a few monolayers. Davies [14] has shown that for a surface with micro-roughness the distribution of heights alone is sufficient for calculating its reflecting and scattering properties. Given a Gaussian distribution of heights where $\sigma \ll \lambda$, an ideal mirror of reflectivity R_0 has its reflectivity at normal incidence reduced to

$$R = R_0 \exp[-(4\pi\sigma/\lambda)^2] \qquad (3)$$

A linewidth of 0.9 Å implies that the FWHM of the height distribution is ~50 Å. From TEM images, the growth is known to be smoother than this. So one can conclude that micro-roughness is a relatively unimportant scattering mechanism at normal incidence in comparison to large scale roughness.

Finally besides surface roughness, large scale thickness variations also influence the Fabry–Perot linewidth. The effective finesse of the cavity indicates that average cavity width fluctuations across the diameter of the probe spot are less than 0.5 Å, which implies that the mirrors are both very flat and highly parallel. Due to the growth conditions the sample is thicker in the middle than at the edges. The narrowest linewidth was measured in the very centre of the sample where the sample variations are minimal. To either side of this region the cavity linewidth increases by 50% to 1.5 Å implying that the difference in the theoretical and measured linewidths is due to the very slight curvature of the sample.

2.4
Conclusion

Resonant microcavity structures require simultaneously alignment of the DBR mirrors, the cavity and the quantum well emission. From the growth point of view, it is fairly easy to obtain the right peak emission wavelength of the quantum well. However, the alignment of the stopband of the DBR is more sensitive than the quantum wells to errors in thickness and composition induced by the growth, while the position of the Fabry–Perot mode is the most sensitive component. Combining information extracted from different characterization techniques such as X-ray diffraction, TEM and detailed analysis of reflectivity measurements including both the finesse and the side-lobes can provide valuable insights to growth related parameters such as growth stability, uniformity of growth, and giving upper limits of long-range interface roughness. This growth related feedback is crucial to the growth of such highly sensitive structures like resonant high finesse microcavity devices. In the end, what really counts for microcavity devices are the optical properties of the structure and how the optical wave sees the surface roughness and thickness variations.

DBR microcavity structures with Fabry–Perot linewidths of 0.84 Å at 930 nm, implying finesse in excess of 5500 and an effective finesse greater than 1450 can be grown by MBE. Detailed characterization of such structures show that: (i) DBRs with reflectivity as high as 99.82% are attainable, (ii) the internal losses are <1 cm^{-1} and are not a limiting factor of the Fabry–Perot linewidth in undoped structures, (iii) the stability of the growth is better than 0.25% over 10 hours, (iv) the variation in the homogeneity of the source flux during growth is the limiting factor of the finesse, and (v) samples show a very slight curvature of $\lambda/10^4$.

References

[1] M. Rattier, T. F. Krauss, J. F. Carlin, R. P. Stanley, U. Oesterle, R. Houdré, C. J. M. Smith, R. M. De La Rue, H. Benisty, and C. Weisbuch, Opt. Quantum Electron. **34**, 79 (2002).
[2] B. Zhang, G. Solomon, M. Pelton, J. Plant, C. Santori, J. Vuckovi, and Y. Yamamoto, J. Appl. Phys. **97**, 73507 (2005).
[3] E. H. C. Parker, The Technology and Physics of Molecular Beam Epitaxy (Plenum Press, New York, 1985).
[4] J. Van Hove and P. I. Cohen, Appl. Phys. Lett. **47**(7), 726 (1985).
[5] K. R. Evans, C. E. Stutz, E. N. Taylor, and J. E. Ehret, J. Vac. Sci. Technol. B **9**(4), 2427 (1991).
[6] S. L. Wright, T. N. Jackson, and R. F. Marks, J. Vac. Sci. Technol. B **8**(2), 288 (1989).
[7] J.-P. Reithmaier, Thesis, München (1990).
[8] G. Bastard, Wave mechanics applied to semiconductor heterostructure (Les Editions de Physique, Paris, 1990).
[9] M. Gailhanou, J. F. Carlin, and U. Oesterle, J. Cryst. Growth **140**, 205 (1994).
[10] B. K. Tanner, A. G. Turnbill, C. R. Stanley, A. H. Kean, and M. McElhinney, Appl. Phys. Lett. **59**, 2272 (1991).
[11] M. S. Goorsky, T. F. Kuech, M. A. Tischler, and R. M. Potemski, Appl. Phys. Lett. **59**, 2269 (1991).
[12] A. Wójcik, T. J. Ochalski, J. Muszalski, E. Kowalczyk, K. Goszczyński, and M. Bugajski, Thin Solid Films **412**, 114 (2002).
[13] H. A. MacLeod, Thin-Film Optical Filters (Adam Hilger, Bristol, 1986).
[14] H. Davies, Proc. Inst. Electr. Eng., Part 3, **101**, 209–214 (1954).

Chapter 3
Early Stages of Continuous Wave Experiments on Cavity-Polaritons

Romuald Houdré

3.1
Introduction (1992)

The aim of this paper is to highlight the historical context of the early stages of the continuous wave experiments performed on cavity-polariton with M. Ilegems during the period 1992–2000. The other key players of this work were C. Weisbuch, R. P. Stanley and U. Oesterle.

The first encounter with what was initially called exciton vacuum field Rabi splitting and the group of M. Ilegems occurred some day in early 1992: C. Weisbuch, with whom we had very close collaboration and who used to spend week (and weekend) long visits in our group, convinced us to work on what he familiarly used to call his "Japanese effect" he had recently observed [1] during a sabbatical stay in Japan. At this time a one to one comparison with the atomic physics case was the best picture we could imagine of the phenomenon and we could somewhat naively describe it the following way: Two regimes for photon–matter interaction can occur when the number of photon modes is reduced by a microcavity, those of weak and strong coupling. In the first one the spectral width of the emission can be much narrower than in free space and the spontaneous emission rate can be altered [2, 3], either enhanced or inhibited. The change in spontaneous emission rate had led to much interest in microcavity structures [4, 5]. The strong coupling regime corresponds to an active medium with atomic-like transitions inside a microcavity so strongly coupled to the photon modes that the position of its energy levels is altered by this coupling. In such a regime, the normal perturbative approach of Fermi's golden rule breaks down [6].

The cavity quantum electrodynamics (CQED) treatment of a two level atomic system resonantly coupled to a single photon mode predicts that the eigenstates of the system are no longer the photon and atomic oscillator states

$$|e\rangle_{at} |0\rangle_{phot} \text{ and } |g\rangle_{at} |1\rangle_{phot} \tag{1}$$

(where e and g denotes the excited and ground states of the atomic oscillator and 0 and 1 the optical cavity mode with 0 and 1 photon) but two mixed symmetric and antisymmetric states:

$$|+\rangle = C_{11} |e\rangle_{at} |0\rangle_{phot} + C_{12} |g\rangle_{at} |1\rangle_{phot} \qquad (2)$$

$$|-\rangle = C_{22} |e\rangle_{at} |0\rangle_{phot} - C_{21} |g\rangle_{at} |1\rangle_{phot} \qquad (3)$$

The energy separation being

$$\Delta_n = \hbar\Omega = g\sqrt{1+n} \qquad (4)$$

where g is a coupling factor that only depends on the dipole matrix element and the cavity volume, and n is the number of photons in the cavity [7]. For zero photons ($n = 0$), a splitting still occurs, which can be regarded as coupling between the atomic oscillator and the vacuum field of the cavity (i.e. in the absence of a driving field). This effect was first called vacuum field Rabi splitting by J. J. Sanchez-Mondragon et al. [7] as it is related to the textbook case of intense field Rabi splitting [8, 9] which, in the present case, is induced by the zero point field fluctuations in the cavity. If several atomic oscillators are present it can be shown [10] that the coupling constant increases as the square root of the number N of atoms:

$$g_{n=0}(N) = g_{N=1}\sqrt{N} \qquad (5)$$

Provided the system is prepared in a pure atomic oscillator or photon oscillator state, it will oscillate between these two states at the Rabi frequency Ω. In a classical description, the overall system exhibits an anticrossing behaviour when both oscillators are resonant, with the two split modes corresponding to the normal modes of the system. In an atomic transition language, one considers the system as undergoing a coherent evolution with a photon being absorbed by an atom, which subsequently emits a photon with the same energy and wave vector k, the photon being reabsorbed, and so on.

The conditions for observation of Rabi splitting obtained from a linear dispersion model or CQED are the following. The Rabi frequency must be larger than the damping rate of both oscillators, i.e. $\Omega > \gamma_{cavity}$ and $\Omega > \gamma_{atom}$. The matching condition of both oscillator linewidths implies that, provided that the conditions for vacuum field Rabi splitting are fulfilled, improving one oscillator only is detrimental. In other words, if the cavity damping rate is too long ("too high" Q) the system is badly coupled to the outside world and the coupled particle decays via nonradiative processes before being emitted out of the cavity. The maximum splitting Ω_{max} is reached when both oscillator linewidths are matched. Ω_{max} is a function of the electric dipole matrix element of the atomic transition (d) (or the oscillator strength (f) in a classical model), N the number of atomic oscillators and the cavity size is V_{cavity} (i.e. the length for a planar cavity or the volume for a 3D cavity):

$$\Omega_{max} \propto \sqrt{\frac{f}{V_{cavity}}} \propto d\sqrt{\frac{N}{V_{cavity}}} \qquad (6)$$

A relevant physical parameter that can be used as experimental evidence of vacuum field Rabi splitting is absorption (i.e. the dielectric susceptibility) [10], neither reflectivity nor transmission are unambiguous parameters [11]. The existence of two split levels in the cavity + atom system is evidenced by transitions that can occur at these energies. Existence of a structure in reflectivity does not make the distinction between absorption and a change in the reflectivity/transmission balance (see below). This important point is, even now, often forgotten and led to repetitive erroneous claim by newcomers in the field.

Vacuum field Rabi splitting had been a domain of interest in atomic physics for many years. In order to observe a similar effect in solids we need the equivalent of both atomic and photon oscillators in a monolithic semiconductor structure. Semiconductor microcavity structures have very simple 1D implementations: planar Fabry–Perot microcavities where the mirrors are multiple quarter-wave stacks distributed Bragg reflectors (DBR's) are used for the optical resonator, whilst for the the atomic transition counterpart, excitons are the obvious choice. Due to the electron–hole interaction the oscillator strength of the electron–hole continuum in the volume of the exciton is concentrated at the exciton energy, making it equivalent to a two level atomic system [12]. Quantum well excitons have the advantages of an increased oscillator strength and an increased binding energy compared to bulk ones and it is also possible to locate them at the exact optical field antinode positions inside the cavity.

With this simple but very fruitful comparison we had in mind all the very exciting concepts developed in quantum optics [6] and we did not worry much about an old paper in atomic physics by Y. F. Zhu et al. [13], arguing that even in atomic physics, this so-called vacuum field Rabi splitting did not require any quantum field theory or even quantum mechanics to be explained and that, probably quantum effects, if any, were to be searched in more subtle properties and experiments.

First practical application that was sought was an exciton emitter at room temperature oriented toward the once fashionable threshold-less laser [4, 5, 14, 15]. This

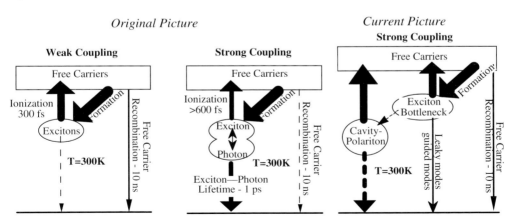

Fig. 3.1 Thermalization and bottleneck effect in the photoluminescence of cavity-polariton, original and current picture. Reprinted from [51].

idea was indeed described in one of the workpart of the Esprit European program named SMILES. The very simple original thoughts were as follows: It was hoped that strong coupling would lead to a new fast and efficient recombination channel in semiconductors at room temperature: whereas in weak coupling excitons are readily ionized after formation before recombining, the strongly coupled exciton can decay radiatively very quickly before re-ionizing with a lifetime of the order of twice the photon lifetime in the empty cavity (Fig. 3.1). As usual things did not turn out as expected and led to the current picture depicted in Fig. 3.1 that will be discussed in Section 3.5.

The first conference or summerschool with long discussion session on Rabi splitting in semiconductors occurred at the Erice summerschool in Sicily entitled "Confined electrons and photons" in 1993.

3.2
First Liquid Nitrogen and Room Temperature Observation (1993)

The first objective to achieve, if any practical application was ever going to exist, was to obtain strong coupling at liquid nitrogen temperature first and then at room temperature. Thanks to the expertise on growth and fabrication of VCSEL acquired in the past, this was an easy task. The design of the first samples was very close to the design of an optically pumped VCSEL.

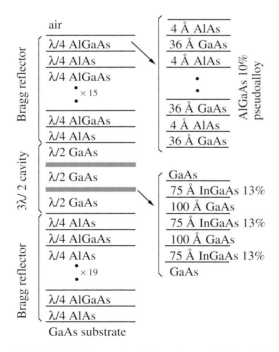

Fig. 3.2 Sketch of a semiconductor microcavity. After [17].

A sketch of a semiconductor microcavity is shown in Fig. 3.2. The low (high) index material is usually AlAs (GaAs) where the refractive index is $n = 2.96$ (3.54). For optical pumping it can be useful to use $Al_xGa_{1-x}As$ ($x \approx 10\%$) instead of GaAs. The alloy is sometimes grown as a pseudo alloy for molecular beam epitaxy technical reasons. The quantum wells are $In_{0.13}Ga_{0.87}As$ and 75 Å thick, allowing experiments in reflection and transmission. In the design of the quantum well, a compromise between the limitations induced by the mismatch strain, the increasing oscillator strength and inhomogeneous linewidth broadening when decreasing the thickness has to be made. In optimizing the structures to achieve a large splitting a compromise also needs to be made because, although Ω increases with $f^{1/2}$, it decreases with the cavity length. In practice all the quantum wells cannot be positioned where the field intensity is maximum. A structure factor arises from the phase difference between the different wells and thus Ω grows more slowly than the square root of the number of quantum wells. The cavity of the structures that have been investigated is $3\lambda/2$ long (i.e. two usable anti node positions) with 3 quantum wells at each anti node. The splitting can then be studied as a function of the detuning, $\delta = E_{FP} - E_X$, between the Fabry–Perot mode and the exciton energy by making use of the fine sample inhomogeneity across the wafer.

Figure 3.3 shows the cavity-exciton absorption spectrum, at 77 K, of the structure described in Fig. 3.2 when both the Fabry–Perot mode and the quantum well exciton are at resonance. The absorption (A) spectrum was deduced from reflectivity (R) and transmittivity (T) measurements. Moreover the structure was deliberately unbalanced (i.e. the reflectivity of the back mirror is greater than the top one) so that at resonance $T \ll R$, then $A = 1 - R - T \approx 1 - R$. The splitting is 8.8 meV. A linear dispersion model was used [13]: In atomic physics it consists of modelling the atomic system by a set of classical Lorentz oscillators and the cavity by the standard Airy description of a Fabry–Perot. The DBR-FP is modelled by a standard transfer matrix method and a 2D exciton is included with a Lorentz oscillator dispersive dielectric constant:

$$\varepsilon(e) = n(e)^2 = \varepsilon_\infty + \frac{f q^2 \hbar^2}{m \varepsilon_0 L_z} \frac{1}{e_0^2 - e^2 - i\gamma e} \tag{7}$$

where f is the oscillator strength per unit area, q (m) the charge (mass) of the electron and L_z the quantum well thickness, e_0 the exciton energy and γ the exciton linewidth. The continuous line in Fig. 3.3 is a fit, where the fitting parameters are the oscillator strength and the excitonic linewidth which is assumed to be the inhomogeneous linewidth. The inhomogeneous linewidth (2.7 meV) and the oscillator strength (4.8×10^{12} cm^{-2} per quantum well) are in good agreement with the literature [16]. It should be noticed that this fitting procedure allowed one of the most accurate and reliable measurements of the oscillator strength of a quantum well exciton. Figure 3.3 shows the absorption spectrum, measured under normal incidence, of the same nominal structure tuned so that the cavity and the quantum wells are resonant at room temperature. The splitting is now 4.6 meV. An excellent fit is obtained with the same oscillator strength that was measured at 77 K, by adding to the 77 K inhomogeneous linewidth the homogeneous linewidth given by LO

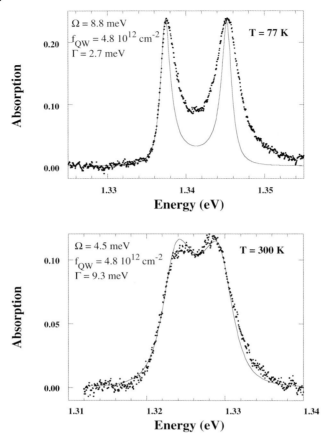

Fig. 3.3 77 K and 300 K absorption spectrum. The solid line is a linear dispersion model fit. The oscillator strength per quantum well is 4.8×10^{12} cm^{-2}. Reprinted from [17].

phonon scattering at 300 K ($\approx$6.5 meV). No broadening due to band to band transitions is observed in this case because of the smaller Rabi splitting.

3.3
Cavity-Polariton Dispersion Curve (1994)

The term cavity-polariton was used for the first time in 1993–1994 in [17] where some remarks implied the claim for the existence of an underlying cavity-polariton dispersion curve resulting from the anti-crossing behaviour between the exciton and Fabry–Perot cavity dispersion curves. It became apparent that we were measuring this dispersion curve when performing angle resolved photoluminescence, but the question why we were measuring it in this case whilst we knew this was never the case for bulk exciton-polariton or exciton in a quantum well, eluded us for a while, although explanation was just in a couple of drawing (thanks to discussions with D. Citrin and L. C. Andreani). However at this time, the field was quite small and sedate and angle resolved measurements remained in a drawer, not-understood for more than a year.

The physical picture is the following: a quantum well exciton state, with a well-defined transverse $\boldsymbol{k}$-momentum ($\boldsymbol{k}_{\parallel}$), is coupled to a single cavity-photon mode with the same in-plane wavevector due to momentum conservation enforced by the in-plane translational invariance of the electronic and photon systems. This selection mechanism explains why the single atomic picture of two coupled levels applies here although we are dealing with a pair of two-dimensional continuum systems. For a given k, a single photon mode and the single coupled exciton interact with an equivalent oscillator strength of about 10^{12} atoms [18]. While experiments previously reported dealt with the optical response of the coupled exciton-cavity photon mode as seen through reflectivity or transmission, it was important to evaluate how much the strong coupling regime modifies the luminescence process, both for fundamental reasons and for applications purposes. The situation now is quite different from atomic physics [19]: an atomic beam, having a monochromatic emission line, can only interact with a single mode of the cavity, while in a semiconductor micro-cavity photons and excitons have an in-plane dispersion. During the photoluminescence process, a number of states with different $\boldsymbol{k}_{\parallel}$ are populated. It is known that this makes the analysis of bulk polariton luminescence somewhat difficult.

Figure 3.4 shows photoluminescence spectrum for series of emission angles. Figure 3.5 is a plot of the positions of the photoluminescence peaks as a function of the in-plane photon wave vector for three different positions on the sample. We have shown that, due to the peculiarity of the emission process of the cavity-polariton, the position of the photoluminescence peaks allows the direct determination of the cavity-polariton dispersion curves [20].

Bulk polariton luminescence line shape is given by the escape rate of polaritons at the surface in the observation solid angle [21]:

$$I(E,\theta_{out})\,d\Omega_{out} = \sum_{i=lp,up} f_i(E,\theta_{in},z=0)\,\rho_i(E)\,v_{g\perp,i}(E)\,T_i(E,\theta_{in})\,d\Omega_{in} \tag{8}$$

where the sum is over the upper and lower branch of the polariton, $f(E, \theta, z)$ is the polariton energy distribution function at the surface, ρ is the polariton density of states, $v_{g\perp}$ is the component of the polariton group velocity perpendicular to the surface, T the transmission coefficient of a polariton incident at the surface into an outside photon, the propagation directions θ_{in} and θ_{out} and solid angles $d\Omega$ are related by Snell–Descartes law. Although the energy and $\boldsymbol{k}_{\parallel}$ selection rule are already included in θ_{in}, θ_{out} and $T(E, \theta)$ it is instructive to include them in the following form:

$$I(E,\theta_{out})\,d\Omega_{out} = \sum_{i=lp,up} f_i(E,\theta_{in},z=0)\,\rho_i(E)\,v_{g,i}(E)\,T_i(E,\theta_{in})\,\delta(E-E(\boldsymbol{k}))$$
$$\times \delta(\boldsymbol{k}_{\parallel}-(\boldsymbol{q}\cdot\boldsymbol{u}_{\parallel})\boldsymbol{u}_{\parallel})\,d\Omega_{in} = \sum_{i=lp,up} I_i(E(\boldsymbol{k}_{\perp}+\boldsymbol{q}\sin\theta_{out}))\,d\Omega_{in} \tag{9}$$

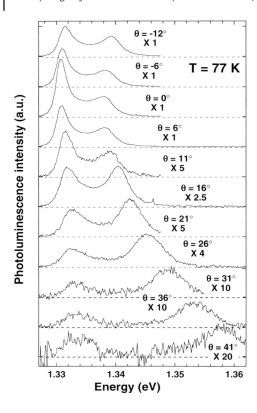

Fig. 3.4 Series of photoluminescence spectrum at $T = 77$ K, for emission angle from $-12°$ to $41°$. The Fabry–Perot at normal incidence is resonant with the quantum well exciton. Reprinted from [20].

where $\boldsymbol{q}$ is the photon wave vector and $\boldsymbol{u}_\parallel$ a unitary vector in the direction of $\boldsymbol{k}$. It follows that the line shape essentially depends on geometrical, static (ρ, $v_{g\perp}$, T, etc.) and dynamic, $f(E, \theta, z)$ factors. Therefore photoluminescence of the bulk polaritons *does not* give direct information on the polariton dispersion curve without an explicit calculation of the energy distribution function.

The cavity-polariton situation is quite different: as cavity-polaritons are a macroscopic property of the microcavity, the luminescence process does not involve any transport of excitation. Therefore, we can describe the luminescence process just by the knowledge of the distribution of the cavity-polariton along its dispersion curve, and by the outside transmission coefficient of such cavity-polaritons. Because of the absence of a perpendicular wave vector the cavity-polariton case is simpler. As both the exciton and the cavity mode energy only depends on $\boldsymbol{k}_\parallel$, the in-plane $\boldsymbol{k}$ selection rule reduces the photoluminescence spectrum to a sum of two delta functions:

$$I(E, \theta_{out}) \, d\Omega_{out} = \sum_{i=lp,up} f_i(E, \theta_{in}) \rho_i(E) T_i(E, \theta_{in}) \delta(E - E(\boldsymbol{k})) \delta(\boldsymbol{k}_\parallel - (\boldsymbol{q} \cdot \boldsymbol{u}_\parallel) \boldsymbol{u}_\parallel) \, d\Omega_{in}$$
$$= \sum_{i=lp,up} I_i(E, \theta_{in}) \delta(E - E(\boldsymbol{q} \sin \theta_{out})) \, d\Omega_{in} \qquad (10)$$

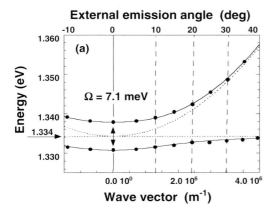
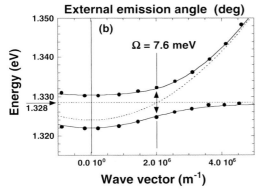
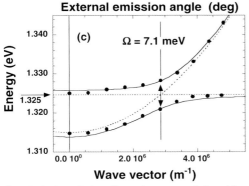

Fig. 3.5 Cavity-polariton dispersion curves, deduced from angle resolved photoluminescence measurements, for different resonance conditions: (a) resonance at $\theta = 0°$, (b) resonance at $\theta = 29°$ and (c) $\theta = 35°$. The continuous lines are theoretical calculations and the dashed lines are the uncoupled exciton and cavity dispersion curves. The interaction energy Ω and exact resonance position are determined from the minimum splitting between both photoluminescence lines. An external emission angle grid is drawn in (a). Reprinted from [20].

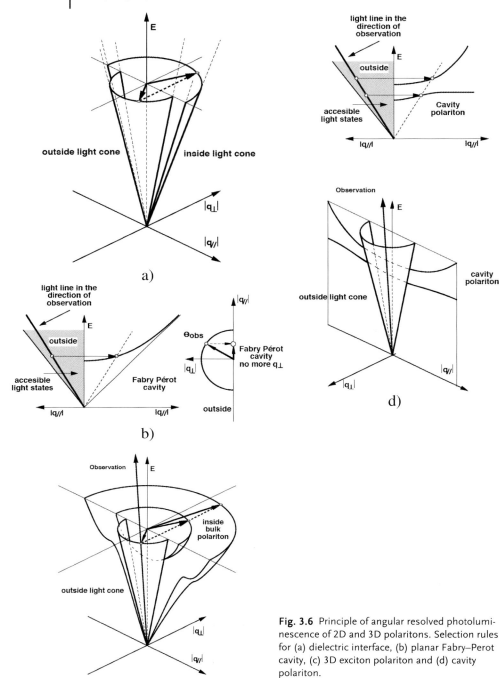

Fig. 3.6 Principle of angular resolved photoluminescence of 2D and 3D polaritons. Selection rules for (a) dielectric interface, (b) planar Fabry–Perot cavity, (c) 3D exciton polariton and (d) cavity polariton.

and therefore the emission spectrum for a fixed incidence angle exhibits two lines whose position are *directly* related to the cavity-polariton dispersion curve and the relative intensities of which are function of static and dynamical factors. In the presence of homogeneous or inhomogeneous broadening the delta functions are replaced by Lorentzian or Gaussian type lines but their energy position is not changed as long as the broadening is smaller than the energy separation of both lines. This allows us to interpret Fig. 3.5 as dispersion curves of the cavity-polariton. These arguments are summarized graphically in Fig. 3.6.

Finally, at this time a more accurate picture started to emerge and it was understood that the so-called vacuum field Rabi splitting effect was more complex in semiconductors than in atomic physics: optical transitions form a continuum and the strong coupling regime cannot be obtained with such excitations because the dephasing time of electron–hole pairs is much shorter than the Rabi frequency. Nevertheless atom-like excitation exist in solids on the form of excitons, which represent the Coulomb-correlated electron–hole pair which is the lowest excited electronic state of the crystal [18, 22]. They can be seen as an electron–hole pair entangled in an hydrogen-like relative motion, the whole pair having a translational motion throughout the entire crystal. This delocalised state leads to an oscillator strength proportional to the crystal volume (bulk) or surface (quantum well). As such the exciton has the same translational symmetry as the crystal, 3D for bulk material and 2D for quantum well and the wavevector remains a good quantum number. An exciton with a well defined in-plane momentum, $k_\parallel$, can only couple to the photon mode of the same momentum. This selection rule explains why, although these states form a continuum, the atomic physics picture remains valid. It should be noted that this strong coupling regime occurs in the case of a bulk material or a quantum well exciton in a planar cavity because both continua have the same dimensionality and therefore a one to one correspondence between each state is possible. The coupled eigenstates are then true eigenstates of the crystal + field system and have an infinite lifetime in the ideal case. Luminescence can only occur through an *extrinsic* scattering or dephasing mechanism (surface …). The case of a 2D exciton not in an optical cavity is fundamentally different because the excitonic state, being coupled to a $k_\perp$ continuum of photons, acquires an *intrinsic* radiative lifetime (Table 3.1).

Table 3.1 Exciton–photon interactions for different class of dimensionalities. (After [51])

Bulk	Exciton 3D	Strong coupling
	Photon 3D	Exciton-polariton
	k_{3D} selection rule	Extrinsic radiative process
Quantum well	Exciton 2D	Weak coupling
	Photon 3D	Intrinsic radiative process
	$k_\perp$ continuum	
Quantum well in planar cavity	Exciton 2D	Strong coupling
	Photon 2D	Cavity-polariton
	$k_\parallel$ selection rule	Extrinsic radiative process

3.4
Bleaching of the Oscillator Strength (1995)

It had been suggested that the strong coupling regime could have an important impact on optoelectronic devices [23]. This point became of particular interest since the strong coupling regime had been observed up to room temperature. New types of lasers and low noise light sources were proposed as well as electrooptic modulators. Lasers in the strong coupling regime were proposed as an approach to the so-called "thresholdless lasers" [14, 15, 24]. From its very definition, as the excitonic state in the strong coupling regime is only coupled to a single photon mode and emission or coupling into other photon modes can be neglected, we exactly are in the $\beta \approx 1$ condition [4]. Although this condition can easily be achieved under resonant optical pumping [25], where only the coupled exciton and photon states are created. The issue of the electrical or more generally nonresonant pumping had not yet been addressed. Optical nonresonant pumping was the most convenient way to investigate such effect. The result showed that for nonresonantly pumped structures the strong coupling regime is destroyed before achieving the proposed effects, at least for the quality of the sample available at this time. Not much later, it became more obvious to us that a "cavity-polariton laser" (if the concept had any meaning) could not be an ordinary laser: As the light–matter interaction is already taken into account in the description of the exciton–photon coupled states, the optical mode of the cavity cannot be exploited again for the realization of a laser consisting of an active gain medium and an optical cavity. Only light amplification could be envisioned with e.g. parametric amplification or cavity-polariton – cavity-polariton scattering (see Section 8).

In fact, these measurements turned out to be a very good tool to study saturation effects: the simple relation between the mode splitting and the exciton oscillator strength enables a direct and precise measure of the oscillator strength. Moreover, because the bandgap renormalisation exactly compensates the decrease of the exciton binding energy [26], the resonance condition is always fulfilled during the bleaching process. Only at much higher excitation intensity does the Fabry–Perot mode become degenerate with the electron–hole pair continuum, whose gap lies at or below the Fabry–Perot mode energy. This results in an important system where a continuous transition from a discrete strongly coupled state to a weakly coupled continuum can be observed (another situation where a transition from weak to strong coupling can be observed is under application of a magnetic field [27]).

Two spectra taken at low (a) and high (b) excitation power densities are shown in Fig. 3.7. The low intensity spectrum exhibits the doublet structure characteristic of the strong coupling regime. The excitation spot on the sample is selected to have exact resonance between the exciton quantum well and the Fabry–Perot mode at normal incidence. Both lines have different relative intensities because of different Boltzmann population factor. The high intensity spectrum exhibits only one single broad line characteristic of a weak coupling regime. Note the symmetric splitting of the strongly coupled lines (a) with respect to the uncoupled energy level as measured from the weakly coupled spectrum (b). This confirms the zero detuning be-

3.4 Bleaching of the Oscillator Strength

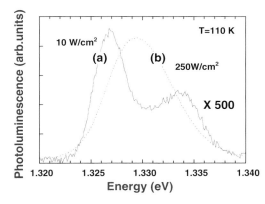

Fig. 3.7 110 K cavity-polariton photoluminescence spectrum under low (a) and high (b) excitation power density under nonresonant excitation. Reprinted from [30].

tween the exciton and Fabry–Perot. The transition from an anti-crossing to a crossing behaviour in the dispersion curves was also verified.

The oscillator strength as a function of the electron–hole density was extracted from the measurements, using a transfer-matrix formalism [11] to calculate absorption spectrum and assuming a Boltzmann distribution along the dispersion curve. The following procedure was used: linewidth of both uncoupled oscillators γ_X and γ_{FP} were measured from a photoluminescence spectrum taken far off resonance. As γ_{FP} is usually found to be larger than the theoretical empty cavity value, losses in the cavity are included to fit the measured value of γ_{FP}. The oscillator strength remains the last free parameter and is used to fit the splitting Ω. No significant broadening was experimentally observed up to the complete bleaching of the strong coupling regime. From lifetime measurements, $\tau \approx 1.5$ ns in the range of excitation power intensity used in this study. The calculated absorption of the pump light is 45% in the GaAs cavity. The plot of the oscillator strength vs. electron–hole density is shown Fig. 3.8, assuming that all the excited carriers are evenly shared between the six quantum wells in the cavity. This last assumption will tend to overestimate the electron–hole pair density, as the capture efficiency is probably less than 100% [28]. The measurements could be well fitted by the standard screening function [29]:

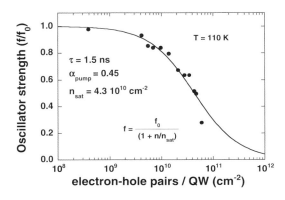

Fig. 3.8 Exciton oscillator strength as a function of electron–hole pairs density. The continuous line is a fit with a usual screening function $f(N_{e-h}) = f_0/(1 + N_{e-h}/N_{sat})$. Reprinted from [30].

$$f(N_{e-h}) = \frac{f_0}{1 + N_{e-h}/N_{sat}} \tag{11}$$

with $N_{sat} = 4.3 \times 10^{10}$ cm^{-2}. This corresponds to a remarkably low incident saturation intensity $I_{sat} = 100$ W cm^{-2} of the cavity-polariton, in agreement with theoretical estimates for an In$_y$Ga$_{1-y}$As/GaAs quantum well [30].

3.5
Continuous Wave Photoluminescence Experiments (1995–1996)

The bleaching results suggested that most of the interesting effects that had been proposed would only be observed under resonant excitation. Nevertheless a detailed investigation of the continuous wave photoluminescence was performed and its analysis provided the ground for novel experiments a couple of years later.

Although the simple atomic picture holds for the optical response, this is not the case for photoluminescence experiments because during the thermalization process a whole set of states with different $k_\parallel$ are involved. Simultaneously, theoretical studies of the thermalization dynamics were extensively studied [31, 32]. One of the major results is the existence of a relaxation bottleneck, as predicted by Tassone et al. [33, 34] and experimentally confirmed [35–37], see "Current Picture" in Fig. 3.1. This makes difficult to take advantages of the dramatic modification of the emission properties under nonresonant excitation [32].

3.5.1
Nonresonant Excitation

Photoluminescence spectra were measured at various positions/detuning on the sample, simultaneously from the front and the edge of the sample. Several features were observed [38] with state of the art samples of this period:

(1) The photoluminescence is thermalized, in that it is identical to the absorption multiplied by a Boltzmann filling factor. The thermal distribution of the photoluminescence (normal incidence, 4° angular aperture) exists for all detuning accessible on the sample and the temperature is independent of δ, this behaviour is still observed at temperature as low as 30 K (Fig. 3.9). This should not be mistaken with a thermal distribution due to a thermodynamic equilibrium, instead what occurs is that both states are populated via phonon scattering from the bottleneck reservoir with rates in a ratio equal to Boltzmann distribution [32].

(2) There is strong cavity pulling, that is the maximum photoluminescence intensity is observed away from resonance and furthermore, it is the photon-like line which has the most intense luminescence in the first sample generation (Fig. 3.10). This effect can be understood in terms of impedance matching of both electronic and photon oscillators [38]. Although this is important for microcavity light emitter devices, it does not have a strong physical relevance.

(3) The spectrally integrated edge photoluminescence shows no variation with detuning implying that the overall lifetime, τ, does not change with detuning. The

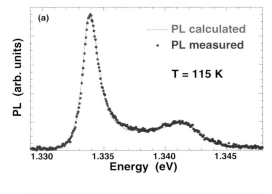

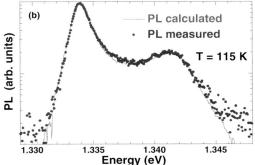

Fig. 3.9 A comparison between the measured photoluminescence (dots) and the spectra calculated by multiplying the absorption spectrum by a Boltzmann filling factor (solid line) using (a) linear, (b) logarithmic scales. Reprinted from [38].

spectrally integrated front surface photoluminescence, has a Lorentzian variation (Fig. 3.11). We showed that this, in conjunction with the thermalized emission and the constant lifetime, is a direct experimental evidence of strong coupling effects as opposed of the picture of uncoupled radiative states within a cavity, acting as a filter. This is an important result as most of the experimental evidence presented before (like time resolved experiments) could not formally make such a distinction (see note 57 of Ref. [38]). The physical argument is that the integrated photoluminescence must have a symmetric variation with respect to the true excitations of the system which are shown to be the cavity-polariton states and not the excitonic states. More precisely, the absorption A of the cavity + exciton object can be described as the product of the density of states times a coupling factor toward the external world, i.e. $A(e - E, \delta) = \rho(e - E, \delta) \cdot C(e - E, \delta)$, where E is the average energy $E = (E_X + E_{FP})/2$, and from symmetry arguments: $A(e - E, \delta) = A(-e - E, -\delta)$. Similarly the emission (which is the reverse process of absorption in the low intensity limit) includes a population factor: $E(e - E, \delta) = r(e - E, \delta) \cdot f_B(e) \cdot C(e - E, \delta)$, where $f_B(e)$ is the Boltzmann factor $\exp(-e/kT)$. The main difference originates from what we include in $\rho(e - E, \delta)$ and $C(e - E, \delta)$. In the filter approach $\rho(e - E, \delta)$ is still the uncoupled exciton density of states $\rho(e - E, \delta) = \rho_X(e - E_X)$ and 'Rabi splitting'-like effects are included in $C(e - E, \delta)$. In the cavity-polariton approach $\rho(e - E, \delta)$ is the density of states of the quantum coupled exciton and photon states,

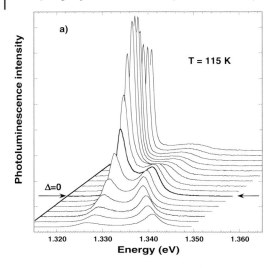

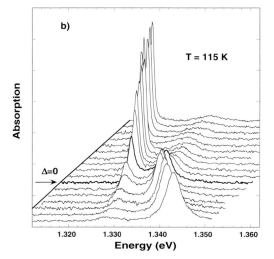

Fig. 3.10 (a) Photoluminescence spectra for a range of Fabry–Perot-exciton detuning, ranging from −10 meV to +12 meV. (b) Absorption spectra for a range of Fabry–Perot-exciton detuning, where the absorption is approximated by 1 − reflectivity, ranging from −10 meV to +12 meV. Reprinted from [38].

$\rho(e - E, \delta) = \rho_{CP}(e - E, \delta)$ and $C(e - E, \delta)$ holds the same symmetry property that $C(e - E, -\delta) = C(-e - E, \delta)$. It has been shown that $A(e - E, \delta)$ is independent of the cavity-polariton or exciton + filter approach [32]. These distinctions become important for the emission. When comparing $E(e, \delta)$ for different values of detuning spectrum must be normalized in such a way that the number of excitations N_0 is kept constant (because $\tau(\delta) = $ const):

$$E(e-E,\delta) = N_0 \cdot \rho(e-E,\delta) \frac{f_B(e)}{\int \rho(e-E,\delta) f_B(e)\,de} C(e-E,\delta) \qquad (12)$$

3.5 Continuous Wave Photoluminescence Experiments

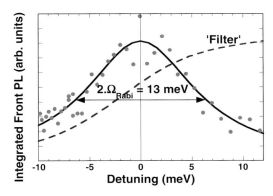

Fig. 3.11 Integrated front surface photoluminescence as a function of detuning (dots). The dashed line labeled 'filter' shows the asymmetry expected from the exciton in a Fabry–Perot filter. Adapted from [38].

As a consequence the integrated emission must be symmetric with respect to δ in the cavity-polariton approach:

$$\int E(e-E,\delta)\,de = N_0 \int C(e-E,\delta)\,de = N_0 \int C(-e-E,-\delta)\,de \tag{13}$$

$$= N_0 \int C(e-E,-\delta)\,de = \int E(e-E,-\delta)\,de \tag{14}$$

while it is not in the exciton + filter approach:

$$E(e-E,\delta) = N_0 \cdot \rho(e-E,\delta) \frac{f_B(e)}{\int \rho_X(e-E_X) f_B(e)\,de} C(e-E,\delta) \tag{15}$$

$$= N_0 \cdot \rho(e-E,\delta) \frac{f_B(e-E_X)}{\int \rho_X(e) f_B(e)\,de} C(e-E,\delta) \tag{16}$$

$$= \text{const} \cdot A(e-E,\delta) f_B(e-E_X) \tag{17}$$

and

$$\int E(e-E,\delta)\,de = \text{const} \cdot \int A(e-E,\delta) f_B(e-E_X)\,de \tag{18}$$

$$= \text{const} \cdot \int A(e-E,\delta) f_B(e-\delta-E)\,de \tag{19}$$

while

$$\int E(e-E,-\delta)\,de = \text{const} \cdot \int A(e-E,-\delta) f_B(e+\delta-E)\,de \tag{20}$$

$$= \text{const} \cdot \int A(-e-E,\delta) f_B(e+\delta-E)\,de \tag{21}$$

$$= \text{const} \cdot \int A(e-E,\delta) f_B(-(e-\delta)-E)\,de > \int E(e-E,\delta)\,de \tag{22}$$

Figure 3.11 is the comparison of the two models with the experimental results, it clearly shows that the asymmetry expected from the exciton + filter model is not observed.

3.5.2
Resonant Excitation

Later on, in continuous wave photoluminescence performed under resonant excitation, we have shown the role of acoustic phonons in the cavity-polariton relaxation [39]. The form of the spectra is almost independent of excitation wavelength at 30 K while the spectra change markedly at 5 K. This difference was attributed to the transition from multiple to single acoustic phonon scattering. An elementary model agrees [39] with the data using only the temperature as a free parameter (Fig. 3.12).

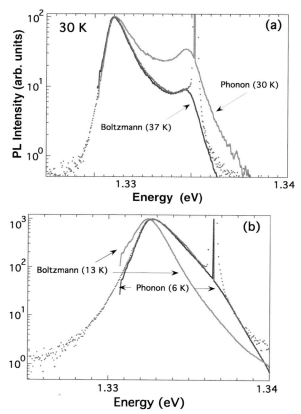

Fig. 3.12 (a) Photoluminescence from a semiconductor microcavity at 30 K. Dots give the experimentally measured photoluminescence, while the solid lines give a fit to the data using the model in [39]. The curve denoted "phonon" uses Eq. (7) while "Boltzmann" uses Eq. (9) of [39]. (b) Same as for (a) except that the temperature is 5.2 K. Reprinted from [39].

The data at 30 K show qualitatively the same trends as that of quantum wells at 4 K, while the 5 K data show that the radiative lifetime has become faster than the acoustic phonon scattering time. These results imply that the acoustic phonon scattering rate is faster than previously estimated for homogeneous cavity polaritons, and gives evidence of the important role that acoustic phonons play in polariton thermalization at low temperatures.

3.6
Linewidth, Disorder Effects and Linear Dispersion Modelling (1995–1997)

The summer of 1995 was also marked by a summerschool in Cargèse (Corsica) entitled "Microcavities and photonic bandgaps: physics and applications". Several new topics in the field of cavity-polariton emerged during the hotly debated discussion sessions, the first one dealt with the strange exciton-polariton ladder of Y. Yamamoto [40] and a preliminary disorder effect calculation from D. Whittaker [41]. The second question asked whether there was anything more than classical solid state and electromagnetism with cavity-polariton also referred to as normal mode splitting, as pioneered by H. Gibbs and S. W. Koch whilst the third question, linked to the previous one, concerned the so-called "Boser" effect form Y. Yamamoto [42] and the potential ability for cavity-polariton to behave as boson and to give rise to Bose–Einstein condensation as proposed in Cargèse by A. Imamoglu. The last topic is fully detailed in a special article of the present issue and deals mainly with non-continuous wave experiments. We concentrated on the first two issues.

3.6.1
More on Linear Dispersion Modelling

As mentioned above the inclusion of a Lorentz oscillator with its associated dispersive dielectric constant (Eq. (7)) within a transfer matrix method reproduces remarkably well the linear optical response of cavity-polariton. Here are a couple of examples:

(1) As pointed out by Savona [43] the necessary condition to observe an anticrossing behaviour differs for the transmission, reflectivity, absorption spectrum and eigenvalues (Eqs. (17), (18), (20) of [43]). For the eigenvalues, the interaction energy must be larger than the difference of the linewidths of both oscillators $(\Omega^2 > (\gamma_{FP} - \gamma_X)^2)$ while this is the sum in the case of absorption $(\Omega^2 > (\gamma_{FP}^2 + \gamma_X^2))$, it is a more complex expression for transmission, see Fig. 3.13. This effect is also well known in atomic physics where J. J. Childs qualifies this intermediate regime as non-perturbative, with no ringing regime [19]. Obviously, only the last condition is meaningful as it is related to the actual existence of two distinct eigenstates with two distinguishable eigenvalues. The observation of a splitting in transmission alone does not allow to distinguish from a genuine normal mode splitting to an absorptive dip created in a broad Fabry–Perot mode. These considerations are perfectly illustrated in a linear dispersion model, as it can be seen in Fig. 3.14. Although this is not an issue anymore for the present generation of samples in GaAs based

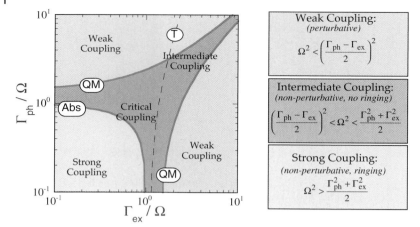

Fig. 3.13 Map of the weak, strong and intermediate coupling as a function of the linewidths of both oscillators. "Abs" denotes the condition to observe a splitting in absorption, "QM" denotes the limit of the perturbative regime. The dashed line denotes the condition to observe a meaningless splitting in the transmission spectrum alone.

materials, this is an important question for samples in new material systems such as nitride.

(2) Any shape of dispersive dielectric constant can be considered, provided care is taken to manipulate Kramers–Krönig conjugated real and imaginary dielectric constants. This can be applied e.g. to excited excitonic transition (2s etc...), collisional broadening, oscillator strength bleaching, modelling of bistabilty effects [44], inhomogeneously broadened system as it will be discussed below and inclusion of band to band transition. Regarding this last point, it should noted that the Kramers–Krönig analytical conjugated real part, n, of a step-like absorption, k, is written as [45, 46]:

$$k(e) = \frac{k_0}{e} \Pi(e - E_0) \quad \text{and} \quad n(e) = 1 + \frac{k_0}{\pi e} \ln\left(\frac{e + E_0}{|e - E_0|}\right) \tag{23}$$

where E_0 is the band edge energy and k_0/E_0 is the step amplitude corresponding to an absorption length $\alpha = 2k_0/(\hbar c)$ which can be rewritten in a more compact form (with the appropriate argument for the complex ln function):

$$n + ik = 1 + \frac{k_0}{\pi e} \ln\left(\frac{e + E_0}{e - E_0}\right) \tag{24}$$

This form is convenient for the introduction of a phenomenological broadening, γ, in using an imaginary component of the band edge energy: $E_0^* = E_0 - i\gamma$, allowing the modelling of the optical response of a crossing/anticrossing behaviour at the continuum band edge.

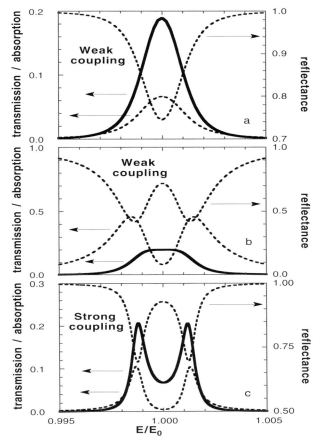

Fig. 3.14 Absorption (thick line), reflectivity and transmission (dashed line) of a Lorentz oscillator in a Fabry–Perot distributed Bragg mirror microcavity. Top: weak coupling regime, center: weak coupling regime with a non-relevant splitting in reflectivity and transmission, bottom: strong coupling regime that exhibits a vacuum field Rabi splitting. Reprinted from [11].

3.6.2
Disorder Effects and Inhomogeneous Broadening (1995)

Disorder effects in quantum wells lead to inhomogeneous broadening in the 1 meV range. In good cavity-polaritons samples disorder effects can be as low as 50 μeV, due to a disorder averaging over a much larger region of the order of the cavity mode size (10 μm) instead of the exciton mode size in quantum well (100 Å). A theoretical description of the effects (initially refereed as motional narrowing [41]) was first given in one-dimension by Savona et al. [47] and then in 2D by Whittaker [48]. The original question that was issued by the exciton-polariton ladder

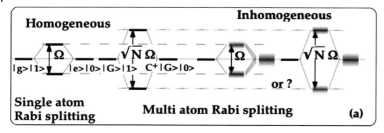
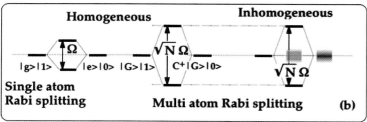

Fig. 3.15 Energy diagrams of the coupled photon–electronic states of the multi-atom Rabi splitting, (a) in the case of homogeneous broadening and possible effects on an inhomogeneous broadening as proposed by several authors. (b) Effect of inhomogeneous broadening from both quantum electrodynamics and classical model. Reprinted from [49].

of Y. Yamamoto [40] was as follows: let Ω be the one atom Rabi splitting, is the Rabi splitting for an inhomogeneous set of n electronic oscillators equal to the collective Rabi splitting of n atoms, inhomogeneously broadened or equal to the incoherent superposition of the one-oscillator Rabi splitting (Fig. 3.15). This last option was proposed to explain the very small Rabi splitting (0.3 meV instead of 3 meV) observed in some semiconductor microcavities [40]. We have shown that none of these hypotheses is entirely correct, the inhomogeneous broadening has no effect on the peak separation of the splitting, and, in general, does not lead to an inhomogeneous broadening of the split states (Fig. 3.15) [49].

The simple physical picture of linear dispersion theory is that in order to form a Fabry–Perot resonance the cavity round-trip phase shift has to be an integer multiple of 2π. Due to the form of the real part of the Lorentz oscillator refractive index, the round-trip phase shift vs. photon energy becomes N-shaped, up to a point where the phase shift conditions are fulfilled three times. This gives rise to the doublet structure because the central solution, which also corresponds to a maximum of absorption, does not create a Fabry–Perot resonance. Extension of this method to a set of non identical oscillator, corresponding to a pure spectral disorder (i.e. no in-plane spatial disorder but a set of oscillator of different energy) is performed in replacing ε by

$$\varepsilon(\nu) = n(\nu)^2 = \varepsilon_\infty + \sum_i \varepsilon_i(\nu) \quad \text{or} \quad \int_{-\infty}^{+\infty} \varepsilon(\nu, \nu_0) g(\nu_0) d\nu_0 \tag{25}$$

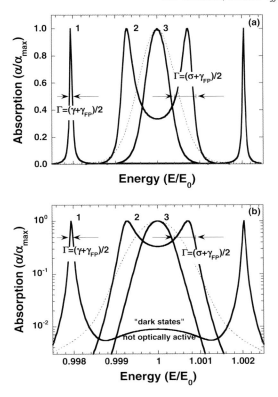

Fig. 3.16 Absorption spectrum of vacuum field Rabi splitting (Ω) for an inhomogeneously broadened system. Dashed line: absorption spectrum of the uncoupled electronic oscillator (plasma dispersion function), $\sigma = \gamma_{inhomog}$, $\gamma = \gamma_{homog}$; (1) strong interaction energy ($\Omega \gg \sigma$), (2) moderate interaction energy ($\Omega = \sigma$), (3) small interaction energy: weak coupling regime; (a) linear and (b) logarithmic scales. Parameters are: $\gamma_c/E_0 = 3 \times 10^{-5}$, $\gamma/E_0 = 1 \times 10^{-4}$, $\sigma/E_0 = 1 \times 10^{-3}$, the relative coupling strengths are 50 (1), 5 (2) and 1 (3). Reprinted from [49].

where $g(\nu_0)$ is the spectral density of oscillators at the frequency ν_0. Simulations are shown in Fig. 3.16. The absorption spectrum exhibits two main lines and residual structures of lower optical activity at the resonance energy. The existence and the peak separation of the splitting in absorption is independent of the homogeneous or inhomogeneous nature of the electronic oscillator. This can easily be understood considering that no specific distinction between inhomogeneous and homogeneous lines is made in this linear dispersion model. The refractive index, n, is only a function of the integrated absorption via the Kramers–Krönig transformation which is linear. Figure 3.16 illustrates that for large splittings the linewidth is given by the homogeneous linewidth. As can be seen when increasing the interaction energy the linewidth $\Delta_\pm$ of the Rabi split lines decreases from $(\sigma + \gamma_{photon})/2$ to $(\gamma + \gamma_{photon})/2$. Here σ refers to an inhomogeneous linewidth and γ to an homogeneous linewidth. This can be understood from a property of the plasma dispersion function (i.e. the convolution of a Gaussian and a Lorentz function) stating that the central energy region has a Gaussian shape while out in the wings ($\nu - \nu_0 \gg \sigma$) the function has a Lorentzian shape (see Fig. 3.16, log. scale). As the linewidth is determined by the slope of the round-trip phase shift vs. energy function, this explains the linewidth reduction. Such models applied with a more realistic (or measured) asymmetric lineshape exciton line have shown to be in perfect agreement with experimental observation and up to now it has not been possible to show quantum effects in the linear response of cavity-polaritons, even if quantum models do sometimes provide pictures that are easier to handle.

Very similar conclusions can be drawn from a simple quantum model with spectral disorder and it can be shown [49] that to the first order in perturbation the $n + 1$ quasi degenerate states (n electronic oscillators and 1 photon mode) gives $n - 1$ uncoupled states that are not optically active (these states often referred as dark states play however an essential role in the relaxation process). At the resonant energy, only two states separated by the same Rabi splitting as in an homogeneous case (i.e. $n^{1/2}\Omega$) and eigenstates are a collective excitation of the whole set of electronic oscillators, analogous of a Dicke state in quantum optics:

$$|\pm, 0\rangle = \frac{1}{\sqrt{2}}\left(|g\ldots g\rangle|1\rangle \pm \frac{1}{\sqrt{n}}\sum_{i=1}^{n}|g\ldots e_i\ldots g\rangle|0\rangle\right)$$
$$= \frac{1}{\sqrt{2}}\left(|g\ldots g\rangle|1\rangle \pm \frac{1}{\sqrt{n}}C^+|g\ldots g\rangle|0\rangle\right) \quad (26)$$

3.6.3
The Second Generation of Samples (1996)

The controversy on the existence of any quantum effect in cavity-polariton augmented over the following years. It was becoming evident that for the present state of the art samples, nothing beyond linear dispersion theory was observed, at least for experiments dealing with the linear response of cavity-polaritons and that if a quantum mechanical description of cavity-polariton was often used, this was because it provided easier to handle conceptual pictures. Observation of any genuine quantum effects required better samples.

We already knew how to fabricate nearly perfect optical oscillators [50]. Figure 3.17 shows a reflectivity spectrum of an "empty" (i.e. no quantum well) cavity fabricated to quantify the performance of the cavity alone. The top and bottom mirrors have 20 and 27 pairs and the cavity is one wavelength long. There is a near perfect fit with the simulated spectrum in the 1.1–1.6 eV range (not shown). The full width at half maximum of the Fabry–Perot mode is 1.4 Å (0.2 meV) which leads to a finesse in excess of 3300. The residual broadening of the Fabry–Perot mode can be modelled by interface roughness and residual sample curvature and shows that the mean square thickness fluctuations is of the order of one monolayer, an intrinsic limit.

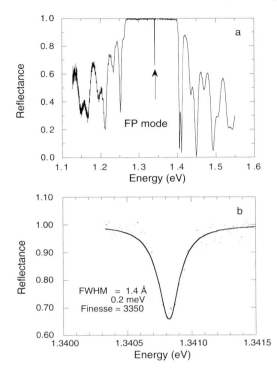

Fig. 3.17 Reflectivity spectrum of a high quality empty cavity. Adapted from [50].

Thus, the next step was how to improve the excitonic oscillator. First idea would be to use a GaAs/Ga$_x$Al$_{1-x}$As quantum well which would not suffer from alloy disorder broadening as it was the case in In$_y$Ga$_{1-y}$As quantum well, however this option complicated the growth of the whole sample because, in order to avoid light absorption, AlAs/Ga$_x$Al$_{1-x}$As had to be used with the drawback of a reduction of the index contrast of the Bragg mirror (and hence a larger number of periods) and last it obliterated the possibility to perform experiment in transmission. A better solution was proposed by the Tucson group (H. Gibbs and G. Khitrova) to use shallow In$_y$Ga$_{1-y}$As quantum well, with the lowest InAs content (as alloy scattering grows as $y(1-y)$) that allows design of quantum well whose energy level confinement and exciton binding energy are large enough compared to the cavity-polariton splitting energy. This optimum occurs for 80 Å wide quantum well with an InAs content around 4 to 5%. Low temperature photoluminescence linewidths as low as 500 µeV could be measured on such a single quantum well structure. Combination of these high quality cavities and narrow linewidth quantum well excitons gave rise to the second generation of samples with linewidths significantly below 1 meV and a cavity-polariton splitting to linewidth ratio exceeding 50.

This second generation of samples turned out to exhibit remarkable temperature dependence broadening with linewidth still extremely low at 100 K (Fig. 3.18). Once again such behaviour could easily be understood with a cavity-polariton picture,

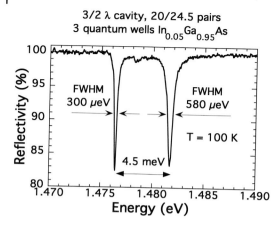

Fig. 3.18 Reflectivity spectrum of a high quality sample, note the 100 K temperature and the extremely narrow linewidths. Reprinted from [103].

however this was still not an evidence of deviation from classical linear dispersion theory, as will be discussed in the next section.

3.6.4
Inhibition of Acoustic Phonon Broadening (1996)

Because of the very peculiar shape of the cavity-polariton dispersion curve homogeneous broadening of the lower cavity-polariton state can be reduced to a value as low as 100 μeV. Photons at these energies do not interact strongly with acoustic phonons so it is the exciton part of the cavity-polariton which determines the scattering and have been calculated in the Born approximation taking into account the confinement of the quantum well exciton. The exciton broadening at $k = 0$ is given by the sum of the transition rates to all possible final states. Assuming the same electron and hole masses, one obtains for the broadening:

$$\Gamma_{ex} = \frac{\hbar |Q_{cv}|^2}{2\pi \rho u} \int_0^\infty \frac{E(k')^2}{(\hbar u)^2} \frac{1}{|\hbar u q_z|} |I_\parallel(k')|^2 |I_\perp(q_z)|^2 \frac{k'}{e^{\beta E(k')} - 1} dk' \qquad (27)$$

where ρ and u are the density and longitudinal sound velocity in GaAs respectively, Q_{cv} is the deformation potential and $E(k')$ is the exciton dispersion relation. Here q_z is given by energy conservation and is found from

$$\hbar u \sqrt{q_z^2 + k'^2} = E(k') \qquad (28)$$

The terms $I_\parallel(k')$ and $I_\perp(q_z^0)$ are the superposition integrals of the exciton envelope function with the phonon wave function in the in-plane and z-directions, respectively. Both superposition integrals introduce cut-offs in wavevector space. The important cut-off for (In,Ga)As is the one in z-direction and corresponds to $q_z \approx 3\pi/L_{qw}$. If we neglect the in-plane phonon dispersion, which is nearly flat compared to the exciton dispersion, we have $\hbar u q_z \approx E(k')$, and the cut-off in q_z becomes equivalent to a cut-off in the energy exchanged in the phonon absorption (or emission), and $|I_\perp(q_z)| \approx \theta(E_{cut} - E)$. The cut-off energy, E_{cut}, comes from the limited overlap of phonons propagating in the z-direction and the quantum well exciton when the phonon wavelength is much smaller than the quantum well width. For a quantum well of width L_{qw}, $E_{cut} = \hbar u \, 3\pi/L_{qw}$. The temperature dependence can be seen by taking the limit of $k_B T > E_{cut}$, then Γ_{ex} reduces to $\Gamma_{ex} \approx bT$, which gives the standard linear temperature dependence of the exciton linewidth. Substituting the standard values for the constants in GaAs and taking $E_{cut} = 3$ meV, then $b \approx 5$ μeV/T for the full width half maximum.

Lower polariton branch (LPB) scattering is shown schematically in Fig. 3.19 (right panel). Polaritons at $k = 0$ are mainly scattered to region B. Scattering to A is re-

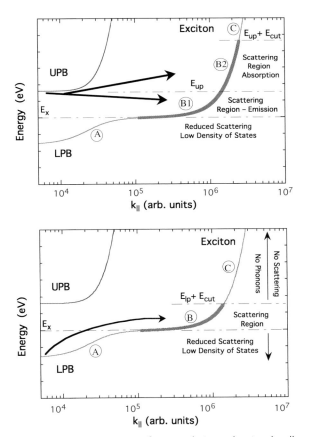

Fig. 3.19 Dispersion curves of cavity-polaritons showing the allowed phonon scattering channels for $k = 0$. Upper panel shows absorption and emission scattering of the upper polariton branch (UPB); lower panel shows scattering for the polariton branch (LPB). Note the log scale for $k_\parallel$. Reprinted from [51].

duced due to the low density of polariton states. There is no scattering to energies greater than E_{cut} (region C). For upper branch polaritons, there is the possibility of both absorption and emission of phonons (see Fig. 3.19, left panel) and no inhibition effects are expected. The cut-off energy directly translates into a detuning cut-off for which the energy separation between the LPB at $k = 0$ and the B region is larger than E_{cut}:

$$\Delta_{cut} = \frac{(\Omega/2)^2 - E_{cut}^2}{E_{cut}} \tag{29}$$

Δ is positive when the cavity is at a higher energy than the exciton. For $\Delta < \Delta_{cut}$ the LPB thermal broadening is negligible. As the Rabi energy increases, the detuning cut-off moves from negative to positive detuning. When $E_{cut} \approx \Omega/2$ the detuning cut-off occurs near resonance.

Figure 3.20 shows the linewidths of both branches as a function of cavity-exciton detuning at 10 and 60 K. At 4 K the linewidth of LPB remains almost constant with

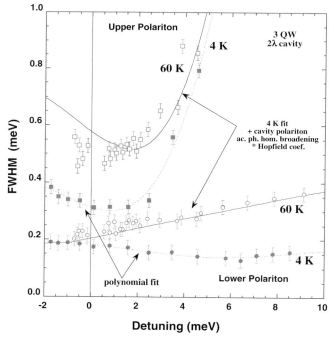

Fig. 3.20 Linewidth of upper (squares) and lower (circles) lines derived from reflectivity spectra. Solid symbols are at 4.2 K and open symbols are for 60 K. The dashed lines are a guide to the eye for the 4.2 K data. The solid lines are the sum of the temperature dependent acoustic phonon scattering broadening calculated at 60 K plus the 4.2 K linewidth. Reprinted from [51].

a slight minimum at 6 meV detuning. In contrast the UPB is larger and its linewidth increases dramatically at around 3 meV detuning due to continuum states which introduce additional absorption. On increasing the temperature, the upper polariton branch (UPB) broadens almost uniformly with detuning. The LPB shows a broadening which decreases towards negative detuning as expected [51].

3.6.5
Test of Linear Dispersion Theory (1997)

Still remained hotly debated the issue whether true quantum effects were observed, not only in non-linear effects ("Boser effect", see below), but also in simple experiments dealing with the linear response of cavity-polariton. The controversy was boosted by the lively participation of the Tucson group in the debate and recent studies of Whittaker et al. [41] and Savona et al. [47] showing that the excitonic response of semiconductor microcavities depends very sensitively on structural disorder [52, 53]. The very fascinating question in this context was whether light-coupling effects are indeed able to modify the influence of structural disorder on the excitonic quasiparticles within their quantum wells. In collaboration with the Tucson group [54], we performed an experimental test using extremely high-quality samples to study the interplay between light-coupling and disorder effects. Even though the concept is rather straightforward, the detailed analysis required substantial accuracy and very good sample characteristics. For this purpose we grew a series of quantum-well structures with and without Bragg mirrors. We then measured the optical response of all these structures such that we knew the energy dependent reflectivity of the bare quantum well in a high-quality microcavity. Clearly, since this is an experiment, all microscopic effects are included consistently. The optical response of all these structures was then measured so that we knew the energy dependent reflectivity of each microcavity containing one or more quantum wells. Next, we measured the optical transmission of our reference quantum wells and used these results to extract the disorder averaged excitonic response. Using this effective measured susceptibility in linear dispersion theory, we computed the microcavity spectrum and compared it with the experimental results of the complete microcavity system. If the linear dispersion theory results were able to reproduce the experimental spectra of the microcavity system, this would demonstrate that polaritonic effects in the exciton-disorder light coupling are negligible.

This was performed with great care on high quality samples form Tucson [54] and EPFL samples. Experimentally measured optical response of the bare quantum wells is shown Fig. 3.21 which allows, via Kramers–Krönig transformation to extract with a good accuracy the real and imaginary part of the susceptibility of the bare quantum well. Plugging such measured susceptibility spectrum into a simple transfer matrices formalism for calculation of the optical response of the cavity-polariton leads to spectrum in perfect agreement with the experimental values as shown in Fig. 3.22.

In studies of cavity-polariton linewidths, surprise had been expressed over two observations. The first surprise was that the upper branch linewidth γ_u is broader

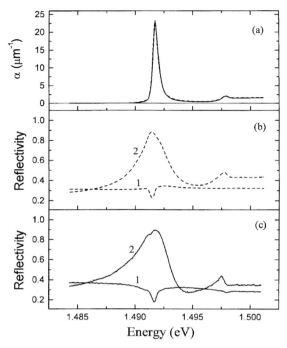

Fig. 3.21 (a) Experimental (solid line) and calculated (dashed line) absorption coefficient as a function of energy. The lines are hard to distinguish. (b) Calculated and (c) measured reflectivity at Bragg resonance (2) and at 0.85 $\lambda/2$ spacing (1) as a function of energy. Reprinted from [54].

than the lower branch linewidth γ_l. The second surprise was that on resonance γ_l is sometimes less than the mean of the bare quantum well exciton linewidth γ_{ex} and the empty cavity linewidth γ_c, which would be the result expected for a homogeneously, symmetrically broadened oscillator. Linear dispersion theory provided an explanation of both of these puzzles by considering the bare quantum well exciton line shape.

The linewidths of the two branches were found to be very sensitive to small local changes in the excitonic absorption coefficient and refractive index in the immediate vicinity of the cavity-polariton peaks, denoted by E_l and E_u. The absorption line, as can be seen in Fig. 3.22a, is affected by disorder, resulting in an asymmetric line shape with higher absorption on the high energy side. Consequently, equality of the reflection dips does not occur at the resonance condition $E_l = E_u$; instead $E_l > E_u$ is required, so that the exciton tail absorptions are equal at E_l and E_u. The linewidths γ_u and γ_l are determined mostly by the locally different slopes of the quantum well refractive index, resulting in $\gamma_u > \gamma_l$, explaining the first surprise. In contrast, at resonance, the smaller low energy tail absorption is the main cause of the smaller linewidth of the lower branch.

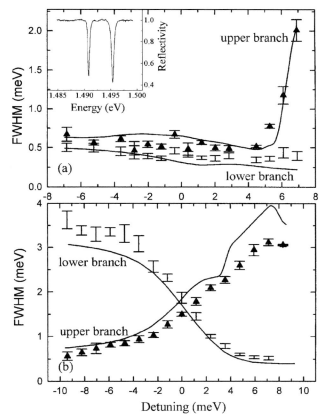

Fig. 3.22 Comparison of measured (symbols) and calculated normal mode coupling linewidths (solid line) for samples with $\gamma_x \approx \gamma_{FP}$ (a) and $\gamma_x \ll \gamma_{FP}$ (b) as a function of the detuning. The inset in (a) shows the reflectivity spectra as a function of energy for approximately equal dips. Reprinted from [54].

The case of $\gamma_{ex} \gg \gamma_c$ where γ_{ex} is strongly inhomogeneously broadened, is the most pronounced example of the second surprise. While in this experiment we have emphasized the importance of line shape asymmetries and spectrally local values of the susceptibility, we had shown (see above) that for symmetrically inhomogeneously broadened lines and an cavity-polariton splitting much larger than γ_{ex}, linewidths become $\gamma_l \approx (\gamma_{ex,hom} + \gamma_c)/2$. This, combined with asymmetric broadening is the basic physics of the second surprise.

This analysis showed clearly that for the currently available top-quality samples it is an excellent approximation to assume that the disorder-averaged excitonic response determines the optical properties. Seven years later this conclusion still holds and we are not aware of high quality samples whose linear optical response cannot be modelled by linear dispersion models. This certainly does not imply that quantum effects do not exist in cavity-polariton but that they have to be searched in more elaborated experiments.

3.7
Rayleigh Scattering (2000)

The second generation of samples also allowed a series of very innovative experiments in the field of elastic or Rayleigh scattering. Interference and coherence effects in elastic light scattering from disordered systems are important in many domains of physics. They have been revived by the analogies that can be made with theories and experiments developed for the propagation of electrons in disordered systems, such as the weak and strong localization regimes.

In such experiments, the light re-emitted by the system consists of: (1) a coherent, elastic part due to *static* disorder, called resonant Rayleigh scattering; (2) an incoherent, elastic part called resonant hot luminescence, involving inelastic scattering events such as phonon scattering; (3) partially or completely thermalized contributions called photoluminescence. These emissions yield information on the energy level scheme and nature of these levels, as well as on their dynamics. This is well demonstrated in 2D semiconductor quantum wells: early work on resonant Rayleigh scattering demonstrated exciton localization effects and the existence of a mobility edge [55] while more recent studies dealt with the dynamics of resonant Rayleigh scattering [56, 57] and the speckle spectroscopy of disorder [58]. Rayleigh

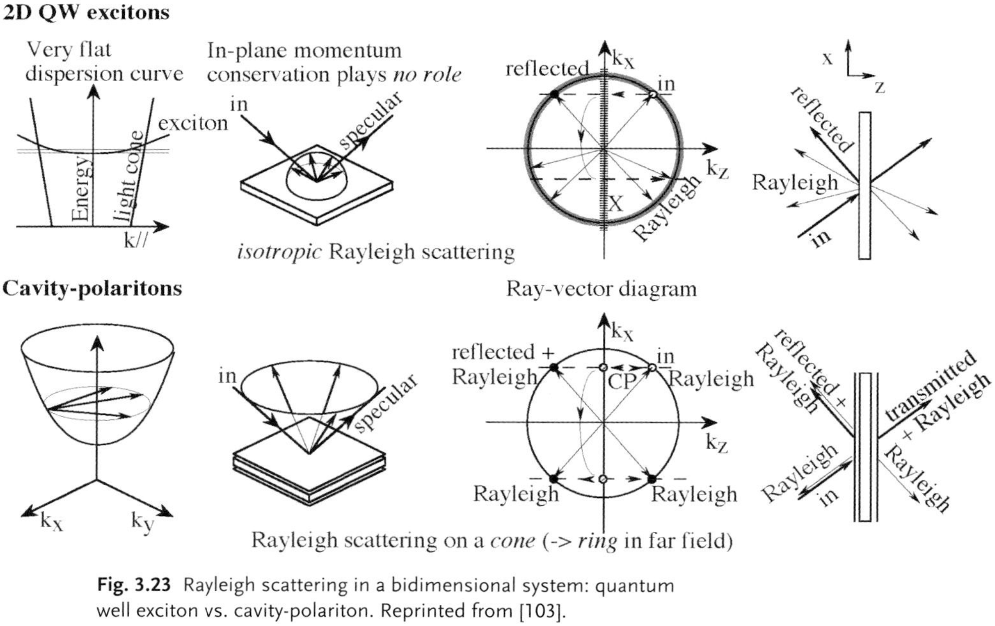

Fig. 3.23 Rayleigh scattering in a bidimensional system: quantum well exciton vs. cavity-polariton. Reprinted from [103].

scattering of cavity-polariton under resonant excitation exhibits remarkable features which impact on light scattering [59].

Let us first consider what happens for *non-resonant* 3D disordered systems in microcavities (the disorder is considered small enough for a cavity mode to develop). In a first approximation, the scattered far-field light is distributed on the Fabry–Perot cones. Now consider resonantly excited excitons in 2D quantum wells, one deals with a quasi-two dimensional electronic system, in which ideally elastically-scattered excitons lie on a constant energy ring in k-space. However, multiple scattering leads to a randomization of the scattered photon wavevector, with energy being conserved only between initial and final photon states (Fig. 3.23). This occurs because of the small exciton dispersion on the scale of the photon wavevector. Therefore the resonant Rayleigh scattering intensity is isotropic and does not reflect the bidimensionality of the excitons [55, 57, 60, 61]. Investigating resonant Rayleigh scattering of a quantum well in a microcavity in the weak-coupling condition, one mainly expects a cavity filtering of the scattered light and therefore a featureless ring of scattered light. In contrast a microcavity in the strong coupling regime is a near ideal 2D photonic system because of the significant curvature of the cavity-polariton dispersion curve (Fig. 3.23). It is therefore expected, and indeed observed, that scattering events occur on the quasi-elastic ring of resonantly excited states. The first preliminary experimental indication of the annular structure of the resonant Rayleigh scattering of cavity-polariton was reported by Freixanet et al. [62].

Figure 3.24 shows the far-field image of the microcavity emission and scattered light as a function of angle. The excitation laser spot is in resonance with the lower polariton branch at an incidence angle of 10°. In this image, a narrow ring of resonantly scattered light can be seen, showing some structure along the ring. The exciting beam covers a finite range of incidence angles larger than the cavity-polariton angular width. A neutral density filter has been used to reduce the intensity of the reflected spot in Fig. 3.24a. A dark arc in the reflected beam can be seen. It coincides with the ring of resonant Rayleigh scattering, and represents the polariton as it would be seen in reflectivity.

Careful investigations show that: (i) The scattered ring is only observed when exciting resonantly the lower polariton branch and for all accessible negative and positive detunings up to $\delta = +2.5$ meV. (ii) At negative detunings, it conserves the linear polarization of the incident photon. (iii) The intensity variations along the ring resemble speckle features. They are very sensitive to any variation in position, angle or wavelength of the laser. (iv) The emission is spectrally identical to the laser. (v) The intensity of the ring is strongly temperature dependent. At higher temperatures (20 K) the intensity vanishes and the ring becomes a featureless ring of *unpolarized* light. (vi) Due to the limited dynamic range of the camera, other effects are not visible in the figure. There is photoluminescence inside the circle which is not observable in the figure because of the weaker intensity. The photoluminescence is non-resonant with the laser, shows the same linewidth as the lower polariton branch, changes wavelength as a function of observation angle and follows the cavity-polariton dispersion curve.

According to the previous discussion on disorder effects in cavity-polariton, it may be surprising to observe a strong Rayleigh scattering corresponding to an in-

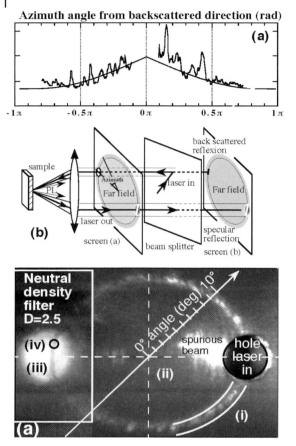

Fig. 3.24 (a) Far field image observed under resonant excitation of the lower cavity polariton branch. Note the elastic Rayleigh arc of circle, its speckle, the attenuated specularly reflected spot, with the absorption beam appearing as a dark arc against an undiminished, nonresonant, reflected beam. Upper part: Rayleigh intensity vs. azimuth scattering angle. (b) Far field emission pattern set-up, screen (a) is used in most measurements, screen (b) is used for backscattering measurements. Reprinted from [59].

plane momentum variation much larger of the reciprocal size of the polariton state. It can be shown that weak disorder first tends to break the vectorial in-plane momentum polariton dispersion, while still preserving the absolute value in-plane momentum polariton dispersion, i.e. while $k_{\parallel} = (k_x, k_y)$ is not a good quantum number anymore, $|k_{\parallel}|$ remains so. The polariton states are then complex coherent superposition, with cylindrical symmetry, of ideal polariton state of different azimuthal orientation. The Rayleigh scattering can then be understood as the excitation of a subset of the whole eigenstate which then re-emits over its whole extension in k-space.

A remarkable feature observed in Fig. 3.24a is that the reflected intensity is stronger in the backscattered direction showing that additional effects do occur. As seen in Fig. 3.24a the line shape shows an enhancement of a factor 2 in the backscattered direction with respect to the forward direction, exactly as is expected from coherent backscattering. However the angular linewidth of coherent backscattering typically reported is in the range of a few mrad (both for 3D and 2D systems). In our observation the angular linewidth is around 90°, leading to λ_{dB}/l^* value close to 0.2 where λ_{dB} is the de Broglie wavelength of the polariton and l^* is the elastic mean free path. It is significantly below the Ioffe-Regel limit $\lambda_{dB}/l^* = l^*$ for optical Anderson strong localization and is a regime never observed before these measurements, where lineshape modification of the coherent backscattering is expected. Quantitative analysis of the present coherent backscattering phenomena was clearly required before making such a strong claim and has not been achieved yet as it requires a new formulation of coherent backscattering in strongly coupled systems. Moreover experimental phenomenology is even more complex due to disorder anisotropy effects along [0 ±1 ±1] crystallographic orientation that are superimposed to coherent backscattering-like lineshape [59]. Finally it should be mentioned that coherent backscattering in 2D system applies only to the in-plane wavevector, i.e. in the $-k_{\parallel \text{exc}}$ propagation direction. This leads to *two* possible directions of observation, one in the usual backscattered direction in *reflection*, as is discussed here, and one in *transmission* which is now in the mirror direction of the incident beam, making it very interesting as this direction is *not* in the specular beam anymore (last sketch of Fig. 3.23). Such enhancement effect in transmission was also observed [63]. Unfortunately, only the surface of this topic has been explored and since then there has not been a lot of activity with the exception of a few but important contributions from W. Langbein et al. [58, 64, 65] and M. Gurioli et al. [66–68] and very recently [69].

3.8
Nonlinear Continuous Wave Effects (1999–2000)

The same high quality samples opened the way to a whole new class of experiment which are mostly non continuous wave experiments and do not belong anymore to the early stages of the cavity-polaritons history. They are detailed in other articles of this issue. Nevertheless some preliminary continuous wave investigations were performed [70] which led to a rich experimental phenomenology [71].

In the high density regime cavity-polaritons exhibit nonlinear interaction, for which remained the unanswered question whether the quasi-particles, being composite bosons, behave as fermions or bosons on the length scale where nonlinear effects manifest. This typically occurs for densities for which the average distance between particles is comparable to the thermal de Broglie wavelength. For quantum well excitons it is well known that on this scale (≈50 nm) excitons behave like fermions.

The large elastic mean-free path and Broglie wavelength implied by their dispersion had led many to search and claim quantum effects [31, 33, 72–77]. Some non-

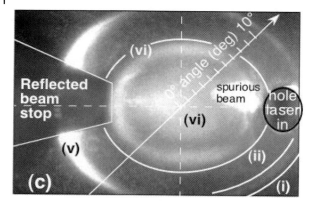

Fig. 3.25 Far field image in the nonlinear regime, above threshold: (i) elastic Rayleigh circle, (ii) photoluminescence from the lower cavity-polariton branch, (v) resonant photoluminescence, (vi) nonlinear structure. Reprinted from [70].

linear phenomena, in particular, the laser-like nonlinear emission observed under both non-resonant and resonant excitation had been the subject of considerable debate. The main mechanisms used to explain such nonlinear emission in this system have been the Boser effect [72], and more recently boson assisted stimulation [77], as well as Bose–Einstein condensation and bi-exciton effects. The Boser effect is the stimulated scattering of bosons into their ground state when the ground state *occupancy* is greater than unity. For cavity-polariton the scattering could be mediated by phonons (Boser) or by the cavity-polariton themselves. Part of the problem was that some experiments dwelt on the energy spectrum of the emission, neglecting the dispersion which is fundamental to the nature of cavity-polariton, other experiments neglect spatial effects.

The experimental set up and sample were identical to the Rayleigh scattering experiment. On increasing the excitation power, a sudden transition occurs and an intense emission occurs as either a disk, a ring or a more complex pattern (depending on conditions) (region (vi) in Fig. 3.25) inside the elastic Rayleigh ring, and a crescent shaped resonant photoluminescence becomes clearly visible (region (v)). Performing angular resolved photoluminescence measurements, three components are found in the emitted spectra, of varying amplitude for different emission angles: a Rayleigh component at the exciting energy and centered on the Rayleigh cone, a cavity-polariton luminescence line the energy of which shifts with angle according to the cavity-polariton dispersion curve, a strong nonlinear emission line which is *dispersionless*, at an energy slightly above the bottom of the cavity-polariton dispersion curve. This shift ruled out an interpretation of the new emission line as a Boser-type effect as it would be unshifted compared to the bottom of the cavity-polariton band.

Key phenomena can be observed under various experimental conditions and are summarized in the following way: (i) At threshold, depending on the detuning, either an angular point or an actual discontinuity can be observed in the "light in – light out" curves. (ii) Around −1 meV detuning, the threshold can reach values as low as 17 mW peak incident power, corresponding to an estimated absorbed photon density of 7×10^{19} cm^{-2} s^{-1}. It is difficult to translate this number into a steady-state population as both lifetime and excited volume depend on excitation intensity *via* changes in relative populations of cavity-polariton states with widely different lifetimes, resonance energy shift and spatial pattern formation. (iii) The nonlinear emission energy, angular spread, even emission pattern depend on the excitation conditions. For a given cavity detuning, the nonlinear emission *follows* the excitation variation, both in energy (i.e. nonlinear emission energy vs. pump energy) and angular extension (i.e. nonlinear emission angular spread vs. incident pump angle), although in a non-trivial manner. (iv) For pump power densities accessible in our experiment, this nonlinear behavior is only observed while exciting the lower polariton branch. (v) From a spatially-filtered image, one observes that the strong nonlinear emission originates from a region which is no greater than half of the excitation spot (measurement limited by our spatial resolution at the large numerical aperture of the experiment). (vi) No clear differences are found between linear and circular polarization excitation. This latter result rules out *a priori* nonlinear mechanisms based on biexcitonic effects as it can be assumed that under resonant excitation little spin depolarization occurs, and therefore circular polarization should inhibit biexciton formation as the total biexciton angular momentum must be 0.

By pumping a strongly-coupled microcavity intensely the system evolves toward the weak coupling regime [30]. Therefore it was important to check whether the cavity was still under strong coupling conditions when reaching threshold. While the persistence of the Rayleigh scattering of the polariton state and observation of a cavity-polariton photoluminescence line would imply that strong coupling was not bleached, we have shown that the experimental features are far more complex [70]. Under high excitation the polariton absorption observed in the reflected beam is bleached at the center of the excitation spot, while absorption continues to occur in the outer part of the spot. The observation of this inhomogeneous situation greatly helps to solve the puzzle of the many apparently incompatible observed facts, the main one being the coexistence of "well-behaved" cavity-polariton photoluminescence with a new emission line incompatible with a cavity-polariton description: they are due to the coexistence of regions in strong coupling (periphery of excited spot) and regions (center of excited spot) where the resonance energy is spatially shifted, by either carrier renormalization, bleaching or by homogeneous broadening.

Below threshold radiative cavity-polaritons states are directly excited through the resonant excitation. The energy is then redistributed in energy, direction, eventually space, through inelastic processes (relaxation) or elastic processes (Rayleigh scattering). Neither the thermalized photoluminescence nor the elastic ring change in intensity at threshold. Therefore there must be another channel invisible in our experiment into which the pump energy is scattered below threshold. Such channels could be leaky and guided modes that remain trapped inside the GaAs material.

Regarding the nonlinear mechanism, cavity-polariton / cavity-polariton scattering, which conserves both energy and momentum, can be viewed as parametric amplification (or non-degenerate four wave mixing) where two cavity-polaritons scatter simultaneously from $k = k'$ to $k = 0$ and $k = 2k'$. Such phenomena were later on demonstrated by Stevenson et al. [78] and Savvidis et al. [76], in pump–probe experiments, where the cavity-polaritons were excited resonantly at $k = k'$ and $k = 0$, with resulting emission at $k = 2k'$. Their results and ours triggered a novel theoretical treatment [79] based on a phase-matched four wave mixing or parametric amplification. In our experiments, the thermal photoluminescence acts as an idler (probe) which seeds the parametric process. After a certain pump intensity there is sufficient power to provide gain at $k = 0$. Many of the features observed in our experiments such as the blue shift of the nonlinear emission with respect to the cavity mode by a quantity of the order of the cavity-polariton homogeneous linewidth are also predicted in this model [79] as well as off branch polaritons multiple scattering [76]. In addition quantum noise measurement shows a phase dependence of the amplification which is typical of coherent multiwave mixing process [80].

3.9
Conclusion

After these initial studies on cavity-polariton until around the year 2000 most of the "easy" and straightforward effects were observed and more elaborated experiments and models started to emerge. They are discussed in the following chapters of the present special issue. The samples discussed in the last sections of the text still represent the state of the art of the present time. Nevertheless this short history of the early stages of the cavity-polaritons would not be complete without the mention of work in other material systems, such as II–VI [81, 82] where the temperature dependence differs sensibly due to a different scale of the exciton binding energy and linewidths, oscillator strength, optical phonon energy, bulk active layer [83, 84], organic materials [85–89], numerous attempts to achieve strong coupling with a collection or single quantum dots in different kind of optical cavities, just recently successful [90, 91], attempts to electrically inject carriers [92], tunability [93], interest for nitrides was mentioned early but up to now results compare at best with the first generation of samples [94–96] and other groups, namely Y. Yamamoto et al. in Stanford [4, 5, 15, 24, 25, 31, 33, 40, 42, 97], H. Gibbs and G. Khitrova et al. from Tucson [98], S. W. Koch et al. from Marburg [99, 100], M. Skolnick et al. from Sheffield [41, 76, 77, 93], R. Planel and now J. Bloch et al. from Paris [75, 101], A. Quattropani et al. from EPFL [32, 43, 47, 79], R. André, Le Si Dang and J. Bleuse et al. in University and CEA Grenoble [81], E. Giacobino et al. from Laboratoire Kastler-Brossel and Ecole Normale Supérieuere, Paris [80, 102].

Acknowledgements

This work was supported by EPFL, Thomson CSF, the Swiss National Priority Program for Optics, the SMILES EEC Esprit program, and would not have existed without numerous contributions and discussion with C. Weisbuch, R. P. Stanley, U. Oesterle, V. Savona, L. C. Andreani, A. Quattropani, C. Ciuti, E. Giaco-bino, D. S. Citrin, H. Gibbs, S. W. Koch, Y. Yamamoto, R. Planel, L. A. Dunbar and many others.

References

[1] C. Weisbuch, M. Nishioka, A. Ishikawa, and Y. Arakawa, Phys. Rev. Lett. **69**, 3314 (1992).
[2] J. P. Wittke, RCA Review **36**, 655 (1975).
[3] D. J. Heinzen, J. J. Childs, J. E. Thomas, and M. S. Feld, Phys. Rev. Lett. **58**, 1320 (1987).
[4] Y. Yamamoto, S. Machida, and G. Björk, Phys. Rev. A **44**, 657 (1991).
[5] G. Björk, S. Machida, Y. Yamamoto, and K. Igeta, Phys. Rev. A **44**, 669 (1991).
[6] S. Haroche, in: New trends in atomic physics, edited by G. Grynberg and R. Stora (North-Holland, Amsterdam, 1983), 193.
[7] J. J. Sanchez-Mondragon, N. B. Narozhny, and J. H. Eberly, Phys. Rev. Lett. **51**, 550 (1983).
[8] I. I. Rabi, Phys. Rev. **51**, 652 (1937).
[9] B. R. Mollow, Phys. Rev. **188**, 1969 (1969).
[10] G. S. Agarwal, Phys. Rev. Lett. **53**, 1732 (1984).
[11] R. Houdré, R. P. Stanley, U. Oesterle, M. Ilegems, and C. Weisbuch, J. Phys. IV (France) **3**, 51 (1993).
[12] J. J. Hopfield, Phys. Rev. **112**, 1555 (1958).
[13] Y. F. Zhu, D. J. Gauthier, S. E. Morin, Q. L. Wu, H. J. Carmichael, and T. W. Mossberg, Phys. Rev. Lett. **64**, 2499 (1990).
[14] H. Yokoyama and S. D. Brorson, J. Appl. Phys. **66**, 4801 (1989).
[15] G. Björk, A. Karlsson, and Y. Yamamoto, Phys. Rev. A **50**, 1675 (1994).
[16] L. C. Andreani and A. Pasquarello, Phys. Rev. B **42**, 8928 (1990).
[17] R. Houdré, R. P. Stanley, U. Oesterle, M. Ilegems, and C. Weisbuch, Phys. Rev. B **49**, 16761 (1994).
[18] J. J. Hopfield, in: Quantum Optics, edited by R. J. Glauber (Academic Press, New York, 1969), 340.
[19] J. J. Childs, K. An, R. R. Dasari, and M. S. Feld, in: Cavity quantum electrodynamics, edited by P. R. Berman (Academic Press, Boston, 1994), p. 123.
[20] R. Houdré, C. Weisbuch, R. P. Stanley, U. Oesterle, P. Pellandini, and M. Ilegems, Phys. Rev. Lett. **73**, 2043 (1994).
[21] A. Bonnot and C. Benoit à la Guillaume, in: Polaritons, edited by E. Burstein and F. D. Martini (Pergamon, New York, 1974), p. 197.
[22] R. S. Knox, Theory of excitons (Academic, New York, 1963).
[23] C. Weisbuch and E. Burstein (eds.), Confined electrons and photons (Plenum, Boston, 1995).
[24] Y. Yamamoto and G. Björk, Jpn. J. Appl. Phys. Pt.2-Lett. **30**, L2039 (1991).
[25] Y. Yamamoto, F. Matinaga, S. Machida, A. Karlsson, J. Jacobson, G. Björk, and T. Mukai, J. Phys. IV (France) **3**, 39 (1993).
[26] S. Schmitt-Rink and C. Ell, J. Lumin. **30**, 585 (1985).
[27] J. Tignon, P. Voisin, C. Delalande, M. Voos, R. Houdré, U. Oesterle, and R. P. Stanley, Phys. Rev. Lett. **74**, 3967 (1995).

[28] H. J. Polland, K. Leo, K. Rother, K. Ploog, J. Feldmann, G. Peter, E. O. Gobel, K. Fujiwara, T. Nakayama, and Y. Ohta, Phys. Rev. B **38**, 7635 (1988).

[29] D. A. B. Miller, D. S. Chemla, D. J. Eilenberger, P. W. Smith, A. C. Gossard, and W. T. Tsang, Appl. Phys. Lett. **41**, 679 (1982).

[30] R. Houdré, J. L. Gibernon, P. Pellandini, R. P. Stanley, U. Oesterle, C. Weisbuch, J. Ogorman, B. Roycroft, and M. Ilegems, Phys. Rev. B **52**, 7810 (1995).

[31] S. Pau, G. Björk, J. Jacobson, H. Cao, and Y. Yamamoto, Phys. Rev. B **51**, 7090 (1995).

[32] V. Savona, F. Tassone, C. Piermarocchi, A. Quattropani, and P. Schwendimann, Phys. Rev. B **53**, 13051 (1996).

[33] F. Tassone and Y. Yamamoto, Phys. Rev. B **59**, 10830 (1999).

[34] F. Tassone, C. Piermarocchi, V. Savona, A. Quattropani, and P. Schwendimann, Phys. Rev. B **56**, 7554 (1997).

[35] A. I. Tartakovskii, M. Emam-Ismail, R. M. Stevenson, M. S. Skolnick, V. N. Astratov, D. M. Whittaker, J. J. Baumberg, and J. S. Roberts, Phys. Rev. B **62**, R2283 (2000).

[36] P. Senellart, J. Bloch, B. Sermage, and J. Y. Marzin, Phys. Rev. B **62**, R16263 (2000).

[37] M. Muller, J. Bleuse, and R. Andre, Phys. Rev. B **62**, 16886 (2000).

[38] R. P. Stanley, R. Houdré, C. Weisbuch, U. Oesterle, and M. Ilegems, Phys. Rev. B **53**, 10995 (1996).

[39] R. P. Stanley, S. Pau, U. Oesterle, R. Houdré, and M. Ilegems, Phys. Rev. B **55**, R4867 (1997).

[40] H. Cao, S. Pau, Y. Yamamoto, and G. Björk, Phys. Rev. B **54**, 8083 (1996).

[41] D. M. Whittaker, P. Kinsler, T. A. Fisher, M. S. Skolnick, A. Armitage, A. M. Afshar, M. D. Sturge, and J. S. Roberts, Phys. Rev. Lett. **77**, 4792 (1996).

[42] S. Pau, H. Cao, J. Jacobson, G. Björk, Y. Yamamoto, and A. Imamoglu, Phys. Rev. A **54**, R1789 (1996).

[43] V. Savona, L. C. Andreani, P. Schwendimann, and A. Quattropani, Solid State Commun. **93**, 733 (1995).

[44] R. Houdré, unpublished.

[45] M. A. Dupertuis, unpublished.

[46] Y. Suzuki and H. Okamoto, J. Electron. Mater. **12**, 397 (1983).

[47] V. Savona, C. Piermarocchi, A. Quattropani, F. Tassone, and P. Schwendimann, Phys. Rev. Lett. **78**, 4470 (1997).

[48] D. M. Whittaker, Phys. Rev. Lett. **80**, 4791 (1998).

[49] R. Houdré, R. P. Stanley, and M. Ilegems, Phys. Rev. A **53**, 2711 (1996).

[50] R. P. Stanley, R. Houdré, U. Oesterle, M. Gailhanou, and M. Ilegems, Appl. Phys. Lett. **65**, 1883 (1994).

[51] R. Houdré, R. P. Stanley, U. Oesterle, and C. Weisbuch, C. R. Phys. (France) **3**, 15 (2002).

[52] S. D. Baranovskii, U. Doerr, P. Thomas, A. Naumov, and W. Gebhardt, Phys. Rev. B **48**, 17149 (1993).

[53] R. Zimmermann and E. Runge, J. Luminesc. **60**(1), 320 (1994).

[54] C. Ell, J. Prineas, T. R. Nelson, S. Park, H. M. Gibbs, G. Khitrova, S. W. Koch, and R. Houdré, Phys. Rev. Lett. **80**, 4795 (1998).

[55] J. Hegarty, M. D. Sturge, C. Weisbuch, A. C. Gossard, and W. Wiegmann, Phys. Rev. Lett. **49**, 930 (1982).

[56] S. Haacke, R. A. Taylor, R. Zimmermann, I. BarJoseph, and B. Deveaud, Phys. Rev. Lett. **78**, 2228 (1997).

[57] G. R. Hayes, S. Haacke, M. Kauer, R. P. Stanley, R. Houdré, U. Oesterle, and B. Deveaud, Phys. Rev. B **58**, R10175 (1998).

[58] W. Langbein, J. M. Hvam, and R. Zimmermann, Phys. Rev. Lett. **82**, 1040 (1999).

[59] R. Houdré, C. Weisbuch, R. P. Stanley, U. Oesterle, and M. Ilegems, Phys. Rev. B **61**, R13333 (2000).

[60] V. Savona and R. Zimmermann, Phys. Rev. B **60**, 4928 (1999).

[61] S. Haacke, G. Hayes, R. A. Taylor, B. Deveaud, R. Zimmermann, and I. BarJoseph, phys. stat. sol. (b) **204**, 35 (1997).
[62] T. Freixanet, B. Sermage, J. Bloch, J. Y. Marzin, and R. Planel, Phys. Rev. B **60**, R8509 (1999).
[63] R. Houdré, unpublished.
[64] W. Langbein, J. Phys.: Condens. Matter **16**, S3645 (2004).
[65] W. Langbein and J. M. Hvam, phys. stat. sol. (a) **190**, 327 (2002).
[66] M. Gurioli, F. Bogani, D. S. Wiersma, P. Roussignol, G. Cassabois, G. Khitrova, and H. Gibbs, Phys. Rev. B **64**, (2001).
[67] M. Gurioli, F. Bogani, D. S. Wiersma, P. Roussignol, G. Cassabois, G. Khitrova, and H. Gibbs, phys. stat. sol. (a) **190**, 363 (2002).
[68] M. Gurioli, L. Cavigli, G. Khitrova, and H. M. Gibbs, Physica E **17**, 463 (2003).
[69] M. Gurioli, F. Bogani, L. Cavigli, H. Gibbs, G. Khitrova, and D. S. Wiersma, Phys. Rev. Lett. **94** (2005).
[70] R. Houdré, C. Weisbuch, R. P. Stanley, U. Oesterle, and M. Ilegems, Phys. Rev. Lett. **85**, 2793 (2000).
[71] R. Houdré, C. Weisbuch, R. P. Stanley, U. Oesterle, and M. Ilegems, Physica E **7**, 625 (2000).
[72] A. Imamoglu and R. J. Ram, Phys. Lett. A **214**, 193 (1996).
[73] M. Kira, F. Jahnke, S. W. Koch, J. D. Berger, D. V. Wick, T. R. Nelson, G. Khitrova, and H. M. Gibbs, Phys. Rev. Lett. **79**, 5170 (1997).
[74] L. S. Dang, D. Heger, R. André, F. Boeuf, and R. Romestain, Phys. Rev. Lett. **81**, 3920 (1998).
[75] P. Senellart and J. Bloch, Phys. Rev. Lett. **82**, 1233 (1999).
[76] P. G. Savvidis, C. Ciuti, J. J. Baumberg, D. M. Whittaker, M. S. Skolnick, and J. S. Roberts, Phys. Rev. B **64**, 075311 (2001).
[77] P. G. Savvidis, J. J. Baumberg, R. M. Stevenson, M. S. Skolnick, D. M. Whittaker, and J. S. Roberts, Phys. Rev. Lett. **84**, 1547 (2000).
[78] R. M. Stevenson, V. N. Astratov, M. S. Skolnick, D. M. Whittaker, M. Emam-Ismail, A. I. Tartakovskii, P. G. Savvidis, J. J. Baumberg, and J. S. Roberts, Phys. Rev. Lett. **85**, 3680 (2000).
[79] C. Ciuti, P. Schwendimann, B. Deveaud, and A. Quattropani, Phys. Rev. B **62**, R4825 (2000).
[80] G. Messin, J. P. Karr, A. Baas, G. Khitrova, R. Houdré, R. P. Stanley, U. Oesterle, and E. Giacobino, Phys. Rev. Lett. **87**12 (2001).
[81] R. André, F. Boeuf, D. Heger, L. Dang, R. Romestain, J. Bleuse, and M. Muller, Acta Phys. Pol. A **96**, 511 (1999).
[82] A. Huynh, J. Tignon, O. Larsson, P. Roussignol, C. Delalande, R. André, R. Romestain, and L. S. Dang, Phys. Rev. Lett. **90** (2003).
[83] Y. Chen, A. Tredicucci, and F. Bassani, Phys. Rev. B **52**, 1800 (1995).
[84] A. Tredicucci, Y. Chen, V. Pellegrini, M. Borger, L. Sorba, F. Beltram, and F. Bassani, Phys. Rev. Lett. **75**, 3906 (1995).
[85] V. Agranovich, H. Benisty, and C. Weisbuch, Solid State Commun. **102**, 631 (1997).
[86] D. G. Lidzey, D. D. C. Bradley, M. S. Skolnick, T. Virgili, S. Walker, and D. M. Whittaker, Nature **395**, 53 (1998).
[87] L. G. Connolly, D. G. Lidzey, R. Butte, A. M. Adawi, D. M. Whittaker, M. S. Skolnick, and R. Airey, Appl. Phys. Lett. **83**, 5377 (2003).
[88] R. Shimada, A. L. Yablonskii, S. G. Tikhodeev, and T. Ishihara, IEEE J. Quantum Electron. **38**, 872 (2002).
[89] J. Ishi-Hayase and T. Ishihara, Semicond. Sci. Technol. **18**, S411 (2003).
[90] J. P. Reithmaier, G. Sek, A. Loffler, C. Hofmann, S. Kuhn, S. Reitzenstein, L. V. Keldysh, V. D. Kulakovskii, T. L. Reinecke, and A. Forchel, Nature **432**, 197 (2004).
[91] T. Yoshie, A. Scherer, J. Hendrickson, G. Khitrova, H. M. Gibbs, G. Rupper, C. Ell, O. B. Shchekin, and D. G. Deppe, Nature **432**, 200 (2004).

[92] G. Klimovitch, G. Bjork, H. Cao, and Y. Yamamoto, Phys. Rev. B **55**, 7078 (1997).
[93] T. A. Fisher, A. M. Afshar, D. M. Whittaker, M. S. Skolnick, J. S. Roberts, G. Hill, and M. A. Pate, Phys. Rev. B **51**, 2600 (1995).
[94] A. Kavokin and B. Gil, Appl. Phys. Lett. **72**, 2880 (1998).
[95] N. Antoine-Vincent, F. Natali, D. Byrne, P. Disseix, A. Vasson, J. Leymarie, F. Semond, and J. Massies, Superlattices Microstruct. **36**, 599 (2004).
[96] T. Tawara, H. Gotoh, T. Akasaka, N. Kobayashi, and T. Saitoh, Phys. Rev. Lett. **92**, (2004).
[97] H. Cao, S. Pau, J. M. Jacobson, G. Björk, Y. Yamamoto, and A. Imamoglu, Phys. Rev. A **55**, 4632 (1997).
[98] G. Khitrova, H. M. Gibbs, F. Jahnke, M. Kira, and S. W. Koch, Rev. Mod. Phys. **71**, 1591 (1999).
[99] F. Jahnke, M. Kira, and S. W. Koch, phys. stat. sol. (b) **206**, 19 (1998).
[100] M. Kira, F. Jahnke, W. Hoyer, and S. W. Koch, Prog. Quantum Electron. **23**, 189 (1999).
[101] J. Bloch, J. Y. Marzin, R. Planel, V. Thierry-Mieg, B. Sermage, I. Abram, and J. M. Gerard, J. Phys. IV (France) **9**, 15 (1999).
[102] E. Glacobino, J. P. Karr, G. Messin, H. Eleuch, and A. Baas, C. R. Phys. (France) **3**, 41 (2002).
[103] R. Houdré, R. P. Stanley, U. Oesterle, and C. Weisbuch, Physica E **11**, 198 (2001).

Chapter 4
Exciton-Polaritons and Nanoscale Cavities in Photonic Crystal Slabs

Lucio Claudio Andreani, Dario Gerace, and Mario Agio

4.1
Introduction

Strong exciton–light coupling and the formation of polariton states is a recurring theme in solid-state physics and optics. The quantum theory of exciton-polaritons in bulk crystals was first formulated by Hopfield [1] and Agranovich [2]. The effect of quantum confinement on exciton-polariton states has been the subject of a number of investigations (for reviews see, e.g., [3–5]). Photon confinement in planar semiconductor microcavities with embedded quantum wells (QWs) leads to a strong-coupling regime of exciton–light coupling and to robust cavity polariton states [6–9]. Exciton–light coupling in microcavities with full three-dimensional photon confinement, like micro-pillars and micro-disks, has also been investigated [10–12].

The field of photonic crystals (PhCs) has become increasingly important since the pioneering works of Yablonovitch [13] and John [14]. In particular, photonic crystals embedded in planar waveguides (also known as photonic crystal slabs) can lead to a full control of light propagation because of a two-dimensional (2D) photonic lattice in the slab plane combined with dielectric confinement in the vertical direction [15–17]. Nanoscale cavities in PhC slabs with extremely high Q-factors and low mode volumes have been recently demonstrated [18, 19]. The performance of these nanocavity structures for optical confinement is, in principle, much better than that of conventional planar microcavities or of micro-pillar and micro-disk structures. Recently, the strong-coupling regime of quantum dot transitions coupled to high-Q cavity modes has been demonstrated for both micro-pillars [20] and PhC nanocavities [21].

In this paper we describe recent theoretical work dealing with exciton-polariton states in PhCs and in nanocavities. In Section 2 we review a few basic concepts related to photonic mode dispersion in PhC slabs. In Section 3 we present a quantum-mechanical formulation of exciton-light coupling in PhC slabs with embedded QWs, leading to the conditions for the occurrence of a strong-coupling regime, and calculations of variable-angle reflectance from the slab surface: the quantum and semiclassical approaches agree with each other and show that radiative exciton-polaritons can be probed by surface reflectivity. In Section 4 we describe a basic

Physics of Semiconductor Microcavities: From Fundamentals to Nanoscale Devices
Edited by Benoit Deveaud
Copyright © 2007 WILEY-VCH Verlag GmbH & Co. KGaA, Weinheim
ISBN: 978-3-527-40561-9

nanocavity structure and present results for the Q-factor as a function of cavity geometry. In Section 5 we treat the coupling of a single quantum-dot transition with a high-Q cavity mode, and quantify the conditions for the occurrence of a strong-coupling regime. Section 6 contains concluding remarks.

4.2
Mode Dispersion and Linewidths in Photonic Crystal Slabs

Photonic crystal slabs consist of planar dielectric waveguides patterned with a one-dimensional (1D) or two-dimensional (2D) lattice. They can have either a weak refractive index contrast between core and claddings (like in the GaAs/AlGaAs or InP/InGaAsP systems) or a strong index contrast like in the self-standing membrane or air-bridge. Electromagnetic eigenmodes in PhC slabs can be either truly guided (if their frequency lies below the light line of the cladding material) or quasi-guided (if the frequency lies above the light line). Truly guided modes are evanescent in the cladding regions and have low propagation losses that are due only to fabrication disorder (in the transparency region of the medium in which absorption losses are absent). Quasi-guided modes, instead, have a radiative component in the cladding regions and suffer from high scattering losses due to diffraction out of the slab plane. For the same reason, however, they couple to an electromagnetic wave incident on the slab surface and represent optically active excitations of the photonic system. Indeed, the dispersion relations of quasi-guided modes have been studied in a number of angle-resolved linear [22–25] and nonlinear [26–29] experiments. Recently, truly-guided modes have also been probed by optical experiments from the slab surface using an attenuated-total-reflectance configuration [30].

In order to calculate the dispersion relations of guided and quasi-guided modes in PhC slabs, we adopt the guided-mode expansion (GME) method recently developed [31]. As conveniently done for photonic crystals, we start from the second-order equation for the magnetic field

$$\nabla \times \left[\frac{1}{\varepsilon(\mathbf{r})} \nabla \times \mathbf{H} \right] = \frac{\omega^2}{c^2} \mathbf{H}, \tag{1}$$

where $\varepsilon(\mathbf{r})$ is the spatially dependent dielectric constant. If the magnetic field is expanded in an orthonormal basis set as $\mathbf{H}(\mathbf{r}) = \sum_\mu c_\mu \mathbf{H}_\mu(\mathbf{r})$, then Eq. (1) is transformed into a linear eigenvalue problem

$$\sum_\nu \mathcal{H}_{\mu\nu} c_\nu = \frac{\omega^2}{c^2} c_\mu, \tag{2}$$

where the matrix $\mathcal{H}_{\mu\nu}$ (which is the analog of a quantum Hamiltonian) is given by

$$\mathcal{H}_{\mu\nu} = \int \frac{1}{\varepsilon(\mathbf{r})} (\nabla \times \mathbf{H}_\mu^*(\mathbf{r})) \cdot (\nabla \times \mathbf{H}_\nu(\mathbf{r})) \, d\mathbf{r}. \tag{3}$$

For the case of a PhC slab we have a waveguide along z and a periodic patterning in the xy plane. The basis states $H_\mu(\mathbf{r})$ are chosen to consist of the guided modes of an effective waveguide, where each layer j has a homogeneous dielectric constant given by the spatial average of $\varepsilon_j(x,y)$. The index μ can be written as $\mu = (\mathbf{k}+\mathbf{G}, \alpha)$, where $\mathbf{k}$ is the 2D Bloch vector, $\mathbf{G}$ is a reciprocal lattice vector and α labels the guided modes at wave vector $\mathbf{k}+\mathbf{G}$. The basis states with the same Bloch vector $\mathbf{k}$ are coupled by the dielectric modulation. The matrix elements (3) can be expressed in terms of the inverse dielectric tensor in each layer $\varepsilon_j^{-1}(\mathbf{G},\mathbf{G}')$, which is evaluated by a numerical inversion of the dielectric matrix $\varepsilon_j(\mathbf{G},\mathbf{G}')$.

The basis set consisting of the guided modes of the effective waveguide is orthonormal but not complete since the leaky modes of the waveguide are not included. Coupling to leaky modes produces a second-order shift of the mode frequency: the neglect of this effect (which is usually small, at least for the low air fractions that are employed here) is the main approximation of the method. When the guided modes are folded in the first Brillouin zone, many of them fall above the light line and become quasi-guided. Indeed, first-order coupling to leaky modes at the same frequency leads to a radiative decay width, which is expressed as twice an imaginary part of the frequency. This can be calculated by time-dependent perturbation theory, like in Fermi Golden Rule for quantum mechanics, and is given by

$$-\mathrm{Im}\left(\frac{\omega_k^2}{c^2}\right) = \pi \left| \mathcal{H}_{\text{leaky, guided}} \right|^2 \rho\left(\mathbf{k}; \frac{\omega_k^2}{c^2}\right), \tag{4}$$

where $\rho(\mathbf{k};\omega_k^2/c^2)$ is the 1D photonic density of states at fixed in-plane wave vector [32, 33]. Notice that the mode Q-factor can be obtained as $Q = \omega/[2\,\mathrm{Im}(\omega)]$.

As an example of photonic mode dispersion, in Fig. 1 we show the photonic bands of a triangular lattice of air holes in a membrane with the dielectric constant of GaAs at optical frequencies, along the Γ–M and Γ–K symmetry directions of the 2D Brillouin zone. In Fig. 4.1(a) a schematic picture of the structure and a definition of its direct and reciprocal lattices is displayed. A high-index membrane supports both guided modes (lying between the cladding and core light lines) and quasi-guided modes (lying above the air light line). It should be noted that the photonic band dispersion is calculated assuming the low temperature value at 1.48 eV for the dielectric constant of the GaAs layer, i.e., $\varepsilon = 12.95$. Although the band dispersion in Fig. 4.1(b) is calculated with a frequency-independent dielectric constant, an exciton level at $E_{\text{exc}} = 1.48$ eV (corresponding to a low-temperature exciton in a typical InGaAs quantum well) is also shown. This will be useful for the study of radiation–matter interaction in the next Section. It can be seen that the exciton energy crosses several photonic modes of the 2D photonic lattice. Only even modes with respect to the horizontal midplane (indicated with $\sigma_{xy} = +1$) are considered here.

An expanded plot of the dispersion and the mode linewidths (twice the imaginary parts of the energies) along the main symmetry directions is shown in Fig. 4.2. Along the Γ M orientation, mode 2 has vanishing radiative linewidth when cross-

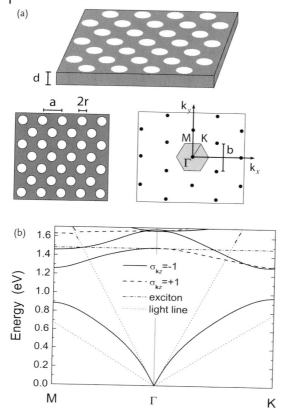

Fig. 4.1 (a) Schematic picture of a high index PhC membrane of thickness d patterned with a triangular lattice of air holes, together with its direct and reciprocal lattices. (b) Photonic band dispersion for the structure in (a) with the following parameters: dielectric constant $\varepsilon = 12.95$, lattice constant $a = 350$ nm, membrane thickness $d = 0.4a$, hole radius $r = 0.3a$. The fundamental exciton level at 1.48 eV is also plotted. Only even modes with respect to the horizontal midplane ($\sigma_{xy} = +1$) are shown, and for each symmetry direction the modes are classified as odd ($\sigma_{kz} = -1$) or even ($\sigma_{kz} = +1$) with respect to the corresponding vertical plane of incidence. The dotted lines represent the light dispersion in the air claddings and in the effective waveguide core.

ing the light line and becoming a truly guided mode, while mode 1 has vanishing linewidth on approaching the Γ point. The latter behavior is determined by symmetry considerations [15, 33] since at Γ only dipole-active, twofold-degenerate modes are coupled to normally incident light. Considering now the ΓK direction, the two photonic modes have different symmetries with respect to the corresponding vertical plane of incidence. In particular, the even mode (indicated with $\sigma_{kz} = +1$) has vanishing linewidth on approaching the Γ point, like the corresponding mode along ΓM. On the contrary, the odd mode ($\sigma_{kz} = -1$) has much higher

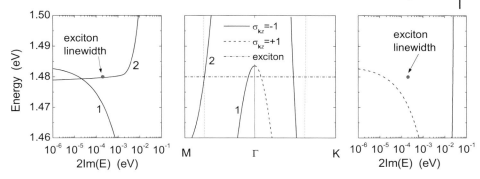

Fig. 4.2 Complex energy dispersion of $\sigma_{xy} = +1$ photon modes for energies around the excitonic resonance, along the main symmetry directions of the triangular lattice (parameters as in Fig. 4.1). Left: mode linewidths along ΓM. Middle: real part of mode energies. Right: mode linewidths along ΓK. The points corresponding to a non-dispersive exciton resonance with $E_{exc} = 1.48$ eV and linewidth $\Gamma_{exc} = 2 \times 10^{-4}$ eV are also shown.

radiation linewidths ($2\,\text{Im}(E) > 20$ meV). The value of the photonic mode (and exciton) linewidth is a crucial parameter when considering the interaction between photon and exciton states, as discussed in the next section.

It is important to notice, in the left panel of Fig. 4.2, that in correspondence to the exciton resonance both photonic eigenmodes have very small linewidths ($2\,\text{Im}(E) < 10^{-3}$ eV). We thus reasonably expect that photonic crystal polaritons should form at two different points in the irreducible Brillouin zone along ΓM, with two distinct anticrossings between exciton center-of-mass levels and photonic bands.

4.3
Exciton-Polaritons in Photonic Crystal Slabs

In order to develop a quantum-mechanical theory of polaritons in PhC slabs, we have to quantize both the photon and exciton states in the dielectric structure (a detailed account of the formalism is presented in Ref. [34]). The electric and magnetic fields are expanded as

$$E(r,t) = \sum_\mu \left(\frac{2\pi\hbar\omega_\mu}{4\pi\varepsilon_0 V}\right)^{1/2} \left[\hat{a}_\mu E_\mu(r)\, e^{-i\omega_\mu t} + \hat{a}_\mu^\dagger E_\mu^*(r)\, e^{i\omega_\mu t}\right], \tag{5}$$

$$H(r,t) = \sum_\mu \left(\frac{2\pi\hbar\omega_\mu}{4\pi\varepsilon_0 V}\right)^{1/2} \left[\hat{a}_\mu H_\mu(r)\, e^{-i\omega_\mu t} + \hat{a}_\mu^\dagger H_\mu^*(r)\, e^{i\omega_\mu t}\right], \tag{6}$$

where $\hat{a}^{\dagger}_{\mu}$ ($\hat{a}_{\mu}$) are creation (destruction) operators of field quanta with energies ω_{μ}. In the above formulas, ε_0 is the vacuum permittivity, V is a quantization volume, and $\mu = (\mathbf{k}, n)$ is a combined index which includes the Bloch vector $\mathbf{k}$ and the photonic band index n. The field eigenmodes $\mathbf{E}_{\mu}(\mathbf{r}), \mathbf{H}_{\mu}(\mathbf{r})$ can be calculated by solving the classical Maxwell equations and are normalized as

$$\int_V \varepsilon(\mathbf{r}) \, \mathbf{E}_{\mu}(\mathbf{r}) \, \mathbf{E}^{*}_{\mu'}(\mathbf{r}) \, d\mathbf{r} = \delta_{\mu\mu'} \, , \tag{7}$$

$$\int_V \mathbf{H}_{\mu}(\mathbf{r}) \, \mathbf{H}^{*}_{\mu'}(\mathbf{r}) \, d\mathbf{r} = \delta_{\mu\mu'} \, . \tag{8}$$

For the exciton part, we consider a membrane containing a thin QW layer that is also patterned with air holes, assume strong electron–hole confinement leading to separability of the exciton wavefunction, and solve the Schrödinger equation for the center-of-mass motion in the QW plane

$$\left[-\frac{\hbar^2 \nabla^2_{\parallel}}{2M_{\text{exc}}} + V(\mathbf{r}_{\parallel}) \right] F_{\text{cm}}(\mathbf{r}_{\parallel}) = \hbar \Omega \, F_{\text{cm}}(\mathbf{r}_{\parallel}) \, , \tag{9}$$

where $\mathbf{r}_{\parallel} = (x, y)$ and the effective potential $V(\mathbf{r}_{\parallel}) = \infty$ in air regions, while $V(\mathbf{r}_{\parallel}) = 0$ in the non-patterned regions of the quantum well. We neglect dead-layer effects (the thickness of the dead layer is usually less than 10 nm, i.e., much smaller than the length scale of the photonic structure) and also image-charge potentials. Equation (9) is solved numerically by plane wave expansion, yielding quantized center-of-mass levels in the periodic potential. By this procedure, the exciton levels are labelled by the same quantum number of the electromagnetic modes: i.e., by a Bloch vector $\mathbf{K}_{\text{exc}}$ and a discrete index ν. This allows introducing exciton creation (destruction) operators $\hat{b}^{\dagger}_{\sigma}$ ($\hat{b}_{\sigma}$) corresponding to the energies $\hbar \Omega_{\sigma}$, where $\sigma = (\mathbf{K}_{\text{exc}}, \nu)$ is again a combined index.

In the interaction with photon states, the Bloch vector is conserved, or $\mathbf{K}_{\text{exc}} \simeq \mathbf{k}$. However, a photonic mode with band index n couples to exciton center-of-mass levels with any ν. The interaction is determined by a matrix element of the full Hamiltonian, as first shown in Refs. [1, 2] for bulk exciton–polaritons and later extended to quantum-confined systems [3–5]. The coupling matrix element between exciton and photon takes the form

$$C_{kn\nu} = \left(\frac{2\pi e^2 \hbar \Omega^2_{k\nu}}{4\pi \varepsilon_0 \omega_{kn}} \right)^{1/2} \langle \Psi^{(\text{exc})}_{k\nu} | \sum_j \mathbf{E}_{kn}(\mathbf{r}_j) \cdot \mathbf{r}_j | 0 \rangle \, , \tag{10}$$

where $\Psi^{(\text{exc})}_{k\nu}$ is the exciton wavefunction, and the sum is over all the electrons in the QW material. If the QW exciton is a heavy-hole state, only the in-plane components of the electric field are involved and $\sigma_{xy} = +1$ modes (often called TE-like modes in the literature) are preferentially coupled. The integral can be expressed in terms the oscillator strength f of the excitonic transition, which is generally defined as

$$f = \frac{2m\omega}{\hbar} \left| \langle \Psi_{kv}^{(\text{exc})} | \hat{e} \cdot \sum_j r_j | 0 \rangle \right|^2 = \frac{2}{m\hbar\omega} \left| \langle \Psi_{kv}^{(\text{exc})} | \hat{e} \cdot \sum_j p_j | 0 \rangle \right|^2, \tag{11}$$

where m is the free-electron mass, $\hat{e}$ is the polarization unit vector of the exciton and r_j (p_j) is the position (momentum) operator of the QW electrons. The matrix element (10) is found to depend on the oscillator strength per unit area, f/S, as well as on the spatial overlap between the exciton center-of-mass wavefunction and the mode electric field in the QW plane:

$$C_{knv} \simeq \left(\frac{\pi e^2 \hbar^2}{4\pi\varepsilon_0 m} \frac{f}{S} \right)^{1/2} \int \hat{e} \cdot E_{kn}(r_\parallel, z_{\text{QW}}) F_{\text{cm}}(r_\parallel) \, dr_\parallel . \tag{12}$$

The full quantum Hamiltonian describing coupled photon and exciton states is finally given by

$$\hat{H} = \sum_{k,n} \hbar\omega_{kn} \hat{a}_{kn}^\dagger \hat{a}_{kn} + \sum_{k,v} \hbar\Omega_{kv} \hat{b}_{kv}^\dagger \hat{b}_{kv} + i \sum_{k,n,v} C_{knv} (\hat{a}_{kn} + \hat{a}_{-kn}^\dagger)(\hat{b}_{kv}^\dagger - \hat{b}_{-kv})$$

$$+ \sum_{k,v} \sum_{n_1,n_2} \frac{C_{kn_1v}^* C_{kn_2v}}{\hbar\Omega_{kv}} (\hat{a}_{-kn_1} + \hat{a}_{kn_1}^\dagger)(\hat{a}_{kn_2} + \hat{a}_{-kn_2}^\dagger) . \tag{13}$$

It should be noticed that both photon and exciton energies are taken to be complex quantities: i.e., the imaginary part of the frequency for quasi-guided photonic modes, as well as the exciton linewidth arising from non-radiative processes, are included in the calculation. Hamiltonian (13) is diagonalized by using a generalized Hopfield transformation [1, 35] to expand new destruction (creation) operators $\hat{P}_k$ ($\hat{P}_k^\dagger$) as linear combinations of $\hat{a}_{kn}$ ($\hat{a}_{kn}^\dagger$) and $\hat{b}_{kv}$ ($\hat{b}_{kv}^\dagger$), with the condition $[\hat{P}_k, \hat{H}] = E_k \hat{P}_k$. The transformation, which leads to a non-hermitian eigenvalue problem, applies also to a Hamiltonian that includes dissipative terms. The new eigenenergies E_k correspond to mixed excitations of radiation and matter, which can be either in a weak or in a strong-coupling regime: in the latter case they give rise to *photonic crystal polaritons*.

For a GaAs membrane containing a typical InGaAs/GaAs quantum well at the field antinode, the coupling matrix element is calculated to be of the order of 6 meV in the present structure. The exciton linewidth in high-quality structures at low temperature can be made lower than 0.6 meV [36], i.e., negligibly small as compared to the energy scale of the interaction. Thus, the eventual regime of the exciton–photon coupling depends critically on the value of the photonic mode linewidth. If the exciton interacts with a truly-guided mode, the (intrinsic) linewidth is zero and the exciton–light coupling is always in a strong-coupling regime. The resulting polaritons are evanescent and non-radiative, as they lie below the air light line. Radiative polariton states are obtained when the exciton interacts with a quasi-guided mode whose linewidth is smaller than the exciton–photon coupling. Several possible situations for the interaction of the exciton with quasi-guided modes are illustrated in Fig. 4.2 above.

Radiative polaritons in PhC slabs can be probed by reflectance (or transmittance) from the slab surface, as done in a pioneering paper where the strong exciton resonance of an organic molecule [bis-(phenethyl-ammonium) tetraiodoplumbate (PE-PI)] with a giant oscillator strength per unit area was employed to observe the strong-coupling at room temperature [22]. Observing the same effect in III–V semiconductors is more difficult and has not been achieved at time of writing. Here we calculate the angle-resolved reflectance from the surface of the PhC slab using the scattering-matrix method [37, 38], which yields an exact numerical solution of Maxwell equations for a stratified medium consisting of patterned layers that are homogeneous in the z-direction. The presence of the exciton resonance is taken into account at a semiclassical level by adding a Lorentz-oscillator term to the dielectric function in the QW layers:

$$\varepsilon_{exc}(\omega) = \varepsilon_\infty \left(1 + \frac{\hbar\omega_{LT}}{\hbar(\Omega_{exc} - \omega) - i\Gamma_{exc}/2}\right), \tag{14}$$

where ε_∞ is the background dielectric constant of the QW material, ω_{LT} is the longitudinal–transverse (LT) splitting and $\hbar\Omega_{exc}$, Γ_{exc} are the bare exciton energy and linewidth as in the quantum calculation. The LT splitting may be related to the quantum-mechanical oscillator strength per unit area by

$$\hbar\omega_{LT} = \frac{2\pi\hbar e^2}{4\pi\varepsilon_0\varepsilon_\infty m\Omega_{exc} L_{QW}} \frac{f}{S}, \tag{15}$$

where L_{QW} is the QW width. This relation is valid for a single quantum well in the strong (electron–hole) confinement regime. A fuller discussion of the relation between semiclassical and quantum descriptions of the light–exciton interaction can be found elsewhere [3–5].

In Fig. 4.3 we show a comparison between quantum and semiclassical treatments of PhC polaritons (results for 1D photonic lattices were previously presented in Ref. [39]), for a structure containing two InGaAs QWs. In order to make the quantum and semiclassical calculations consistent with each other, an oscillator strength per unit area $f/S = 4.2 \times 10^{12}$ cm^{-2} is assumed for each InGaAs QW, corresponding to a quantum well of width $L_{QW} = 8$ nm and a LT splitting $\hbar\omega_{LT} = 0.2$ meV in the semiclassical calculation. Panels (a), (c) display angle-resolved reflectance spectra along the Γ–M and Γ–K orientations, respectively, while panel (b) reports the dispersion of coupled exciton-photon modes as calculated from the quantum theory (small circles) and extracted from the spectral structures in reflectance taking into account parallel momentum conservation (square points). Along the Γ–M orientation (TE polarization of incident light) there are two anticrossing points, i.e., exci-

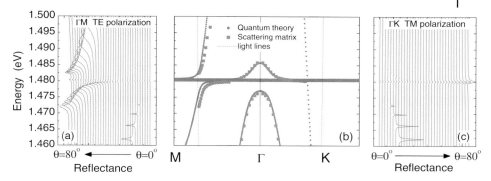

Fig. 4.3 Photonic crystal polaritons in a 2D triangular lattice along the main symmetry directions (parameters as in Fig. 4.1). Scattering matrix calculations of reflectance spectra along (a) ΓM (TE incident light) and (c) ΓK (TM incident light) are compared to quantum calculations of mode dispersion in (b): small circles are from the quantum theory results, while square points are extracted from reflectance spectra in (a) and (c).

ton–light coupling is in a strong-coupling regime for both photonic modes 1 and 2 previously shown in Fig. 4.2. Along the Γ–K orientation (TM incident polarization) there is only one anticrossing point. It is worth noting that odd modes ($\sigma_{k_z} = -1$) are excited by TE incident radiation along ΓK, whilst even modes ($\sigma_{k_z} = +1$) are excited by TM incident beams along ΓM. The results of Fig. 4.3 demonstrate that radiative polaritons can be observed by angle-resolved reflectance, provided the photonic mode linewidth (and, of course, the exciton linewidth) be smaller than the exciton–photon coupling. They also demonstrate that the semiclassical and quantum treatments of photonic-crystal polaritons yield results that are in very good agreement with each other – the expected results for *linear* optical properties.

Notice that the polariton splitting at resonance is of the order of 10 meV with two embedded QWs. This is larger than common values for III–V microcavities which are typically of the order of 4 meV with two QWs [40], and no larger than 6–8 meV even with several quantum wells close to the field antinodes [7–9]. This increase of the polariton splitting has little to do with the x, y dependence of the electric field and of the exciton envelope function: since the exciton center-of-mass levels are nearly degenerate, for a given photonic mode there is always a linear combination of exciton states which has the proper spatial dependence to yield an overlap matrix element (10) close to unity. In this respect, the physics of exciton-light coupling in photonic crystals is similar to the case of pillar microcavities [35] where a cavity mode couples to several exciton states and the polariton splitting has only a slight dependence on the pillar radius. The reason for the increased polariton splitting in PhCs lies in a better field confinement along the vertical direction: in a microcavity with dielectric mirrors the penetration of the electric field in the distributed Bragg reflectors reduces the overlap of the cavity mode with the exciton state [5, 40], while in a PhC slab the fundamental waveguide mode is almost perfectly confined within the slab, thus yielding optimal coupling with the QW exciton.

4.4
Nanoscale Cavities in Photonic Crystal Slabs

Point defects in PhC slabs behave as 0D cavities and support localized modes in the photonic gap. Cavity modes are always subject to radiation losses, as they have no wave vector and are coupled to the continuum of leaky slab modes by the dielectric modulation. Still, photonic cavities with large quality factor Q and small mode volumes can be defined. The quality factor can be increased by a momentum-space design, which allows to reduce the radiative component of the confined photonic mode [41]. In real space, this corresponds to changes of the position or size of the nearby holes. One of the best performing cavity structures consists of three missing holes along the ΓK direction of the triangular lattice: by using the principle of "gentle confinement", which consists of shifting the positions of the holes close to the defect, Q-factors as high as 1.5×10^5 have been demonstrated [42]. The very high Q-factors can also be interpreted with a Fabry–Perot model [43]. Using a conceptually different design approach, based on PhC slab heterostructures with varying lattice constants, measured Q-factors of the order of 6×10^5 have been reported [44].

Within the present method, the quality factor is calculated as $Q = \omega/[2\,\mathrm{Im}(\omega)]$ by introducing a supercell in two directions and evaluating $\mathrm{Im}(\omega)$ in perturbation theory with the use of Eq. (4). Notice that by using a supercell, all photonic modes that fall above the light line upon folding in a reduced Brillouin zone become radiative: by this procedure, the determination of the Q-factor of cavity modes is similar to the calculation of propagation losses of extended modes. We focus on cavities with one, two or three missing holes in the triangular lattice (L1, L2, L3 defect) and consider a displacement or a shift of the nearby holes in ΓK direction, as illustrated in Fig. 4.4. We calculate only intrinsic losses, i.e., we do not include the effect of disorder which is left for further analysis.

In Fig. 4.5 we show the quality factor as a function of (a) hole displacement and (b) hole shrinking. All curves have a pronounced maximum, confirming that the Q-factor is indeed increased by gentle confinement. For the case of the L3 defect

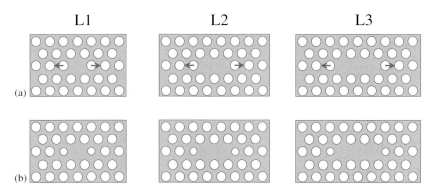

Fig. 4.4 Schematic structure of L1, L2, L3 point defects with (a) hole displacement and (b) hole shrinking.

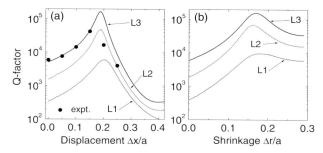

Fig. 4.5 Quality factor for L1, L2, L3 defects in a silicon membrane with $\varepsilon_r = 12$, $a = 420$ nm, $d/a = 0.6$, $r/a = 0.29$ as a function of (a) displacement and (b) shrinking of the two holes close to the point defect along the ΓK direction. The experimental points are taken from Ref. [18]. The mode considered has symmetry $\sigma_{xy} = +1$.

with hole displacement, we find $Q = 4.5 \times 10^4$ for $\Delta x/a = 0.15$, in agreement with the experimental results [18] obtained on nanocavities in Silicon membranes with a cavity mode around $\lambda = 1.55$ μm. The maximum calculated value is $Q \simeq 1.5 \times 10^5$ at $\Delta x/a = 0.18$. The experimental values for $\Delta x/a = 0.2$ and 0.25 are lower than the theoretical ones. Turning now to the case of hole shrinking, we notice that the maximum of the Q-factor as a function of $\Delta r/a$ is broader, implying that the structure may be more tolerant to small imperfections in fabrication. When the two nearby holes are shrunk to zero radius, the curve relative to the Ln defect tends to the value for the L(n+2) defect at $\Delta r = 0$. The results of Fig. 4.5 show clearly that an L3 cavity with optimal hole displacement or shrinking (maximum of L3 curves) has a higher Q than a bare L5 cavity (end point of L3 curve in Fig. 4.5b).

In Fig. 4.6 we show the electric field profile along the main symmetry directions for the ground mode of the L3 cavity (for null displacement and shrinking of nearby holes). The square modulus of the electric field has a maximum at the cavity center, but it oscillates along the Γ–K direction (and, to a lesser extent, along the Γ–M

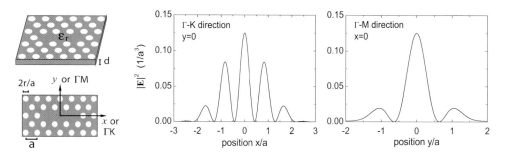

Fig. 4.6 Schematice picture of a L3 nanocavity and field profile of the ground cavity mode along the Γ–K and Γ–M symmetry directions for structure parameters as in Fig. 4.5. The field is calculated for the structure with no shift or shrinking of nearby holes.

direction) with secondary maxima. The oscillations along Γ–K result from the physical origin of the L3 cavity mode, which results from quantization of the line-defect mode corresponding to a missing row of holes along the Γ–K direction or *W*1 *waveguide* [45]: the quantized wave vector in the extended zone scheme is of the order of $4\pi/(3a)$, which explains the period of the oscillations. Figure 4.6 implies that there are three favorable positions for placing a dipole emitter at field antinodes: however, the extension of the region where the field is close to its maximum value is of the order of $\pm 0.1a$, i.e., much smaller than the lattice constant. From Fig. 4.6 it is also possible to deduce the *mode volume*, which is generally defined as [46, 47]

$$\widetilde{V} = \frac{\int \varepsilon(r)|E(r)|^2 \, dr}{\varepsilon(r_{\text{peak}})|E(r_{\text{peak}})|^2}, \qquad (16)$$

where r_{peak} is the peak position of the product $\varepsilon(r)|E(r)|^2$ and the integral is the normalization of the electric field. The mode volume for the field shown in Fig. 4.6 is $\widetilde{V} = (\varepsilon_r |E(r_{\text{peak}})|^2)^{-1} \simeq 0.67 a^3$, where the electric field is normalized according to Eq. (7), and since the dimensionless frequency of the cavity mode is $a/\lambda \simeq 0.27$ we get $\widetilde{V} \simeq 0.56(\lambda/n_r)^3$ ($n_r = \sqrt{\varepsilon_r}$ is the dielectric constant at the peak position). The mode volume of this kind of PhC nanocavity is smaller than a wavelength cubed, i.e., the electromagnetic field is very well confined in all three spatial directions.

4.5
Strong Exciton-Light Coupling in Nanocavities

In this section we consider a single InAs quantum dot (QD) coupled to a PhC nanocavity realized in a GaAs membrane. While quantum-well excitons are described by bosonic operators, the exciton transition in a single quantum dot can be modelled to a first approximation by a two-level system. The theory of radiation-matter coupling in this case relies on the Jaynes–Cummings model. If the QD interacts with a single cavity mode, the Hamiltonian is

$$H = \hbar \Omega_{\text{exc}} \hat{\sigma}_3 + \hbar \omega_\mu (\hat{a}^\dagger_\mu \hat{a}_\mu + \tfrac{1}{2}) + i\hbar \Omega_0 (\hat{\sigma}_- \hat{a}^\dagger_\mu - \hat{\sigma}_+ \hat{a}_\mu), \qquad (17)$$

where Ω_{exc} is the fundamental exciton frequency, $\hat{\sigma}_+, \hat{\sigma}_-, \hat{\sigma}_3$ are pseudo-spin operators for the two-level system with ground (excited) state $|g\rangle$ ($|e\rangle$) and $a^\dagger_\mu, a_\mu$ are creation/destruction operators for the cavity mode μ. The coupling constant $\Omega_0 = \langle \mathbf{d} \cdot \mathbf{E} \rangle / \hbar$ of the quantum dot–cavity interaction is

$$\Omega_0 = \left(\frac{1}{4\pi\varepsilon_0} \frac{\pi e^2 f}{\varepsilon_r m \widetilde{V}_\mu} \right)^{1/2}, \qquad (18)$$

where f is the oscillator strength of the transition, $\widetilde{V}_\mu$ is the mode volume defined in Eq. (16), $\varepsilon_r (\varepsilon_0)$ is the relative (vacuum) permittivity and m is the free-electron

4.5 Strong Exciton-Light Coupling in Nanocavities

mass. We are assuming that the quantum dot is located at the peak position of the electric field. The condition of spatial overlap between the QD and the cavity mode can be met by aligning the cavity around a chosen quantum dot, as demonstrated in Refs. [48–50].

The quantum Hamiltonian (17) has a discrete spectrum consisting in a ladder of dressed states, in which each excited state is split into two levels separated by $2\hbar\Omega_0\sqrt{n+1}$, where n is the number of photons in the cavity mode [51]. In the weak excitation regime, we can consider only the transition between the ground state and the first excited doublet, whose splitting $2\hbar\Omega_0$ corresponds to the vacuum-field Rabi splitting between the QD transition and the single cavity mode. In order to take into account the finite linewidth of both the QD exciton (Γ_{exc}) and the cavity mode ($2\,\mathrm{Im}(\hbar\omega_\mu) = \hbar\omega_\mu/Q_\mu$), a master-equation approach has been used, which allows calculating the spontaneous emission spectrum. This leads to an analytical expression for the complex energy splitting of the two oscillators [52, 53]

$$\hbar\Omega_\pm = \hbar\Omega_{exc} \pm \sqrt{\hbar^2\Omega_0^2 - \left(\frac{\Gamma_{exc}-(\hbar\omega_\mu/Q)}{4}\right)^2} - i\left(\frac{\Gamma_{exc}+(\hbar\omega_\mu/Q)}{4}\right). \quad (19)$$

We assume the quantum dot to be in resonance with the cavity mode, i.e., the QD has to be not only *spatially* but also *spectrally* resonant. Achieving spectral overlap is made difficult by the size distribution of self-assembled QDs. Spectral resonance is imposed here by properly designing the GaAs PhC nanocavity to have the ground cavity mode at energy $\hbar\omega_\mu \sim 1.3$ eV ($\lambda_\mu \sim 950$ nm), which is a typical value for the fundamental exciton resonance of InAs QDs.

The solutions of Eq. (19) are plotted in Fig. 4.7 as a function of the Q-factor. We take a mode volume $\tilde{V} \simeq 1.1 \times 10^{-14}$ cm^3, estimated from $\sim 0.56(\lambda/n_r)^3$ at a wavelength $\lambda = 950$ nm and $n_r = \sqrt{\varepsilon_r} \simeq 3.54$ (low temperature value for GaAs at 1.3 eV). The oscillator strength $f \simeq 10.7$ is a typical value for self-assembled InAs QDs corresponding to the measured lifetime $\tau \sim 1$ ns [10]. The crossover from weak to strong

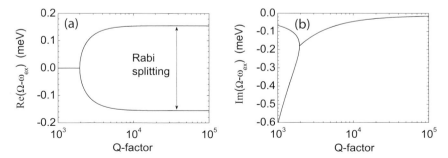

Fig. 4.7 (a) Real and (b) imaginary parts of Eq. (19) as a function of Q-factor, for QD parameters $\Gamma_{exc} = 0.05$ meV, $f = 10.7$, and effective cavity mode volume $\tilde{V} = 1.1 \times 10^{-14}$ cm^3.

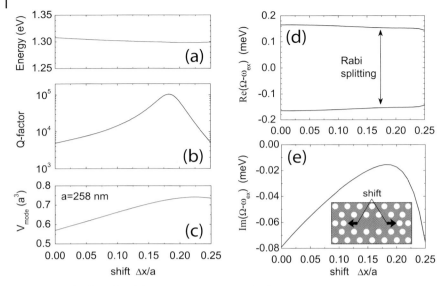

Fig. 4.8 Results for a PhC slab nanocavity in GaAs air bridge. Parameters of the structure are: $\varepsilon_r = 12.53$, $d = 126$ nm, $a = 258$ nm, $r/a = 0.3$. (a) Energy, (b) Q-factor, and (c) mode volume for the cavity mode as a function of the shift of two holes along the x-direction. (d) Real and (e) imaginary parts of Eq. (19), plotted as a function of the holes' shift by using the quantities calculated in (a), (b), and (c), and QD parameters $\Gamma_{exc} = 0.05$ meV, $f = 10.7$.

coupling is seen to appear at $Q \sim 2000$, even if the corresponding imaginary part is still larger than the Rabi splitting. The maximum Rabi splitting for this kind of systems is seen to be reached already for $Q \sim 10\,000$. Such values of Q are well within the reach of present-day fabrication technology, even for cavities in GaAs slabs for which the Q-factor is limited by absorption at the GaAs/native oxide interfaces at the hole sidewalls [50].

In Fig. 4.8(a–c) the calculated mode energy, Q-factor and effective volume are plotted as a function of the holes' shift, $\Delta x/a$, for proper design parameters of the GaAs PhC membrane. The resonance energy and the mode volume do not change appreciably, while the Q-factor has a dramatic increase with a maximum $Q > 10^5$ for $\Delta x/a \sim 0.18$. The latter results are employed to calculate the complex splitting as a function of $\Delta x/a$, which is shown in Fig. 4.8(d) and (e). It is evident that the system is always in the strong coupling regime, regardless of the displacement of the nearby holes. This result is in agreement with the calculations of Fig. 4.7, because the Q-factor is always higher than 2000 for the present nanocavity. The imaginary part of the complex splitting has a minimum for a value of $\Delta x/a$ corresponding to the maximum Q-factor. It is arguable that, in order to observe the strong coupling, shifting the holes in the PhC slab nanocavity could be of importance for reducing the emission linewidth of the two peaks. These results agree well with recent experimental findings [21] as well as with a theoretical study of the strong-coupling based on a Green's function approach [54].

Notice that the physics of the Jaynes–Cummings model (17) is very different from that of the Hamiltonian (13) describing the interaction between photonic modes and quantum-well excitons. The point is that the QW exciton is a delocalized excitation that represents a collection of excited unit cells and has a bosonic character, while the quantum-dot transition is localized and has to be modelled by a two-level system which cannot be excited more than once. The two systems behave in a similar way under weak excitation conditions, but the differences become manifest on increasing the excitation, as the quantumdot transition coupled to the nanocavity mode gives a Rabi splitting that increases like $\sqrt{n+1}$. Indeed, the coupled QD–cavity system is expected to display a Mollow-type spectrum [55] with a classical Rabi splitting at high excitation intensity. Of course, a more complete model should take into account bi-exciton and multi-exciton states of the quantum dot [56, 57] with their complex many-body interactions.

4.6
Conclusions

Photonic crystal slabs are very suitable systems for the control of light propagation and confinement in all spatial directions. A recently developed theory of photonic crystal slabs has been reviewed. The main conclusions of the present work are as follows.

Exciton-polaritons can form in PhC slabs with embedded quantum wells when a narrow excitonic transition is in resonance with either a truly-guided or a quasi-guided photonic mode: in the latter case, the intrinsic linewidths of the exciton and of the photonic mode need to be smaller than the coupling matrix elements. When these conditions are met, the polariton splitting can be larger than for polariton in microcavities, due to the tighter field confinement in a high-index planar waveguide. Polaritons arising from excitons in interaction with quasi-guided modes are radiative and can be probed by reflectance from the slab surface. The results of quantum and semiclassical treatments of photonic crystal polaritons are in very good agreement with each other.

Nanoscale cavities realized in photonic crystal slabs have very high Q-factors and low mode volumes: on these respects they are more performing than usual 3D microcavities like micro-pillars and micro-disks. On the other hand, their electric field profile is rapidly varying, making spatial alignment of a single quantum dot more difficult to achieve. Starting from a Jaynes–Cummings model for a two-level system coupled to a cavity mode, we have quantified the conditions for a single QD transition interacting with a PhC nanocavity to be in a strong-coupling regime with a vacuum-field Rabi splitting. Quality factors larger than 2000 are already sufficient to achieve strong coupling. This makes PhC nanocavities very promising systems for quantum-electrodynamics applications at a nanoscale level.

Acknowledgements

This paper is dedicated to Professor Marc Ilegems on the occasion of his sixty-fifth birthday. Useful discussions with R. Ferrini, K. Hennessy and R. Houdré are gratefully acknowledged. This work was supported by the Italian Ministry for Education, University and Research (MIUR) through Cofin and FIRB programs as well as by the National Institute for the Physics of Matter (INFM) through PRA PHOTONIC.

References

[1] J. J. Hopfield, Phys. Rev. **112**, 1555 (1958).
[2] V. M. Agranovich, J. Exp. Theor. Phys. **37**, 430 (1959) [Sov. Phys. JETP **37**, 307 (1960)].
[3] L. C. Andreani, in: Confined Electrons and Photons – New Physics and Devices, edited by E. Burstein and C. Weisbuch (Plenum Press, New York, 1995), p. 57.
[4] V. Savona, in: Confined Photon Systems – Fundamentals and Applications, edited by H. Benisty, J.-M. Gérard, R. Houdré, J. Rarity, and C. Weisbuch (Springer, Berlin, 1999), p. 173.
[5] L. C. Andreani, in: Electron and Photon Confinement in Semiconductor Nanostructures, edited by B. Deveaud, A. Quattropani, and P. Schwendimann (IOS Press, Amsterdam, 2003), p. 105.
[6] C. Weisbuch, M. Nishioka, A. Ishikawa, and Y. Arakawa, Phys. Rev. Lett. **69**, 3314 (1992).
[7] R. Houdré, C. Weisbuch, R. P. Stanley, U. Oesterle, P. Pellandini, and M. Ilegems, Phys. Rev. Lett **73**, 1043 (1994).
[8] R. Houdré, R. P. Stanley, U. Oesterle, M. Ilegems, and C. Weisbuch, Phys. Rev. B **49**, 16761 (1994).
[9] M. S. Skolnick, T. A. Fisher, and D. M. Whittaker, Semicond. Sci. Technol. **13**, 645 (1998).
[10] J.-M. Gérard, B. Sermage, B. Gayral, B. Legrand, E. Costard, and V. Thierry-Mieg, Phys. Rev. Lett. **81**, 1110 (1998).
[11] J. Bloch, F. Boeuf, J.-M. Gérard, B. Legrand, J.-Y. Marzin, R. Planel, V. Thierry-Mieg, and E. Costard, Physica E **2**, 915 (1998).
[12] B. Gayral, J.-M. Gérard, B. Sermage, A. Lamaitre, and C. Dupuis, Appl. Phys. Lett. **78**, 2828 (2001).
[13] E. Yablonovitch, Phys. Rev. Lett. **58**, 2059 (1987).
[14] S. John, Phys. Rev. Lett. **58**, 2486 (1987).
[15] K. Sakoda, Optical Properties of Photonic Crystals (Springer, Berlin, 2001).
[16] S. G. Johnson and J. D. Joannopoulos, Photonic Crystals: the Road from Theory to Practice (Kluwer, Dordrecht, 2002).
[17] For recent reviews, see e.g. papers in IEEE J. Quantum Electron. **38**, Feature Section on Photonic Crystal Structures and Applications, edited by T. F. Krauss and T. Baba (2002), pp. 724–963.
[18] T. Akahane, T. Asane, B.-S. Song, and S. Noda, Nature **425**, 944 (2003).
[19] M. Notomi, A. Shinya, S. Mitsugi, E. Kuramochi, and H.-Y. Ryu, Opt. Express **12**, 1551 (2004).
[20] J. P. Reithmaier, G. Sȩk, A. Löffler, C. Hofmann, S. Kuhn, S. Reitzenstein, L. V. Keldysh, V. D. Kulakovskii, T. L. Reinecke, and A. Forchel, Nature **432**, 197 (2004).
[21] T. Yoshie, A. Scherer, J. Hendrickson, G. Khitrova, H. M. Gibbs, G. Rupper, C. Ell, O. B. Shchekin, and D. G. Deppe, Nature **432**, 200 (2004).
[22] T. Fujita, Y. Sato, T. Kuitani, and T. Ishihara, Phys. Rev. B **57**, 12428 (1998).
[23] V. N. Astratov, D. M. Whittaker, I. S. Culshaw, R. M. Stevenson, M. S. Skolnick, T. F. Krauss, and R. M. De La Rue, Phys. Rev. B **60**, 16255 (1999).
[24] V. Pacradouni, W. J. Mandeville, A. R. Cowan, P. Paddon, J. F. Young, and S. R. Johnson, Phys. Rev. B **62**, 4204 (2000).
[25] M. Patrini, M. Galli, F. Marabelli, M. Agio, L. C. Andreani, D. Peyrade, and Y. Chen, IEEE J. Quantum Electron. **38**, 885 (2002).

[26] A. M. Malvezzi, G. Vecchi, M. Patrini, G. Guizzetti, L. C. Andreani, F. Romanato, L. Businaro, E. Di Fabrizio, A. Passaseo, and M. De Vittorio, Phys. Rev. B **68**, 161306 (R) (2003).
[27] J. P. Mondia, H. M. Van Driel, W. Jiang, A. R. Cowan, and J. F. Young, Opt. Lett. **28**, 2500 (2003).
[28] J. Torres, D. Coquillat, R. Legros, J. P. Lascaray, F. Teppe, D. Scalbert, D. Peyrade, Y. Chen, O. Briot, M. Le Vassor d'Yerville, E. Centeno, D. Cassagne, and J. P. Albert, Phys. Rev. B **69**, 085105 (2004).
[29] G. Vecchi, J. Torres, D. Coquillat, M. Le Vassor D'Yerville, and A. M. Malvezzi, Appl. Phys. Lett. **84**, 1245 (2004).
[30] M. Galli, M. Belotti, D. Bajoni, M. Patrini, G. Guizzetti, D. Gerace, M. Agio, L. C. Andreani, and Y. Chen, Phys. Rev. B **70**, 081307(R) (2004).
[31] L. C. Andreani and M. Agio, IEEE J. Quantum Electron. **38**, 891 (2002).
[32] T. Ochiai and K. Sakoda, Phys. Rev. B **64**, 045108 (2001).
[33] L. C. Andreani, phys. stat. sol. (b) **234**, 139 (2002).
[34] D. Gerace, Photonic Modes and Radiation–Matter Interaction in Photonic Crystal Slabs, Ph.D. thesis, University of Pavia (2004).
[35] G. Panzarini and L. C. Andreani, Phys. Rev. B **60**, 16799 (1999).
[36] L. A. Dunbar, R. P. Stanley, M. Lynch, J. Hegarty, U. Oesterle, R. Houdré, and M. Ilegems, Phys. Rev. B **66**, 195307 (2002).
[37] D. M. Whittaker and I. S. Culshaw, Phys. Rev. B **60**, 2610 (1999).
[38] A. L. Yablonskii, E. A. Muljarov, N. A. Gippius, S. G. Tikhodeev, T. Fujita, and T. Ishihara, J. Phys. Soc. Jpn. **70**, 1137 (2001).
[39] D. Gerace, M. Agio, and L. C. Andreani, phys. stat. sol. (c) **1**, 446 (2004).
[40] V. Savona, L. C. Andreani, P. Schwendimann, and A. Quattropani, Solid State Commun. **93**, 733 (1995).
[41] K. Srinivasan and O. Painter, Opt. Express **10**, 670 (2002).
[42] Y. Akahane, T. Asano, B.-S. Song, and S. Noda, Opt. Express **13**, 1202 (2005).
[43] C. Sauvan, Ph. Lalanne, and J. P. Hugonin, Phys. Rev. B **71**, 165118 (2005).
[44] B.-S. Song, S. Noda, T. Asano, and Y. Akahane, Nature Mater. **4**, 207 (2005).
[45] L. C. Andreani, D. Gerace, and M. Agio, Photon. Nanostruct. **2**, 103 (2004).
[46] O. Painter, R. K. Lee, A. Scherer, A. Yariv, J. D. O'Brien, P. D. Dapkus, and I. Kim, Science **284**, 1819 (1999).
[47] S. T. Ho, L. Wang, and S. Park, in: Confined Photon Systems – Fundamentals and Applications, edited by H. Benisty, J.-M. Gérard, R. Houdré, J. Rarity, and C. Weisbuch (Springer, Berlin, 1999), p. 243.
[48] K. Hennessy, C. Reese, A. Badolato, C. F. Wang, A. Imamoğlu, P. M. Petroff, E. Hu, G. Jin, S. Shi, and D. W. Prather, Appl. Phys. Lett. **83**, 3650 (2003).
[49] K. Hennessy, A. Badolato, P. M. Petroff, and E. Hu, Photon. Nanostruct. **2**, 65 (2004).
[50] A. Badolato, K. Hennessy, M. Atatüre, J. Dreiser, E. Hu, P. M. Petroff, and A. Imamoğlu, Science **308**, 1158 (2005).
[51] S. Haroche in: Fundamental Systems in Quantum Optics, edited by J. Dalibard, J. M. Raimond, and J. Zinn-Justin (Elsevier, Amsterdam, 1992).
[52] H. J. Carmichael, R. J. Brecha, M. G. Raizen, H. J. Kimble, and P. R. Rice, Phys. Rev. A **40**, 5516 (1989).
[53] L. C. Andreani, G. Panzarini, and J.-M. Gérard, Phys. Rev. B **60**, 13276 (1999).
[54] S. Hughes and H. Kamada, Phys. Rev. B **70**, 195313 (2004).
[55] B. R. Mollow, Phys. Rev. **188**, 1969 (1969).
[56] E. Moreau, I. Robert, L. Manin, V. Thierry-Mieg, J.-M. Gérard, and I. Abram, Phys. Rev. Lett. **87**, 183601 (2001).
[57] D. V. Regelman, U. Mizrahi, D. Gershoni, E. Ehrenfreund, W. V. Schoenfeld, and P. M. Petroff, Phys. Rev. Lett. **87**, 257401 (2001).

Chapter 5
Parametric Amplification and Polariton Liquids in Semiconductor Microcavities

Jeremy J. Baumberg and Pavlos G. Lagoudakis

5.1
Introduction

The discovery of parametric amplification in semiconductor microcavities in 2000 has opened up a new highly-nonlinear optical regime to explore [1, 2]. Prior to this, optically-induced changes in the response of excitons in a semiconductor were at the level of a few per cent or less, before the photo-injected carriers screened the excitons into ionization. In semiconductor microcavities, the induced changes can be several thousand percent while still retaining the bound exciton [3]. This opens the way to exploring in some detail the nonlinear dynamics of excitons, and in particular of exciton-polaritons. Because polaritons with small kinetic energy live on a distorted dispersion relation, their dynamics is rather different to excitons, and can be directly observed [4].

Here, we explore some of the rich possibilities that emerge from the dynamics of polaritons. We aim to capture the broad feel of the phenomena through time resolved studies, which have been essential to clarify and disentangle the different scattering process that can occur. The predominant process is the pair scattering of two polaritons to different final states along the dispersion relation, which is constrained by the shape of the lower polariton dispersion [5]. We first review the simplest pair scattering, which occurs with the pump pulse incident at a 'magic' angle (in the language of non-linear optics, we would call this a triply phase-matched angle). We also explore the relationship between nonlinear optics and semiconductor quasiparticle scattering. We then show that multiple scattering plays a role and new 'nonlinear' polariton modes can appear, further distorting the dispersion [6]. Moving to a geometry in which the pump pulse is normally incident, we explore the way that the polariton dispersion can transiently distort, and propose a model in which the nonlinear interaction of light (of particular in-plane k) with polaritons can be thought of in terms of a k-dependent oscillator strength. We assume the reader is familiar with the strong-coupling regime of exciton-polaritons in a microcavity, and refer to previous reviews in this field [7].

Physics of Semiconductor Microcavities: From Fundamentals to Nanoscale Devices
Edited by Benoit Deveaud
Copyright © 2007 WILEY-VCH Verlag GmbH & Co. KGaA, Weinheim
ISBN: 978-3-527-40561-9

5.2
Parametric Scattering at the Magic Angle

5.2.1
Ultrafast Experiments on Semiconductor Microcavities

Measuring the dynamics of polaritons in semiconductor microcavities requires an additional expansion of the tools of ultrafast spectroscopy. Undertaking angle-dependent measurements is vital to identify the different mechanisms involved in the scattering of these composite particles. Here we use two different techniques: angle-dependent pump–probe differential reflection/transmission, and angle-dependent luminescence analysed in a spectrometer or a streak camera. This required us to develop the first ultrafast goniometer in which several laser pulses can be adjusted in their incident angle on a sample without changing their time-delay. A number of implementations are possible in which the total path length travelled by each pulse remains constant while the incident angle (and hence k) is tuned [8]. The microcavity samples here are cooled to 4 K using a wide field-of-view, cold-finger cryostat, allowing us to collect the light emission emerging at a range of angles.

Parametric scattering has been observed in a number of samples in our group, including InGaAs Multiple Quantum Wells (MQWs) in $3\lambda/2$ cavities, and GaAs Single Quantum Wells (SQWs) in λ and 2λ cavities. Many groups have also observed the effects in alternative strongly-coupled heterostructures, including II–VI microcavities [9, 10]. A number of questions remain open on the different efficiency of the scattering in different samples, including the effects of the number and placing of the quantum wells, the role of disorder both in the QW and the mirror stacks, as well as the composition. While the standard coupled mode theory [11] gives a good account of the angle- and detuning-dependence of the gain, it cannot account for the temperature-, heterostructure-, and disorder-dependence. To account for these effects it is then needed to consider scattering with all the exciton and polariton states in the microcavity. However, there still does not exist a good understanding of which states become occupied after excitation on the lower polariton branch. For instance, upper branch emission is also seen after lower branch excitation. Nor do we have a good understanding of how occupation of other states on the polariton dispersion affects pair scattering at the magic angle. An alternative treatment including all 2-exciton states has highlighted the role of nonlinear absorption at the idler in controlling the gain possible in the parametric amplification process [12].

5.2.2
Simple Pair Scattering

The simplest pair scattering process is one in which two polaritons at k_p mutually scatter to polaritons at $k = 0$, $2k_p$ (Fig. 5.1a). This process can only occur near the region of the lower polariton dispersion relation, which satisfies energy and momentum conservation between initial and final states. This polariton region is given

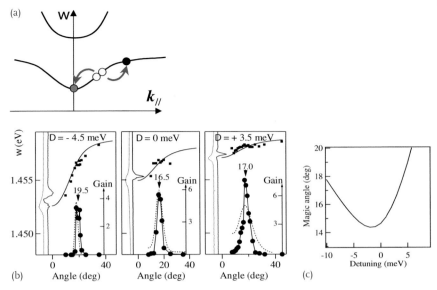

Fig. 5.1 (a) Schematic magic angle parametric amplification, for final states at $k = 0, 2\,k_p$. (b) Measured gain of a weak probe at normal incidence as a function of pump angle, for various detunings. (c) Optimum angle for gain as a function of detuning, for a typical InGaAs/GaAs microcavity.

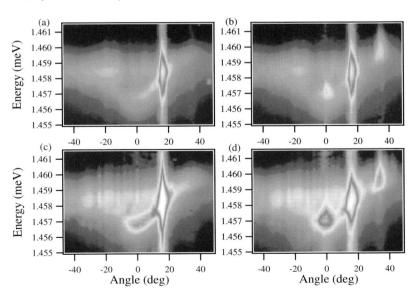

Fig. 5.2 Time-integrated emission (log scale) from the lower polariton branch with incident pump pulse $\theta_{pump} = 16°$, when the normally-incident probe pulse is absent (a, c) or present (b, d), at low (a, b) 10 W/cm^2 or high (c, d) 30 W/cm^2 pump power.

by the solution of $2E(\mathbf{k}_p) = E(0) + E(2\mathbf{k}_p)$. This can be clearly seen experimentally [1] when the gain of the normally incident probe pulse is measured as a function of the incident pump angle, at several detuning conditions between the cavity and the exciton resonance, $\Delta = E_c - E_x$ (Fig. 5.1b).

The optimum gain occurs at a magic angle which reaches a minimum near zero detuning (Fig. 5.1b, c), due to the way the shape of the dispersion changes with detuning. The strength of the pair scattering depends on the exciton fraction in the initial and final states and the strength of the dephasing of these states, and hence large positive or negative detuning conditions reduce the gain. Pair scattering is virtually absent on the upper polariton dispersion, or on the bare exciton dispersion, since the dispersion for these does not favour energy-momentum conservation.

To track the pair scattering process in more detail, the light emitted from the lower polariton branch is recorded over a wide range of angles, both in the spontaneous (no probe) and stimulated (with probe) regime (Fig. 5.2).

When the probe pulse is absent, polaritons scatter in pairs from the pump into a wide range of final states along the lower polariton dispersion [Fig. 5.2(a, c)]. The higher $\mathbf{k}$ states appear weaker because their photon fraction is smaller, so they escape more slowly from the microcavity, while the bigger exciton fraction of higher $\mathbf{k}$ states increases the probability of scattering to states outside the light cone. When the probe pulse is present, the pair scattering preferentially picks out the $\mathbf{k} = 0$ 'signal' state. This language of pump, signal (low energy) and idler (high energy) modes originates from optical parametric oscillator (OPO) theory. Occupation of a polariton state *increases* the probability of scattering into that state; there is a tendency for polaritons to accumulate in particular positions along the dispersion. The transition of the pair scattering from spontaneous in the absence of the probe to stimulated when the probe is present is a manifestation of the bosonic symmetry of the polariton wavefunction. The absence of such transitions in bare quantum wells implies that polaritons are 'better' bosons (the lower energy polaritons have

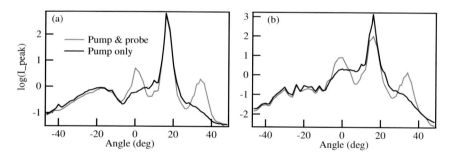

Fig. 5.3 Maximum emission intensity at each angle (log scale) from Fig. 5.2, with and without the probe pulse at low (a) 10 W/cm² and high (b) 30 W/cm² pump power. Regions of resonant Rayleigh pump scattering, spontaneous parametric scattering and stimulated pair scattering are observed.

less multi-particle scattering mixed into their states) and paves the way for the study of polariton Bose condensation related phenomena [13–15]. The corresponding 'idler' polaritons are clearly visible at $2k_p$ (Figs. 5.2 and 5.3). However it is possible to note already a consistent feature of the idler polaritons, in that their energy spectrum and their wavevector distribution are considerably broader than that of the signal polaritons.

The other universal feature observed is the blue shift of the entire polariton dispersion when the strong pump injects significant polariton densities at the magic angle. This blue shift matches very well the simple theory, which is normally used to quantitatively describe the parametric scattering [16, 17]. Following this, it is possible to cast the equations for the coupling of the signal, idler and pump polaritons into a form which shows that new mixed states composed of both signal and idler experience the gain (Section 5.2.3). These mixed states are the eigenstates of the perturbed system and contain components from different in-plane wavevectors [18].

5.2.3
Quasimode Theory of Parametric Amplification

In this section, we extend the theories developed for CW parametric scattering to the dynamic regime. We aim to find the transient eigenstates of the pair polaritons at each time, which independently experience the gain/loss. We assume a slowly varying polariton amplitude (which is a reasonable approximation for these narrow spectral linewidth cavities), and also work in the limit of negligible pump depletion (i.e. at low probe powers). In this case the equations governing the slowly-varying envelope of signal (S) and idler (I) can be written

$$\frac{\partial S}{\partial t} = -\gamma_S S - \Lambda I^*$$

$$\frac{\partial I^*}{\partial t} = -\gamma_I I^* - \Lambda^* S \qquad (1)$$

where $\Lambda(t) = iVP(t)^2 e^{i\nu t}$ accounts for the coupling. Here V is the exchange interaction between polaritons, P is the dynamic pump polariton occupation, and $\nu = 2\omega_P - \omega_S - \omega_I$ is the frequency mismatch from the magic angle condition. We look for solutions corresponding to gain: $S, I^* \propto e^{qt}$. Solving the determinant of Eq. (1) produces the two solutions for the damping:

$$q_\pm = -\left(\frac{\gamma_S + \gamma_I}{2}\right) \pm \sqrt{\alpha^2 + |\Lambda|^2} \qquad (2)$$

with $\alpha = (\gamma_S - \gamma_I)/2$. These solutions are time dependent, with $q_\pm < 0$ away from the pump pulse corresponding to the individual damping of signal and idler. They

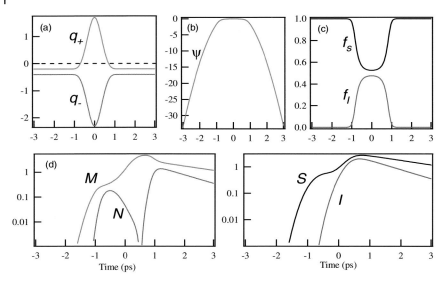

Fig. 5.4 Quasimode calculations as a function of real time (ps) for (a) eigenvalues of *M, N*, (b) mixing parameter ψ, (c) fractional amount of signal and idler components in *M*, (d) dynamics of eigenmodes *M, N* and (e) of signal and idler, when the probe pulse is at $t = -1$ ps, pump at $t = 0$ ps.

repel strongly when the pump arrives, to produce transient gain ($q_+ > 0$, Fig. 5.4a). The eigenvectors of these solutions correspond to the two mixed modes (*M, N*) which experience these gains.

$$M = \frac{1}{\sqrt{1+e^{2\psi}}}[-e^{i\varphi} S + e^{\psi} I^*]$$

$$N = \frac{1}{\sqrt{1+e^{2\psi}}}[e^{\psi} S + e^{i\varphi} I^*] \tag{3}$$

where we have defined $\sinh \psi = \alpha/|A|$ and $|A|e^{i\varphi} = A$. This mixed complex transformation of the signal and idler is controlled by the phase mismatch, $\varphi = vt$, and a mixing parameter, $\psi(t)$. The mode *M* is amplified when the pump pulse arrives, while the mode *N* is de-amplified. The gain of these modes is given by $q_\pm = \bar{\gamma} \pm |A| \cosh \psi$, with the average damping, $\bar{\gamma} = (\gamma_S + \gamma_I)/2$. The incident probe couples into both modes, giving new instantaneously decoupled dynamical equations:

$$\frac{\partial M}{\partial t} = q_+ M - \frac{e^{i\varphi}}{\sqrt{1+e^{2\psi}}} S_{\text{probe}}(t)$$

$$\frac{\partial N}{\partial t} = q_- N + \frac{e^\psi}{\sqrt{1+e^{2\psi}}} S_{\text{probe}}(t) \tag{4}$$

In the vicinity of the pump pulse, the modes M, N contain roughly equal admixtures of the signal and idler (Fig. 5.4c): in other words, when the pump is present, the true modes of the system are not S, I but M, N. The dynamics of the quasi-uncoupled modes and the signal and idler are shown in Fig. 5.4d, e for a probe pulse which is 1 ps before the pump, and with damping of signal and idler, $\gamma_{s,i} = 0.2, 0.4$ meV corresponding to the experiments.

From these equations it can be seen that the amplification of the population of polaritons in the M-mode is roughly given by:

$$\left|\frac{M_{\text{out}}}{M_{\text{in}}}\right|^2 \approx \exp(2q_+ T) \approx \exp(2|\Lambda|T) \approx \exp(2VI_{\text{pump}}T) \tag{5}$$

where T is the pulselength and I_{pump} is the pump power. This recovers the experimental result. It is also not what might be at first expected from a pair scattering process which in an uncoupled system would have a gain proportional to the *square* of the pump intensity. The completely mixed nature of signal and idler polaritons is what makes the parametric amplification so sensitive to dephasing of the idler component.

5.2.4
Multiple Scattering at the Magic Angle

The above analysis shows that the parametric process creates new mixed pair states linked by the scattering with the strong pump. Effectively the strong pump acts to tie together polariton states in a pair-wise fashion. However, in pulsed experiments it is possible to transiently inject very large pump polariton densities, which in turn can lead to very large polariton densities at the signal and idler. These in turn act to tie together new pairs of polariton states, further distorting the polariton dispersion relation. We have previously analysed this case in some detail and summarise the results here to draw together the theme of parametric amplification.

If the data of Fig. 5.2(d) are analysed carefully it is possible to see additional polariton emission at positions which do not lie on the blue-shifted dispersion relation (Fig. 5.5) [6]. These directly correspond to the expected positions of new polariton branches created by 'dressing' the dispersion with signal, idler and pump polaritons. A number of off-branch modes have also now been observed under CW excitation [19].

Hence the meaning of the dispersion relation becomes more difficult to ascertain in the highly-nonlinear regime of parametric scattering in semiconductor microavities. At any moment in time, a range of parametric interactions are available,

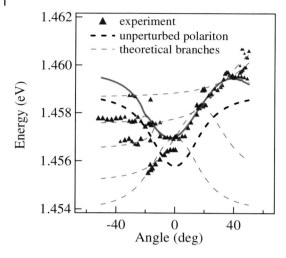

Fig. 5.5 Extracted peak positions of emitted modes, with both pump and probe present (from Fig. 5.2d). Off-branch polaritons can be seen produced by renormalisation of the polariton dispersion under multiple scattering.

whose strengths also control the *k*-dependent self-energy shifts of the dispersion-relation. In turn, this modifies the allowed parametric interactions. This new regime of nonlinear optics more closely corresponds to the strong-coupling of He atoms in superfluidity, and could be termed a strongly-correlated '*polariton liquid*' [20, 21].

5.2.5
Double Resonant On-Branch Multiple Scattering

The experiments presented so far have concentrated on the simplest pair scattering situation with the probe pulse arriving at normal incidence. However, it is clear from the spontaneous scattering regime (e.g. Fig. 5.2c) that many final pair states can satisfy energy-momentum conservation along the lower polariton dispersion. We describe now the results when the probe pulse is shifted to an angle of incidence of 6°, while the pump pulse remains at the magic angle of 17°. In this case we see not just the signal (S1) (as an amplified probe) and idler (I1) at 28° but a second signal beam (S2) emerging at an angle of −4° from the sample. Complementing these two signals is a second idler pulse (I2) at 42°. These results are summarized in Fig. 5.6.

Two possible routes exist for the pair scattering process to populate S2 and I2: either directly from the pump (P + P → S2 + I2), or from the coupled pair of signal and idler (S1 + I1 → S2 + I2). This process can be seeded either from the S2 (which can be populated by Rayleigh scattering of the initial signal) or from I2 (which is near degenerate with the exciton). This multiple scattering of coherent bosons can also be accounted for as the four-wave-mixing of pump and probe pulses, producing S2 and I1. In this equally valid description it is however less clear how to account for the strength of I2, which must arise from higher mixing. Both descriptions similarly treat the mixing of coherent bosonic fields, however four-wave-mixing

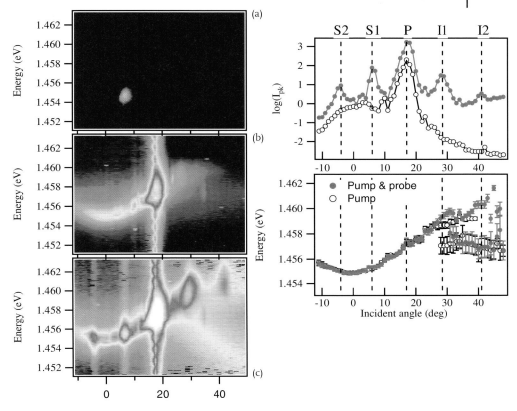

Fig. 5.6 Lower polariton branch emission (on a log scale), for the three cases (a) probe alone at 6°, (b) pump alone at 17°, (c) both pump and probe incident. With the probe arriving away from $k = 0$, new multiple scattering processes take place, giving rise to a second signal at −4° and a second idler at 42°. The gain for the signal is ~400 in these conditions.

retains coherence in the photon fields and polarizations, while the parametric scattering uses the polariton basis which more naturally accounts for the different properties of polaritons along the lower branch. The double-resonant pair scattering process seen here depends on all four states (S1, S2, I1, S2) being resonant on the polariton branches for its strength. This is only true for certain selected conditions of injected pump and probe polaritons, and for example does not work when the probe pulse arrives at −6°. In general, as the probe in-plane wavevector approaches that of the pump polaritons, the number of multiple scattering processes increases strongly, allowing many orders to be observed.

Another interesting question is the extent to which the modes S1, S2, I1, I2 form a new mixed polariton state with four components. It should be possible to probe intensity and phase correlations in the photons emitted from these states. In the Fourier domain, these correspond to spatial correlations in the polaritons inside the microcavity. However it is noticeable that the second idler is predominantly built of exciton states, and thus suffers very strong scattering processes. This may account for the large increase in emission from a broad range of lower energy states at high angles ($\theta > 40°$) when the multiple scattering process turns on. It is also possible that such broadband emission observed at large angles originates from localised exciton states (at smaller angles the strong emission from the pump and the polariton states in the trap hinders detection of light emitted from localised states that exist is small numbers). Under non-resonant excitation, emission from localised states can easily be observed and even lead to the appearance of a lasing mode due to the achievable population inversion in these low density states [22].

5.3
Local Deformations of the Dispersion: Beyond Pair Scattering

The complete admixture of the idler in the mixed polariton state which is amplified explains the limitations of the parametric amplification. A number of careful experiments have compared how the gain changes with temperature, and number and type of quantum wells [23]. Two models have been advanced to account for this data, the first linking the increase in idler decoherence to the exciton binding energy, the second ascribing this to excitation induced dephasing of the pair-state to higher-lying continuum states [11, 12]. The common theme is clearly the proximity of the idler component of this mixed state to the very large density of excitons states (both localised and not) and unbound electronic states [24, 25]. One way to progress is thus to identify processes in which the idler can be energetically lower down inside the polariton trap, and thus better protected from these dephasing mechanisms.

A possible approach uses pump and probe beams at angles much closer to normal incidence, with all the scattering polaritons thus caught in the trap and so more stable against ionization and scattering. The problem is the small gain in this configuration due to the less advantageous shape of the dispersion for energy conservation in the pair scattering, and the short lifetime of polaritons before they escape as photons emitted from the sample [26, 27]. Only the broadening of the polariton mode in both energy and in-plane momentum caused by the finite lifetime of the cavity photons and diffraction in the planar microcavity allows the process to occur at all. This low gain is not predicted by the current models, which provide estimates for the polariton gain which are comparable or exceed those at the magic angle.

5.3.1
Polariton Liquids at the Bottom of the Polariton Trap

To explore the transient energy shifts during the pair scattering process, we collect the time-integrated transmission of pulses at different angles on the far side of the sample, and spectrally analyse it. We compare the results both with, and without, a probe pulse incident at $\theta_{probe} = 7°$ arriving simultaneously with the pump pulse

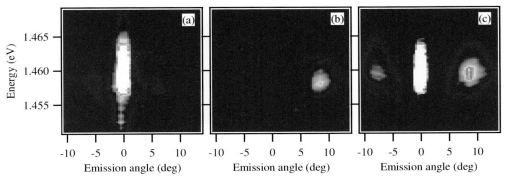

Fig. 5.7 Time-integrated emitted light in transmission from the lower polariton, with the pump pulse normally incident, and the probe arriving at +7°. Parametric scattering is observed when both pulses arrive (c), producing an idler pulse at −7°. For comparison, pump alone (a) and probe alone (b).

which is at normal incidence. The polarization of the pump beam is linear while that of the probe beam is circular, in order to maximize the parametric amplification process [28, 29]. When the pump and probe arrive together, the scattering produces a gain of up to 15 of the probe beam, and a similarly intense backscattered beam in the momentum-matched direction of $-\theta_{probe}$, which is the idler. The strength of the pair scattering driven by the pump depends on the cavity detuning from the exciton and is maximised for $\Delta = -1$ meV. To achieve this gain also depends on increasing the pump bandwidth (to several times the lower polariton linewidth) to allow resonance throughout the transient blue-shifting of the lower polariton branch. The angle dependent emission of pump, probe and phase-conjugate idler are clearly seen in Fig. 5.7 demonstrating the expected energy selectivity in the pair scattering process near the bottom of the polariton trap.

The parametric scattering process at the bottom of the trap nominally does not conserve energy-momentum, with a mismatch of 0.2 meV expected from the unperturbed dispersion. However this should be compared to the lower polariton linewidth for this sample at $k = 0$, which is 0.5 meV. To gain an insight into the effects of the population-induced shifts in the dispersion, we extract the positions of peak emission and the emission strength as the pump power is varied by a factor of twenty (Fig. 5.8). Initially, it can be seen that increasing the pump power increases the signal and idler strength, however they rapidly saturate in intensity. On the other hand, the energy of the signal and idler continually blue shift with pump intensity. What is more surprising is that the the lower polariton branch at the bottom of the trap is rapidly deformed to a "w" shaped non-linear dispersion. Strong blue shifting appears where the pump, signal and idler are located. In the next section we discuss a model that tries to give some insight into these results.

It is also clear that these radical energy shifts can have a dramatic effect on the maximum possible gain observed (which is much less than expected). The gain can be clamped when the energy shifts pass beyond the condition for allowed energy-

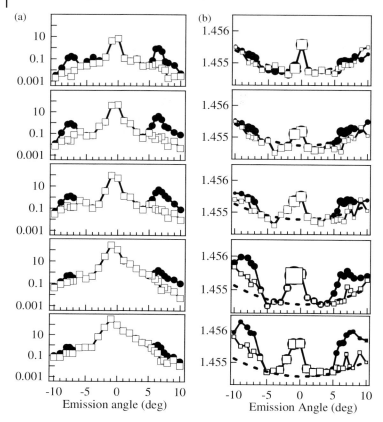

Fig. 5.8 Extracted peak emission strength (a) and peak emission energy (b) across the bottom of the lower polariton trap. The pump is at 0°, and probe at 7°, (open squares = pump only, solid circles = pump & probe), while the dashed line is the unperturbed lower polariton dispersion. The pump power increases from top to bottom, from 0.5 W cm^{-2} to 10 W cm^{-2}, while the probe power is kept constant.

momentum scattering. At higher pump powers, it is clear that pair scattering at other points of the dispersion enables more states on the bottom of the polariton trap to become occupied. Once again this appears to be a situation more like that of the strongly-correlated polariton liquid.

5.3.2
Local Oscillator Strength Model

One way to account for the data observed in the previous section is to treat the blue shifts as arising from the reduction in oscillator strength of the exciton component of the polariton. Since the Rabi splitting depends on the oscillator strength, the non-bosonic residual component of the polaritons which is sensitive to exciton densities can cause a reduction in the Rabi splitting, Ω. However we observe k-dependent energy shifts which change with the polariton populations at each k. We postulate that the admixture of exciton states in the polariton state at a particular k, can correspondingly lead to the change of oscillator strength for *only that admixture of excitons*. In other words, the oscillator strength which causes the Rabi splitting can be k-dependent when it is the polaritons that are populated, $\Omega(k) \propto \sqrt{f[I(k)]}$.

The lower polariton energy is given by

$$\omega_p(k) = \tfrac{1}{2}[\omega_c(k) + \omega_x(k)] - \tfrac{1}{2}\sqrt{[\omega_c(k) - \omega_x(k)]^2 + \Omega(k)^2} \qquad (6)$$

From this, the resulting fractional energy shifts are found to be comparable to the fractional change in oscillator strength $\Delta f/f$. This mechanism would imply that the upper polariton should also show an equivalent k-dependent reduction in energy. However such experiments are hampered by the difficulty of observing the upper polariton simultaneously with the pump–probe parametric scattering – no emission is observed at the upper polariton, and injecting a probe here can radically modify the polariton scattering processes which occur.

Another possible way to account for the observations is that the emission we observe is time-integrated, and that the signal and idler polaritons emerge promptly when the pump polaritons are still in the sample (and thus the dispersion is blueshifted), while the emission at other k occurs much more slowly at much later times after the pump polaritons have decayed from the sample (hence they emerge from the unshifted dispersion). The difficulty with this explanation is that all polaritons should show the same time dependence of emission – there are no non-resonant states excited which would contribute to delayed emission. However to verify this behaviour, it is necessary to directly time-resolve the emission during the parametric scattering as we show in the next section.

5.3.3
Direct Time-Resolved Emission During Parametric Amplification

To directly resolve the dynamical energy-shifts during the parametric scattering process, we use a streak camera to track all the $k = 0$ emission as a function of time together with a broadband 150 fs probe pulse which monitors the lower polariton occupation. We revert to the magic angle geometry (Section 5.2.2) in these preliminary experiments, and vary the time delay between the pump and probe pulses on the microcavity (Fig. 5.9). The signal light emerging from the microcavity at normal incidence after the probe pulse is reflected is directly resolved in time – the dip in the reflected spectrum at $t = 0$ ps is the lower polariton branch. When the pump pulse arrives close in time, a portion of the probe pulse becomes amplified. Clearly visible is the chirp in the signal (or parametrically-amplified probe): the blue-shift is initially large and reduces as time progresses as the pump polaritons are depleted and escape from the sample. The chirp is reduced as the pump pulse arrives at later

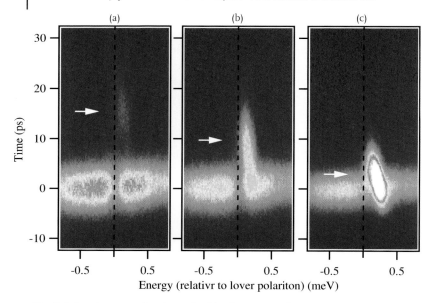

Fig. 5.9 Streak camera directly-resolved $k = 0$ emission when a normally-incident probe pulse arrives at time delays ($t = 0$ ps) close to the pump pulse (delayed at time marked by the arrow) which is at the magic angle. The dip in reflection is the lower polariton. The energy shifts of the $k = 0$ polariton can be directly resolved. In (a) the pump pulse arrives after the probe pulse, and only weak gain is seen, at energies close to the polariton. In (c) the pump pulse is nearly simultaneous, and the gain is large, as is the energy shift.

times. Further experiments are underway to explore the parametric scattering described above, in the polariton liquid regime, to try and disentangle the different processes occurring.

5.4
Historical Perspective (JJB)

For this special issue reviewing our research on semiconductor microcavities, it is perhaps worthwhile to reflect on how we stumbled upon the enormously-strong parametric scattering process in semiconductor microcavities. Since the pioneering work of Arakawa and Weisbuch in identifying the strong coupling regime in these structures, much excellent work in the 1990s at EPFL, Sheffield and elsewhere had gone into understanding how luminescence emerges from the two polariton branches. From my background in ultrafast spectroscopy of semiconductors, it was natural to try and understand the dynamics of polaritons. However one of the most frequent assumptions always made previously in the study of *any* semiconductor heterostructure was that the angle of incident and emitted light is decoupled from any dynamics in the sample, essentially because the lowest energy exciton states observed are relatively localised in space.

Initial experiments that I undertook at Hitachi Cambridge in 1997 in collaboration with the Sheffield group thus used a conventional geometry in which incident pump and probe beams were a few degrees from normal incidence thus observing polaritons only around $k = 0$. The results from these experiments were truly puzzling. Among the crucial observations was that the precise pump spectrum used made a huge difference to both the time-resolved reflectivity, and the four-wave mixing dynamical signatures. By using coherent pulse shaping with real-time spectral filtering [30], we were able to deduce that not only were there drastic differences between pumping the upper polariton alone, and pumping the lower polariton alone, but also when pumping both branches together. Even more surprising, was when pumping the lower polariton simultaneously with states at $k = 0$ at higher energy up to the exciton energy (even though there is no direct absorption into polaritons) produced a markedly different signal. This work signalled that something remarkably odd was sensitive in the polariton system, and on moving the group to Southampton in 1998, I set out to try and understand what this could be.

Research at Sheffield beautifully depicted the angle-dependence of the polariton branches, and I began to wonder about k-selection rules in the system. In particular, the identification of a bottleneck in relaxation of lower polaritons which reduces the rate at which higher-k excitons relax into the lower branch $k = 0$ polaritons suggested that the dynamics would be interesting. Hence I decided to attempt angularly-sensitive pump–probe experiments. To do this required development of a new piece of apparatus for femtosecond spectroscopy since normal beamline geometries had the unfortunate effect of changing the time delay every time the angle of incidence was changed. The femtosecond goniometer that we developed had a nice property that the temporal overlap of pump and probe pulses was preserved (within about 100 fs) as the pump and probe angles were freely adjusted, allowing simple comparison of dynamics at different angles, and hence different polariton in-plane k. Pavlos Savvidis, my first PhD student at Southampton took on the job of measuring the response when pumping higher-k polaritons to see how quickly they could relax to $k = 0$, where they were probed by a second weak pulse. It was during these first initial runs that Pavlos came to me saying there was something peculiar ('wrong') with the data. At certain pump incident angles, the dynamical response changed sign from giving a small pump-induced decrease in the probe reflectivity to giving a vast increase. More puzzling, under investigation, it transpired that more probe light could be reflected from the sample when the pump arrived, than was actually incident on the sample. (This is rather easy to measure since the microcavity reflectivity off the polariton branches is close to 100%). At this point it was clear that some new process which was particularly angular-resonant was occurring, and we rapidly mapped out the angular-, spectral- and ultrafast-spectroscopy to reveal the classic magic-angle parametric scattering process.

One remaining note in this story is salutary. The original paper that we submitted to Physical Review Letters [1] also discussed the spin-selection rules for the parametric scattering. Experimentally we had observed that seeding the stimulated scattering required the *opposite* probe circular polarisation to the pump. The incredulity of the referee to this physics forced us to check repeatedly this fact. However the point about the stimulated scattering is that it is so very strong that an extremely weak probe pulse is necessary, on the order of microwatts. Hence the process of checking directly the circular polarisation helicity of the probe was always done by increasing the probe power and rotating a half-wave plate before a polarisation beam splitter which split the pump–probe beamlines. Typically such optics are imperfect at the level of 10^{-4} and, having confirmed the helicity of the probe, we reduced the probe power assuming that the helicity remained unchanged. The severe scepticism of Benoit Deveaud to our findings eventually forced us to track down the background amount of opposite helicity only present at low probe powers, and confirm the expected intuition of spin-conserving parametric scattering. Interestingly, subsequent research by Pavlos Lagoudakis in the group uncovered a far more complicated story of spin in the parametric scattering process which was partly triggered by the ideas discussed when trying to reconcile such peculiar spin observations.

There are several observations to be taken from the semiconductor microcavity field. Firstly, just because a research sub-topic slows down does not mean that there is not huge life still to be teased out of it. Too quick an opportunistic redirecting of research towards the latest sexy topic will miss much of the physics. Secondly, I realise how little I understood about excitonic states in quantum wells until forced to confront the issues when they are embedded in microcavities. The direct access to states of different k which are here separated clearly in energy provides a spectacular window on the excitonic state. I still understand less about the quantum well electronic states than about the coupled exciton–photons in the microcavity, but I think much progress has been made, visible throughout the papers collected here. Thirdly, the physics opened up by exploring parametric scattering is an excellent paradigm of research in action, and an excellent demonstrator of the breadth of quantum optoelectronics: the connection between Bose-condensation of quasiparticles in the solid state, and the technological application of ultrafast ultra-high-gain micro-parametric oscillators and amplifiers is not an obvious one. Currently these are still vibrant and open questions.

Acknowledgements

We would like to thank Delores Martin-Fernandez, Cristiano Ciuti, and Pavlos Savvidis for the results and ideas they contributed, and Maurice Skolnick and David Whittaker for long-term collaboration, samples and many fruitful discussions.

References

[1] P. G. Savvidis, J. J. Baumberg, R. M. Stevenson, M. S. Skolnick, D. M. Whittaker, and J. S. Roberts, Phys. Rev. Lett. **84**, 1547 (2000).

[2] R. M. Stevenson, V. N. Astratov, M. S. Skolnick, D. M. Whittaker, M. Emam-Ismail, A. I. Tartakovskii, P. G. Savvidis, J. J. Baumberg, and J. S. Roberts, Phys. Rev. Lett. **85**, 3680 (2000).

[3] M. Saba, C. Cuiti, J. Bloch, V. Thierry-Mieg, R. Andre, Le Si Dang, S. Kundermann, A. Mura, G. Bongiovanni, J. L. Staehli, and B. Deveaud, Nature **414**, 731 (2001).

[4] R. Houdré, C. Weisbuch, R. P. Stanley, U. Oesterle, P. Pellandini, and M. Ilegems, Phys. Rev. Lett. **73**, 2043 (1994).

[5] P. G. Savvidis, J. J. Baumberg, R. M. Stevenson, M. S. Skolnick, J. S. Roberts, and D. M. Whittaker, Phys. Rev. B **62**, R13278 (2000).

[6] P. G. Savvidis, C. Ciuti, J. J. Baumberg, D. M. Whittaker, M. S. Skolnick, and J. S. Roberts, Phys. Rev. B **64**, 075311 (2001).

[7] "Special Issue on Microcavities", edited by J. J. Baumberg and L. Vina, Semicond. Sci. Technol. **18**, S279 (2003).
G. Khitrova et al., Rev. Mod. Phys. **71**, 1591 (1999).
M. S. Skolnick, T. A. Fisher, and D. M. Whittaker, Semicond. Sci. Technol. **13**, 645 (1998).
A. Kavokin and G. Malpuech, Cavity Polaritons (Elsevier, Amsterdam, 2003)

[8] P. G. Savvidis and P. G. Lagoudakis, Semicond. Sci. Technol. **18**, S311 (2003).

[9] A. Huynh, J. Tignon, O. Larsson, P. Roussignol, C. Delalande, R. Andre, R. Romestain, and L. S. Dang, Phys. Rev. Lett. **90**, 106401 (2003).

[10] M. Muller, R. Andre, J. Bleuse, R. Romestain, L. S. Dang, A. Huynh, J. Tignon, P. Roussignol, and C. Delalande, Semicond. Sci. Technol. **18**, S319 (2003).

[11] C. Ciuti, P. Schwendimann, and A. Quattropani, Semicond. Sci. Technol. **18**, S279 (2003).

[12] S. Savasta, O. Di Stefano, and R. Girlanda, Semicond. Sci. Technol. **18**, S294 (2003).

[13] D. Porras, C. Ciuti, J. J. Baumberg, and C. Tejedor, Phys. Rev. B **66**, 085304 (2002).

[14] H. Deng, G. Weihs, C. Santori, J. Bloch, and Y. Yamamoto, Science **298**, 199 (2002).

[15] G. Weihs, H. Deng, R. Huang, M. Sugita, F. Tassone and Y. Yamamoto, Semicond. Sci. Technol. **18**, S386 (2003).

[16] C. Ciuti, P. Schwendimann, B. Deveaud, and A. Quattropani, Phys. Rev. B **62**, R4825 (2000).

[17] C. Ciuti, P. Schwendimann, and A. Quattropani, Phys. Rev. B **63**, 041303 (2001).

[18] C. Ciuti, Phys. Rev. B **69** 245304 (2004).

[19] A. I. Tartakovskii, D. N. Krizhanovskii, D. A. Kurysh, V. D. Kulakovskii, M. S. Skolnick, and J. S. Roberts, Phys. Rev. B **65**, 081308 (2002).

[20] L. Tisza, Phys. Rev. **72**, 838 (1947).

[21] O. Penrose and L. Onsager, Phys. Rev. **104**, 576 (1956).

[22] P. G. Lagoudakis, M. D. Martin, J. J. Baumberg, G. Malpuech, and A. Kavokin, J. Appl. Phys. **95**, 8979 (2004).

[23] M. Saba, C. Ciuti, S. Kundermann, J. L. Staehli, and B. Deveaud, Semicond. Sci. Technol. **18**, S325 (2003).

[24] P. G. Lagoudakis, M. D. Martin, J. J. Baumberg, A. Qarry, E. Cohen, and L. N. Pfeiffer, Phys. Rev. Lett. **90**, 206401 (2003).

[25] A. Qarry, G. Ramon, R. Rapaport, E. Cohen, A. Ron, A. Mann, E. Linder, and L. N. Pfeiffer, Phys. Rev. B **67**, 115320 (2003).

[26] F. Quochi, C. Ciuti, G. Bongiovanni, A. Mura, M. Saba, U. Oesterle, M. A. Dupertuis, J. L. Staehli, and B. Deveaud, Phys. Rev. B **59**, R15594 (1999).

[27] G. Messin, J. P. Karr, A. Baas, G. Khitrova, R. Houdre, R. P. Stanley, U. Oesterle, and E. Giacobino, Phys. Rev. Lett. **87**, 127403 (2001).

[28] P. G. Lagoudakis, P. G. Savvidis, J. J. Baumberg, D. M. Whittaker, P. R. Eastham, M. S. Skolnick, and J. S. Roberts, Phys. Rev. B **65**, R161310 (2002).

[29] A. Kavokin, P. G. Lagoudakis, G. Malpuech, and J. J. Baumberg, Phys. Rev. B **67**, 195321 (2003).
[30] J. J. Baumberg, A. Armitage, M. S. Skolnick, and J. Roberts, Phys. Rev. Lett. **81**, 661 (1998).

Chapter 6
Quantum Fluid Effects and Parametric Instabilities in Microcavities

Cristiano Ciuti and Iacopo Carusotto

6.1
Preface

In the last decade, the research group of Professor Marc Ilegems at EPFL has been working intensively and enthusiastically on the physics and device applications of artificial photonic systems, such as semiconductor microcavities and photonic crystals. In this *Festschrift* paper, we are going to present a theory of some exotic physical properties of coherently excited semiconductor microstructures in the strong exciton–photon coupling regime. We hope that the rich phenomenology here described will contribute to a very pleasant celebration of his 65th birthday.

6.2
Introduction

The behavior of a quantum fluid has played an important role in many fields of condensed matter and atomic physics, ranging from superconductors to Helium fluids [1] and, during the last decade, Bose–Einstein condensates of cold trapped atoms [2]. One of the most dramatic manifestations of macroscopic coherence of an interacting many-body system is superfluidity [3].

In this paper, we will provide a comprehensive theoretical analysis of the predicted non-equilibrium propagation properties of a two-dimensional gas of polaritons in a semiconductor microcavity in the strong exciton–photon coupling regime [4, 5]. Thanks to their photonic component, polaritons can be coherently excited by an applied laser field and detected through the emitted light. Thanks to their excitonic component, polaritons have strong binary interactions, which have been shown to produce spectacular and rich polariton amplification effects through matter-wave stimulated collisions [6–11], as well as spontaneous parametric instabilities [12–18]. Recently, a significant amount of research has been also focusing on the quantum optical properties of the polariton emission in the parametric regime with the possibility of observing polariton squeezing and polariton pair entanglement [19–26].

Physics of Semiconductor Microcavities: From Fundamentals to Nanoscale Devices
Edited by Benoit Deveaud
Copyright © 2007 WILEY-VCH Verlag GmbH & Co. KGaA, Weinheim
ISBN: 978-3-527-40561-9

Here, we are going to present a detailed discussion of the interplay between the peculiar polariton collective excitations and the rich variety of parametric instabilities, which occur in presence of a nearly-resonant continuous wave pump laser. In addition, we are going to discuss the impact on the propagation properties of a moving polariton fluid and analyze in detail the superfluid regime, which we predicted in a very recent Letter [27]. As our system is a strongly non-equilibrium one, the polariton field oscillation frequency is not fixed by any equation of state relating the chemical potential to the polariton density, but it can be tuned by the frequency of the exciting pump laser. This opens the possibility of having a collective excitation spectrum which has no counterpart in usual systems close to thermal equilibrium. In particular, we will analyze the propagation in presence of a static potential (either for the photonic or exciton component), which is known to produce resonant Rayleigh scattering (RRS) of the exciting laser field [14, 28–30]. Superfluidity of the polariton fluid manifests itself as a dramatic collapse of the resonant Rayleigh scattering intensity when the flow velocity imprinted by the exciting laser beam is slower than the interaction-induced sound velocity in the polariton fluid. Furthermore, a dramatic reshaping of the RRS pattern due to polariton–polariton interactions can be observed in both momentum and real space even at higher flow velocities, e.g. with the appearance of a variety of Cherenkov-like patterns. We will present a rich set of predicted far-field and near-field images of the resonant Rayleigh scattering emission.

6.3
Hamiltonian and Polariton Mean-Field Equations

In order to describe a planar microcavity containing a quantum well with an excitonic resonance strongly coupled to a cavity mode, we will consider the following model Hamiltonian [31]:

$$\mathcal{H} = \int d\mathbf{x} \sum_{i,j \in \{X,C\}} \hat{\Psi}_i^\dagger(\mathbf{x}) [\mathbf{h}_{ij}^0(-i\nabla) + V_i(\mathbf{x}) \, \delta_{ij}] \hat{\Psi}_j(\mathbf{x})$$
$$+ \frac{\hbar g}{2} \int d\mathbf{x} \, \hat{\Psi}_X^\dagger(\mathbf{x}) \hat{\Psi}_X^\dagger(\mathbf{x}) \hat{\Psi}_X(\mathbf{x}) \hat{\Psi}_X(\mathbf{x})$$
$$+ \int d\mathbf{x} \, \hbar F_p(\mathbf{x}, t) \hat{\Psi}_C^\dagger(\mathbf{x}) + \text{h.c.}, \quad (1)$$

where $\mathbf{x}$ is the in-plane spatial position and the indexes i, j run over the exciton (X) and cavity photon (C) indexes corresponding to the field operators $\Psi_X(\mathbf{x})$ and $\Psi_C(\mathbf{x})$ respectively. They satisfy Bose commutation rules, $[\hat{\Psi}_i(\mathbf{x}), \hat{\Psi}_j^\dagger(\mathbf{x}')] = \delta^2(\mathbf{x} - \mathbf{x}') \delta_{ij}$. Note that, for simplicity, we will limit our treatment to the case of polariton modes with the same circular polarization, which can be excited by a circularly polarized pump. The approach here presented can be generalized to the spin-dependent case by considering appropriate spin-dependent exciton–exciton collisional potentials.

In the k-space representation, the single-particle Hamiltonian h^0 has the simple form

$$h^0(k) = \hbar \begin{pmatrix} \omega_X(k) & \Omega_R \\ \Omega_R & \omega_C(k) \end{pmatrix}, \qquad (2)$$

where $\omega_C(k) = \omega_C^0 \sqrt{1 + k^2/k_z^2}$ is the cavity mode energy dispersion as a function of the in-plane wave-vector k and k_z is the quantized photon wavevector along the growth direction. Ω_R is the Rabi frequency of the exciton-cavity photon coupling. In the following, we will consider a flat exciton dispersion $\omega_X(k) = \omega_X$. In this framework of coupled harmonic oscillators, polaritons simply arise as the eigenstates of the linear Hamiltonian (2). $\omega_{\text{LP(UP)}}(k)$ denotes the dispersion of the lower (upper) polariton branch Fig. 6.1(a). From Eq. (2), one obtains the real-space form which appears in Eq. (1) by simply replacing the wavevector k with the operator $-i\nabla$.

The term proportional to $F_p(x,t)$ in Eq. (1) represents an applied coherent laser pump spot, which drives the cavity and injects polaritons. In the following, we will consider the case of a monochromatic laser field of frequency ω_p and plane-wave profile with in-plane wave-vector k_p. The in-plane wave-vector is linked to the incident direction by the simple relationship $k_p = \sin\theta_p \omega_p/c$, θ_p being the pump incidence angle with respect to the growth direction. This means that an oblique pump incidence generates a polariton fluid with a non-zero flow velocity along the cavity

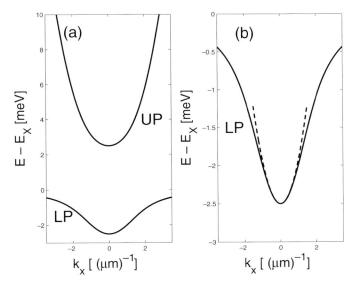

Fig. 6.1 (a) Linear dispersion of the Lower (LP) and Upper (UP) Polariton branches as a function of k_x ($k_y = 0$). Cavity parameters: $\hbar\omega_X = \hbar\omega_C^0 = 1.4$ eV, $2\hbar\Omega_R = 5$ meV. (b) Zoom of the LP branch. The dashed line depicts the parabolic approximation around the bottom of the dispersion.

plane. For $\hbar\omega_p = 1400$ meV, an in-plane wavevector $\mathbf{k}_p = 1$ $(\mu m)^{-1}$ corresponds to a pump incidence angle of $8.1°$.

The nonlinear interaction term in Eq. (1) is due to exciton–exciton collisional interactions and, as usual, is modelled by a repulsive $(g > 0)$ contact potential. For the excitation parameters considered in the present work, the anharmonic exciton–photon coupling [31] has a negligible effect. As we are considering the case of a cavity photon on resonance with the exciton $(\omega_C(0) = \omega_X)$, the wavevector-dependent component of the lower polariton branch is larger than the photonic one. At the level of the present knowledge [31], the collisional exciton–exciton interaction is therefore expected to be an order of magnitude stronger than the anharmonic exciton–photon term, which will be therefore neglected for simplicity in the following of the discussion.

Finally, $V_{X,C}(\mathbf{x})$ are the single particle potentials acting on the exciton and photon fields respectively. These potentials break the translational symmetry along the cavity plane. The exciton potential $V_X(\mathbf{x})$ can be due to natural interface or alloy disorder in the semiconductor quantum wells due to unavoidable growth imperfections. The photonic potential $V_C(\mathbf{x})$ can be due to fluctuations of the cavity length or imperfections in the Bragg reflectors (photonic disorder [24]). More interestingly, $V_C(\mathbf{x})$ can be designed and created deliberately by means of lithographic techniques.

Within the mean-field approximation, the time-evolution of the mean fields $\psi_{X,C}(\mathbf{x}) = \langle \hat{\Psi}_{X,C}(\mathbf{x}) \rangle$ under the Hamiltonian (1) is given by:

$$i\hbar \frac{d}{dt} \begin{pmatrix} \psi_X(\mathbf{x}) \\ \psi_C(\mathbf{x}) \end{pmatrix} = \begin{pmatrix} 0 \\ \hbar F_p e^{i(\mathbf{k}_p \mathbf{x} - \omega_p t)} \end{pmatrix} + \left[h^0 + \begin{pmatrix} V_X(\mathbf{x}) + \hbar g |\psi_X(\mathbf{x})|^2 - \frac{i}{2}\hbar \gamma_X & 0 \\ 0 & V_C(\mathbf{x}) - \frac{i}{2}\hbar \gamma_C \end{pmatrix} \right] \begin{pmatrix} \psi_X(\mathbf{x}) \\ \psi_C(\mathbf{x}) \end{pmatrix}. \quad (3)$$

Using the language of the quantum fluid community, these are the Gross–Pitaevskii equations [2] for our cavity-polariton system. The quantities γ_X and γ_C represent the homogoneous broadening of the exciton and photon modes respectively. In the present work, we will be concerned with an excitation close to the bottom of the LP dispersion, i.e. the region most protected [5] from the exciton reservoir, which may be otherwise responsible for excitation-induced decoherence [32].

6.4
Stationary Solutions in the Homogeneous Case

In the homogeneous case (i.e., $V_{X,C}(\mathbf{x}) = 0$ and translational invariance along the plane), we can look for spatially homogeneous stationary states of the system in which the field has the same plane-wave structure $\psi_{X,C}(\mathbf{x}, t) = \exp[i(\mathbf{k}_p \mathbf{x} - \omega_p t)] \psi_{X,C}^{ss}$ as the incident laser pump field. The mean-field equations

$$\left(\omega_X(\boldsymbol{k}_p) - \omega_p - \frac{i}{2}\gamma_X + g|\psi_X^{ss}|^2\right)\psi_X^{ss} + \Omega_R \psi_C^{ss} = 0, \tag{4}$$

$$\left(\omega_C(\boldsymbol{k}_p) - \omega_p - \frac{i}{2}\gamma_C\right)\psi_C^{ss} + \Omega_R \psi_X^{ss} = -F_p, \tag{5}$$

are the non-equilibrium analogous of the state equation, which in equilibrium systems links the chemical potential to the particle density. Importantly, we stress that while the oscillation frequency of the condensate wavefunction in an isolated gas is equal to the chemical potential μ and therefore it is fixed by the equation of state, in the present driven-dissipative system it is equal to the frequency ω_p of the driving laser and therefore it is an experimentally tunable parameter. Hence, the microcavity polariton system allows us to explore regimes, which are not accessible in systems close to thermal equilibrium, such as the ultracold trapped atoms.

6.5
Linearized Bogoliubov-Like Theory

As usual in the theory of nonlinear systems, stability of the solutions of Eqs. (4, 5) with respect to fluctuations has to be checked by linearizing Eq. (3) around the stationary state. Perturbations can be produced by classical fluctuations of the pump field, quantum noise of the exciton and photon fields as well as the presence of perturbing potentials $V_{C,X}(\boldsymbol{x})$, which have not been considered by the plane-wave solutions in Eqs. (4, 5).

In the stability region, the linearized response of the system to a weak perturbation is analogous to the celebrated Bogoliubov theory of the weakly interacting Bose gas [2]. Let us define the slowly varying fields with respect to the pump frequency as

$$\delta\phi_i(\boldsymbol{x}, t) = \delta\psi_i(\boldsymbol{x}, t)\exp(i\omega_p t), \tag{6}$$

and let us consider the four-component displacement vector

$$\delta\phi(\boldsymbol{x}, t) = (\delta\phi_X(\boldsymbol{x}, t), \delta\phi_C(\boldsymbol{x}, t), \delta\phi_X^*(\boldsymbol{x}, t), \delta\phi_C^*(\boldsymbol{x}, t))^T. \tag{7}$$

The equation of motion for the four-component displacement vector reads

$$i\frac{d}{dt}\delta\phi = \mathcal{L}\cdot\delta\phi + \boldsymbol{f}_{\text{pert}}, \tag{8}$$

where $\boldsymbol{f}_{\text{pert}}$ is the inhomogeneous source term produced by the perturbation. The expression for $\boldsymbol{f}_{\text{pert}}$ depends on which kind of perturbation is considered and will be given explicitly later for the case of a perturbation induced by the single particle potentials $V_{C,X}(\mathbf{x})$. The linear operator $\mathcal{L}$ is

$$\mathcal{L} = \begin{pmatrix} \omega_X + 2g|\psi_X^{ss}|^2 - \omega_p - \frac{i\gamma_X}{2} & \Omega_R & g\psi_X^{ss\,2} e^{2i\mathbf{k}_p \mathbf{x}} & 0 \\ \Omega_R & \omega_C(-i\nabla) - \omega_p - \frac{i\gamma_C}{2} & 0 & 0 \\ -g\psi_X^{ss*2} e^{-2i\mathbf{k}_p \mathbf{x}} & 0 & -(\omega_X + 2g|\psi_X^{ss}|^2) + \omega_p - \frac{i\gamma_X}{2} & -\Omega_R \\ 0 & 0 & -\Omega_R & -\omega_C(-i\nabla) + \omega_p - \frac{i\gamma_C}{2} \end{pmatrix}$$

(9)

6.5.1
Stability of the Stationary Solutions

The stability of the solutions of Eqs. (4, 5) can be determined by calculating the imaginary parts of the eigenvalues of the operator $\mathcal{L}$. If all the eigenvalues have negative imaginary parts (i.e., as it happens in the non-interacting case), then the

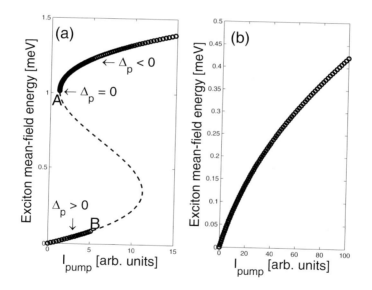

Fig. 6.2 Exciton mean-field energy $\hbar g|\Psi_X^{ss}|^2$ (meV) as a function of the incident pump intensity (arb. units). Cavity parameters: $\hbar\gamma_C = \hbar\gamma_X = 0.1$ meV, $\hbar\omega_X = \hbar\omega_C^0 = 1.4$ eV, $2\hbar\Omega_R = 5$ meV. (a) Bistability curve obtained with the excitation parameters: $k_p = 0.314\,\mu\text{m}^{-1}$ (well in the parabolic region near the bottom of the LP dispersion), $\hbar\omega_p - \hbar\omega_{LP}(k_p) = 0.47$ meV. Circles depict the calculated stable points, while the dashed line represents the unstable branch. The threshold points A and B are, respectively, due to a single-mode (Kerr) or a multi-mode (parametric) instability. (b) Optical limiter curve obtained with the same k_p, but with $\hbar\omega_p - \hbar\omega_{LP}(k_p) = -0.47$ meV. In this case, all stationary solutions are stable.

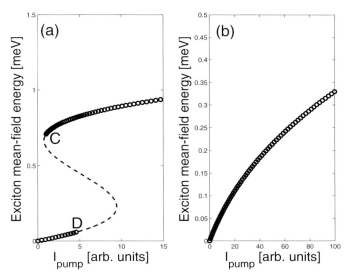

Fig. 6.3 Same plot as in Fig. 6.2, but with pump in-plane wave-vector $k_p = 1.5\,\mu m^{-1}$ (close to the inflection point of the LP dispersion). Pump frequency: (a) $\hbar\omega_p - \hbar\omega_{LP}(k_p) = 0.47$ meV, (b) $\hbar\omega_p - \hbar\omega_{LP}(k_p) = -0.47$ meV. In contrast with the case shown in Fig. 6.2, both threshold points C and D correspond here to the onset of instabilities of parametric type. This explains why the point C is appreciably shifted to the right of the inversion point of the hysteresis curve.

solutions are stable. Otherwise, an instability occurs. If the polariton instability involves only the pump mode, we have the analogous of a *Kerr* instability. If the instability is due to pairs of modes (formation of the so-called signal-idler pairs), we have a *parametric* instability (in the field of quantum fluids, this kind of dynamical instabilities are generally known as *modulational instabilities* [33]). In Fig. 6.2, we have plotted the stationary solutions for the exciton mean-field energy $\hbar g |\psi_X^{ss}|^2$ (meV) as a function of the incident pump intensity (arb. units) for realistic microcavity param-eters and with a small pump wave-vector ($k_p = 0.314\,\mu m^{-1}$), close to the bottom of the LP dispersion.

In Fig. 6.2(a), we have considered the case of a pump frequency, which is blue-detuned with respect to the unperturbed lower polariton energy ($\omega_p > \omega_{LP}(k_p)$). In this case, there is a clear S-shaped bistability curve [34–37]. The unstable branch, determined through the eigenvalues of the linear operator $\mathcal{L}$, have been depicted with a dashed line, while the stable points are represented by circles. The threshold points A and B are due to a Kerr and to parametric instability respectively. By comparison, in Fig. 6.2(b), we have shown the same quantity, but for a red-detuned laser frequency ($\omega_p < \omega_{LP}(k_p)$). In this case, the polariton system behaves as an optical limiter [37], the absorption is highly sublinear and all the points are stable. To give a more complete picture, we report in Fig. 6.3, the analogous calculations, but with a larger wavevector ($k_p = 1.5\,\mu m^{-1}$), close to the inflection point of the LP dispersion. It is apparent that, while the shape is analogous, the boundary between the stable and unstable branches is modified. In particular, the threshold points C and D are both due to parametric instabilities. Note that nice hysteresis loops due to polariton bistability have been recently experimentally demonstrated in the case $k_p = 0$ [34] and for a pump wavevector close to the inflection point of the LP dispersion [36].

6.5.2
Complex Energy of the Collective Excitations

The spectrum of the collective excitations (Bogoliubov modes in the quantum fluid terminology) can be obtained from the eigenvalues of the operator $\mathcal{L}$. As the system is translationally invariant along the plane (we are considering the homogeneous case $V_X = V_C = 0$), the wavevector k is a good quantum number and the eigenvectors of $\mathcal{L}$ have a plane-wave form

$$\delta\phi^{\pm}_{j,k}(x) = \begin{pmatrix} u^{\pm}_{j,X,k}\,e^{ikx} \\ u^{\pm}_{j,C,k}\,e^{ikx} \\ v^{\pm}_{j,X,k}\,e^{i(k-2k_p)x} \\ v^{\pm}_{j,C,k}\,e^{i(k-2k_p)x} \end{pmatrix}, \qquad (10)$$

satisfying the reduced eigenvalue equation

$$\left((\omega^{\pm}_j(k) - \omega_p) \pm - \tilde{\mathcal{L}}(k,k_p)\right) \begin{pmatrix} u^{\pm}_{j,X,k} \\ u^{\pm}_{j,C,k} \\ v^{\pm}_{j,X,k} \\ v^{\pm}_{j,C,k} \end{pmatrix} = 0, \qquad (11)$$

where

$$\tilde{\mathcal{L}}(k,k_p) = \begin{pmatrix} \omega_X + 2g|\psi^{ss}_X|^2 - \dfrac{i\gamma_X}{2} & \Omega_R & g\psi^{ss\,2}_X & 0 \\ \Omega_R & \omega_C(k) - \dfrac{i\gamma_C}{2} & 0 & 0 \\ -g\psi^{ss\,*2}_X & 0 & 2\omega_p - (\omega_X + 2g|\psi^{ss}_X|^2) - \dfrac{i\gamma_X}{2} & -\Omega_R \\ 0 & 0 & -\Omega_R & 2\omega_p - \omega_C(2k_p - k) - \dfrac{i\gamma_C}{2} \end{pmatrix}$$

$$(12)$$

For each $\boldsymbol{k}$, the spectrum is composed by four branches. For each polariton branch $j \in \{LP, UP\}$, two $\pm$ branches exist, which are related by the symmetry

$$\omega_j^-(\boldsymbol{k}) = 2\omega_p - \omega_j^+(2\boldsymbol{k}_p - \boldsymbol{k}). \tag{13}$$

Now, we wish to point out and list clearly the relevant properties and symmetries in this problem. These properties will be later discussed in detail through a set of comprehensive examples and elucidations.

(i) The collective excitations are characterized by the pump-induced coherent coupling between a generic mode with wavevector $\boldsymbol{k}$ and the "idler" wavevector $2\boldsymbol{k}_p - \boldsymbol{k}$. This corresponds to the elementary process $\{\boldsymbol{k}_p, \boldsymbol{k}_p\} \to \{\boldsymbol{k}, 2\boldsymbol{k}_p - \boldsymbol{k}\}$, i.e. the conversion of two pump excitations into a signal-idler pair (to use the quantum optics terminology of parametric oscillators) with the same total momentum.

(ii) The "idler" branch $\omega_{LP}^-(\boldsymbol{k})$ is the "image" of the ordinary branch $\omega_{LP}^+(\boldsymbol{k})$ under the simultaneous transformations [38] $\boldsymbol{k} \to 2\boldsymbol{k}_p - \boldsymbol{k}$ and $\omega \to 2\omega_p - \omega$. The same relationship holds for the UP branches $\omega_{UP}^\pm(\boldsymbol{k})$.

(iii) The matrix $\tilde{\mathcal{L}}(\boldsymbol{k}, \boldsymbol{k}_p)$ is characterized by an *anti-hermitian* coupling between $\boldsymbol{k}$ and $2\boldsymbol{k}_p - \boldsymbol{k}$. This feature is typical of parametric wave-mixing coupling (in the quantum optics language) or Bogoliubov theory (using the quantum fluid literature terminology).

(iv) The four branches of eigenvalues $\omega_{UP,LP}^\pm(\boldsymbol{k})$ are *complex*. As the real parts, the imaginary parts of the eigenvalues depend both on $\boldsymbol{k}$ and on the pump parameters. In the stability region, all imaginary parts are negative.

(v) In case of excitation close to resonance with the lower polariton branch, provided that the interaction energy $g|\psi_X^{ss}|^2$ is much smaller than the polaritonic splitting $\omega_{UP} - \omega_{LP}$, there is no significant mixing between the LP and UP branches. Hence, a simplified approach consists in neglecting the contribution of the upper branch. With this approximation, the branches $\omega_{LP}^\pm(\boldsymbol{k})$ are the eigenvalues of the simplified matrix

$$\tilde{\mathcal{L}}_{LP}(\boldsymbol{k}, \boldsymbol{k}_p) = \begin{pmatrix} \omega_{LP}(\boldsymbol{k}) + 2g_{LP}|\psi_{LP}^{ss}|^2 - \dfrac{i\gamma_{LP}(\boldsymbol{k})}{2} & g_{LP}\psi_{LP}^{ss\,2} \\ -g_{LP}\psi_{LP}^{ss*2} & 2\omega_p - (\omega_{LP}(2\boldsymbol{k}_p - \boldsymbol{k}) + 2g_{LP}|\psi_{LP}^{ss}|^2) - \dfrac{i\gamma_{LP}(2\boldsymbol{k}_p - \boldsymbol{k})}{2} \end{pmatrix} \tag{14}$$

where the stationary lower polariton field is written as a linear superposition of the exciton and cavity photon fields, namely

$$\psi_{LP}^{ss} = X_{LP}(\boldsymbol{k}_p)\psi_X^{ss} + C_{LP}(\boldsymbol{k}_p)\psi_C^{ss}, \tag{15}$$

being $|X_{LP}(\boldsymbol{k}_p)|^2$ and $|C_{LP}(\boldsymbol{k}_p)|^2$ the exciton and photon fractions of the lower polariton mode with the pump wavevector. The lower polariton linewidth is given by $\gamma_{LP}(\boldsymbol{k}) = |X_{LP}(\boldsymbol{k})|^2 \gamma_X + |C_{LP}(\boldsymbol{k})|^2 \gamma_C$. The effective interaction strength

$$g_{LP}(\boldsymbol{k}, \boldsymbol{k}_p) = g |X_{LP}(\boldsymbol{k}_p)|^2 X_{LP}(\boldsymbol{k}) X_{LP}(2\boldsymbol{k}_p - \boldsymbol{k}) \tag{16}$$

takes into account for the exciton fraction of the involved lower polariton modes (pump, signal and idler).

(vi) When the diagonal elements of $\tilde{\mathcal{L}}_{LP}(\boldsymbol{k}, \boldsymbol{k}_p)$ are equal, it is easy to verify that $\Re[\omega_{LP}^+(\boldsymbol{k})] = \Re[\omega_{LP}^-(\boldsymbol{k})]$, while $\Im[\omega_{LP}^+(\boldsymbol{k})] \neq \Im[\omega_{LP}^-(\boldsymbol{k})]$. This means that the parametric coupling produces a splitting of the imaginary parts of the two LP branches, while the real parts are the same. If the difference between the diagonal elements of $\tilde{\mathcal{L}}_{LP}(\boldsymbol{k}, \boldsymbol{k}_p)$ is small compared to the coupling $g_{LP}|\psi_{LP}^{ss}|^2$, then the same property holds. In other words, the dispersions of the two branches $\omega_{LP}^+(\boldsymbol{k})$ and $\omega_{LP}^-(\boldsymbol{k})$ stick together, while their imaginary parts are split. One branch is narrowed with respect to the linear regime, while the other is overdamped. Analogous properties occur for the exact eigenvalues of the 4×4 matrix $\tilde{\mathcal{L}}(\boldsymbol{k}, \boldsymbol{k}_p)$, which are reported in all the figures of this paper.

6.5.2.1 Excitation Near the Inflection Point of the LP Dispersion

In the following, we will show the exact eigenvalues (obtained by numerical calculations) of the matrix $\tilde{\mathcal{L}}(\boldsymbol{k}, \boldsymbol{k}_p)$ in Eq. (12) as a function of the excitation parameters (pump frequency, wavevector and intensity). We will focus on the subtle interplay between the dramatic modification of the energy dispersions (depending on the real part of the eigenvalues) and the onset of the parametric instabilities (depending on the imaginary part).

As a first example, we consider the case of nearly-resonant excitation close to the inflection point of the LP dispersion (see Fig. 6.4). The pump frequency has been taken slightly blue-detuned with respect to the polariton energy in the linear regime. In Fig. 6.4(a) the exact dispersions of the four polariton Bogoliubov branches is shown. The upper polariton branches are energetically far away and play a negligible role, while the relevant physics concerns the lower polariton branches $\omega_{LP}^\pm$ only. The corresponding imaginary parts are shown in Fig. 6.4(b). It is apparent that there is a dramatic modification of the imaginary part around the wave-vectors $k_x = 0$ and $k_x = 2k_p$. Although the stationary solutions are here stable (negative imaginary parts), we can see that we are close to a parametric instability. In fact, there is one branch, whose imaginary part is not far from zero. Note that the imaginary parts are split at $\boldsymbol{k}_p$ as well, even if in a much weaker way. This is a precursor of a Kerr (or single-mode) instability.

In Fig. 6.5, we give another example, with the same excitation parameters as in Fig. 6.3(a) and with an exciton mean-field energy $\hbar g|\Psi_X^{ss}|^2 = 0.699$ meV. In this case, we have an unstable solution, because, as shown in Fig. 6.5(b), there are modes with positive imaginary parts. The parameters of this plot correspond to a point of the hysteresis curve of Fig. 6.3(a) which is just to the left of the threshold point C.

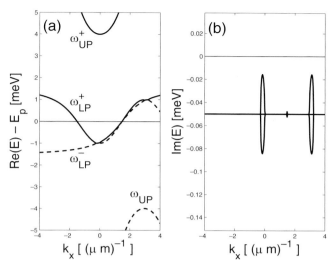

Fig. 6.4 (a) Exact energy dispersions $\Re[\hbar\omega^{\pm}_{LP,UP}]$ of the four polariton Bogoliubov branches measured with respect to the pump photon energy $\hbar\omega_p$ (meV). (b) Corresponding imaginary parts (meV). Note that negative imaginary parts imply stability. Excitation parameters: $k_p = 1.5\,\mu m^{-1}$ (along the x-axis), $\hbar\omega_p - \hbar\omega_{LP}(k_p) = 0.107$ meV, $\hbar g|\Psi^{ss}_X|^2 = 0.05$ meV. Cavity parameters as in the previous figures.

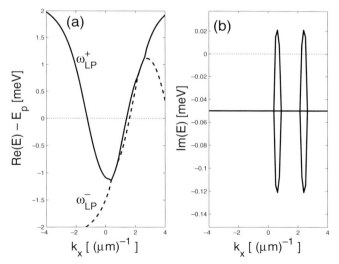

Fig. 6.5 (a) Exact energy dispersions $\Re[\hbar\omega^{\pm}_{LP}]$ of the lower polariton Bogoliubov branches measured with respect to the pump photon energy $\hbar\omega_p$ (meV). (b) Corresponding imaginary parts (meV). Excitation parameters: $k_p = 1.5\,\mu m^{-1}$ (along the x-axis), $\hbar\omega_p - \hbar\omega_{LP}(k_p) = 0.47$ meV, $\hbar g|\Psi^{ss}_X|^2 = 0.699$ meV. Note that here the stationary solution is unstable, because there are modes with positive imaginary parts. This unstable point is close to the point C in Fig. 6.3(a).

The fact that the instability is of parametric (or multi-mode) kind explains why C is close but does not coincide with the inversion point of the hysteresis curve. In fact, this point would instead correspond to the onset of a single mode instability [as it happens at the point A in Fig. 6.2(a)]. In Fig. 6.5(a), we can see that the branches $\omega_{LP}^{\pm}$ stick together in the wavevector region where the parametric instability takes place.

6.5.2.2 Excitation Near the Bottom of the LP Dispersion

Here, we consider the case of a smaller pump excitation wavevector and energy, such as to excite the LP branch close to the bottom of its dispersion. In this region, as shown by Fig. 6.1(b), the dispersion of the unperturbed lower polariton branch is parabolic. In order to stress the non-trivial effects here predicted, we start by showing a spectacular case, depicted in Fig. 6.6. The excitation parameters correspond to Fig. 6.2(a) with the exciton mean-field energy $\hbar g |\Psi_X^{ss}|^2 = 1.02$ meV, i.e., a point close to the threshold point A for the Kerr instability. The dispersions of the branches $\omega_{LP}^{\pm}$ in Fig. 6.6(a) have a corner at the pump wavevector. This dispersion is reminiscent of the celebrated Bogoliubov linear dispersion in superfluid helium and in the atomic condensates. If we look at Fig. 6.6(b), we realize that the modification of the imaginary part is peaked around the pump wavevector itself. Using the quan-

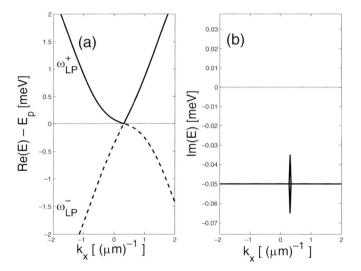

Fig. 6.6 (a) Exact energy dispersions $\Re[\hbar\omega_{LP}^{\pm}]$ of the lower polariton Bogoliubov branches measured with respect to the pump photon energy $\hbar\omega_p$ (meV). (b) Corresponding imaginary parts (meV). Excitation parameters: $k_p = 0.314\,\mu m^{-1}$ (along the x-axis), $\hbar\omega_p - \hbar\omega_{LP}(k_p) = 0.47$ meV, $\hbar g |\Psi_X^{ss}|^2 = 1.02$ meV. Note that this case is the precursor of a Kerr instability, because the imaginary parts are modified at the pumped mode only. This stable point is close to the inversion point A in Fig. 6.2(a).

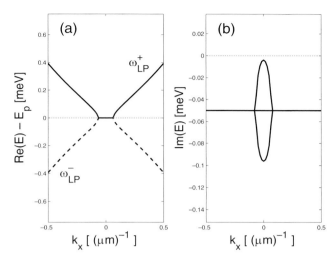

Fig. 6.7 (a) Exact energy dispersions $\Re[\hbar\omega_{LP}^{\pm}]$ of the lower polariton Bogoliubov branches measured with respect to the pump photon energy $\hbar\omega_p$ (meV). (b) Corresponding imaginary parts (meV). Excitation parameters: $\boldsymbol{k}_p = 0$, $\hbar\omega_p - \hbar\omega_{LP}(\boldsymbol{k}_p) = 0.532$ meV, $\hbar g |\Psi_X^{ss}|^2 = 1.2$ meV.

tum optics language, this is the precursor of a Kerr instability, because it involves the pumped mode only.

In Fig. 6.7, we give another example, which has no analog in equilibrium systems. Here, we have a blue-detuned pump at normal incidence, namely $\boldsymbol{k}_p = 0$, $\hbar\omega_p - \hbar\omega_{LP}(\boldsymbol{k}_p) = 0.53$ meV and $\hbar g |\Psi_X^{ss}|^2 = 1.2$ meV. In Fig. 6.7(a), we can clearly see that the dispersion of the polariton collective excitations is flat around the pump wavevector. For the parameters of Fig. 6.7(b), the imaginary parts are all negative, which implies stability. The peak in the imaginary part at the pump wavevector in Fig. 6.6(b) is the precursor of a Kerr instability. In Fig. 6.8, we have an analogous situation, but with a finite pump wavevector. As shown by Fig. 6.8(a), the branches $\omega_{LP}^{\pm}$ stick together around the pump wavevector, with a dispersion exactly linear.

In Fig. 6.9(a), we show the dispersions for a stable point, which is close to the threshold point B in Fig. 6.2(a). Note that here the exciton mean-field energy is considerably smaller than the pump detuning. Hence, the branch sticking occurs in a limited portion of momentum space, where the imaginary parts are affected, as shown in Fig. 6.9(b).

Finally, in Fig. 6.10(a), we give another different example, with a full gap between the branch ω_{LP}^{+} and the branch ω_{LP}^{-}. In Fig. 6.10(b), we can see that the imaginary parts of the eigenvalues are unchanged with respect to the linear regime. Indeed, for these excitation parameters (see the caption of Fig. 6.10), we are on the upper branch of the bistability curve of Fig. 6.2(b) and no instability occurs if the pump intensity is further increased.

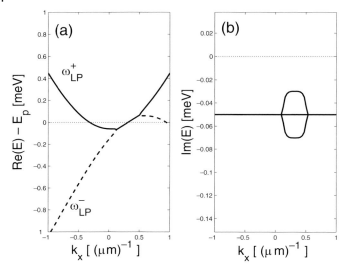

Fig. 6.8 (a) Exact energy dispersions $\Re[\hbar\omega_{LP}^{\pm}]$ of the lower polariton Bogoliubov branches measured with respect to the pump photon energy $\hbar\omega_p$ (meV). (b) Corresponding imaginary parts (meV). Excitation parameters: $\mathbf{k}_p = 0.314\ \mu m^{-1}$ (along the x-axis), $\hbar\omega_p - \hbar\omega_{LP}(\mathbf{k}_p) = 0.04$ meV, $\hbar g |\Psi_x^{ss}|^2 = 0.04$ meV.

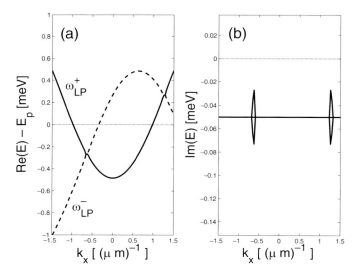

Fig. 6.9 (a) Exact energy dispersions $\Re[\hbar\omega_{LP}^{\pm}]$ of the lower polariton Bogoliubov branches measured with respect to the pump photon energy $\hbar\omega_p$ (meV). (b) Corresponding imaginary parts (meV). Excitation parameters: $\mathbf{k}_p = 0.314\ \mu m^{-1}$ (along the x-axis), $\hbar\omega_p - \hbar\omega_{LP}(\mathbf{k}_p) = 0.47$ meV, $\hbar g |\Psi_x^{ss}|^2 = 0.04$ meV.

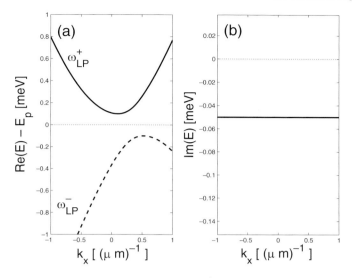

Fig. 6.10 (a) Exact energy dispersions $\Re[\hbar\omega_{LP}^{\pm}]$ of the lower polariton Bogoliubov branches measured with respect to the pump photon energy $\hbar\omega_p$ (meV). (b) Corresponding imaginary parts (meV). Excitation parameters: $\boldsymbol{k}_p = 0.314\,\mu\text{m}^{-1}$ (along the x-axis), $\hbar\omega_p - \hbar\omega_{LP}(\boldsymbol{k}_p) = 0.2\,\text{meV}$, $\hbar g|\Psi_X^{ss}|^2 = 0.6\,\text{meV}$.

6.5.2.3 Simplified Analytical Model for Excitation Close to the Bottom of the LP Dispersion

Now, after having shown a few examples of the rich spectra of non-equilibrium collective excitations and the variety of interaction-induced polariton instabilities, we present here a simple approximated approach, which allows us to grasp effectively the physics contained by the eigenvalues of the matrix in Eq. (12). In particular, we consider the case of negligible mixing with the UP branches, which allows us to focus our analysis on the simpler 2×2 matrix in Eq. (14) instead of the 4×4 matrix in Eq. (12). Moreover, we consider a pump excitation close to the bottom of the LP dispersion, where the dispersion is approximately parabolic (see Fig. 6.1(b)), i.e.,

$$\omega_{LP}(\boldsymbol{k}) \simeq \omega_{LP}(0) + \frac{\hbar \boldsymbol{k}^2}{2m_{LP}}, \tag{17}$$

where m_{LP} is the effective mass of the LP dispersion. Under these assumptions, the spectrum of the LP Bogoliubov excitations can be approximated by the simple expression

$$\omega_{LP}^{\pm} \simeq \omega_p + \delta\boldsymbol{k}\cdot\boldsymbol{v}_p - \frac{i\gamma_{LP}}{2} \pm \sqrt{\left(2g_{LP}|\psi_{LP}^{ss}|^2 + \eta_{\delta k} - \Delta_p\right)\left(\eta_{\delta k} - \Delta_p\right)}, \tag{18}$$

where $\delta \boldsymbol{k} = \boldsymbol{k} - \boldsymbol{k}_p$,

$$\eta_{\delta k} = \frac{\hbar \delta k^2}{2m_{LP}}, \qquad (19)$$

the pump mode flow velocity is $\boldsymbol{v}_p = \hbar \boldsymbol{k}_p / m_{LP}$ and the interaction-renormalized pump detuning

$$\Delta_p = \omega_p - \omega_{LP}(\boldsymbol{k}_p) - g_{LP} |\psi_{LP}^{ss}|^2. \qquad (20)$$

In the case of resonant exciton–photon coupling (i.e. $\omega_C(0) = \omega_X$) and small wavevectors, the excitonic fraction of the lower polariton mode is approximately 0.5. Under these assumptions, we have $|\psi_{LP}^{ss}|^2 \approx 2|\psi_X^{ss}|^2$ and $g_{LP} \approx g/4$ [see Eqs. (15, 16)]. Therefore, $g_{LP}|\psi_{LP}^{ss}|^2 \approx 0.5 g |\psi_X^{ss}|^2$, i.e. the mean-field interaction energy "felt" by the lower polariton is half of the mean-field energy for the exciton field. Note that there are three different cases, according to the value of Δ_p defined in Eq. (20). For given values of $\boldsymbol{k}_p$ and ω_p, one can go from one case to another by moving along the hysteresis curve of Fig. 6.2(a) by varying the pump intensity.

(1) $\Delta_p = 0$. In this resonant situation, the $\pm$ branches touch at $\boldsymbol{k} = \boldsymbol{k}_p$. The effect of the finite flow velocity $\boldsymbol{v}_p$ is to tilt the standard Bogoliubov dispersion [2] via the term $\delta \boldsymbol{k} \cdot \boldsymbol{v}_p$. While in the non-interacting case the dispersion is parabolic, in the presence of interactions [Fig. 6.6(a)] its slope has a discontinuity at $\boldsymbol{k} = \boldsymbol{k}_p$. On each side of the corner, the + branch starts linearly with group velocities respectively given by $v_g^{r,l} = c_s \pm v_p$, c_s being the usual sound velocity of the interacting Bose gas

$$c_s = \sqrt{\hbar g_{LP} |\Psi_{LP}^{ss}|^2 / m_{LP}}. \qquad (21)$$

On the hysteresis curve of Fig. 6.2(a), the condition $\Delta_p = 0$ almost coincides with the inversion point A, where a single-mode Kerr instability appears.

(2) $\Delta_p > 0$. In this case, the argument of the square root in (18) is negative for the wavevectors $\boldsymbol{k}$ such that $\Delta_p > \eta_{\delta k} > \Delta_p - 2g|\Psi_X^{ss}|^2$. In this region, the $\pm$ branches stick together [31] (i.e. $\Re[\omega_{LP}^+] = \Re[\omega_{LP}^-]$) and have an exactly linear dispersion of slope v_p as in Fig. 6.8(a) and in Fig. 6.7(a). The imaginary parts are instead split, with one branch being narrowed and the other broadened [31, 38]. Increasing further the pump intensity, the multi-mode parametric instability [38] sets in, corresponding to the point B in Fig. 6.2(a).

(3) $\Delta_p < 0$. In this case, as it is shown in Fig. 6.10, the branches no longer touch each other at $\boldsymbol{k}_p$ and a full gap between them opens up for sufficiently large values of $|\Delta_p|$. In Fig. 6.2(a), the region $\Delta_p < 0$ is indicated.

6.6
Response to a Static Potential: Resonant Rayleigh Scattering

The dispersion of the polariton elementary excitations is the starting point for a study of the microcavity response to a perturbation. In particular, we shall consider here a moderate static disorder as described by the potential $V_{C,X}(\boldsymbol{x})$. In this case the perturbation source term for the equations of the linearized theory (see Eq. (8)) is the time-independent quantity

$$f_d(\mathbf{x}) = \begin{pmatrix} V_X(\mathbf{x})\phi_X^{ss} \\ V_C(\mathbf{x})\phi_C^{ss} \\ -V_X(\mathbf{x})\phi_X^{ss*} \\ -V_C(\mathbf{x})\phi_C^{ss*} \end{pmatrix}. \qquad (22)$$

The induced perturbation of the exciton and photon fields is given by the expression

$$\delta\phi_d(\mathbf{x}) = -\mathcal{L}^{-1} \cdot f_d(\mathbf{x}). \qquad (23)$$

We remind you that, as shown by Eq. (9), $\mathcal{L}$ is an operator depending on the two-dimensional spatial gradient ∇. It is convenient to perform a spatial Fourier transform, which leads to the algebraic result

$$\begin{pmatrix} \delta\tilde\phi_X(\mathbf{k}) \\ \delta\tilde\phi_C(\mathbf{k}) \\ \delta\tilde\phi_X^*(2\mathbf{k}_p - \mathbf{k}) \\ \delta\tilde\phi_C^*(2\mathbf{k}_p - \mathbf{k}) \end{pmatrix} = -(\tilde{\mathcal{L}}(\mathbf{k},\mathbf{k}_p) - \hbar\omega_p)^{-1} \cdot \begin{pmatrix} \tilde V_X(\mathbf{k})\phi_X^{ss} \\ \tilde V_C(\mathbf{k})\phi_C^{ss} \\ -\tilde V_X(\mathbf{k}-2\mathbf{k}_p)\phi_X^{ss*} \\ -\tilde V_X(\mathbf{k}-2\mathbf{k}_p)\phi_C^{ss*} \end{pmatrix}, \qquad (24)$$

where the eigenvalues of the matrix $\tilde{\mathcal{L}}(\mathbf{k},\mathbf{k}_p)$, defined in Eq. (12), are the 4 branches of polariton Bogoliubov excitations. The perturbation potentials $V_{C,X}(\mathbf{x})$ break the planar translational symmetry of the microcavity system, thus exciting polariton modes with in-plane wavevectors different from the pump wavevector $\mathbf{k}_p$. However, as the perturbation $V_{C,X}(\mathbf{x})$ is static, it can resonantly excite only those Bogoliubov modes whose frequency is equal to ω_p (within the polariton homogeneous linewidth). The observable quantity is

$$I_{RRS}(\mathbf{k}) \propto |\delta\tilde\phi_C(\mathbf{k})|^2, \qquad (25)$$

i.e., the perturbation-induced intensity of the photonic field, which is proportional to the far-field images of the resonant Rayleigh scattering signal [14, 24, 29].

6.6.1
Weak Excitation Regime and Elastic RRS Ring

In the following, we will show a few applications of Eq. (24), using the perturbation potentials depicted in Fig. 6.11. Here, we have considered a single photonic defect (depth 1 meV, width 1.5 µm) for the in-plane photonic potential $V_C(\mathbf{x})$ [see Fig. 6.11(a)] and a disordered excitonic potential [see Fig. 6.11(b)]. We point out that these potentials are just an example and that Eq. (24) can be readily applied for an arbitrary set of perturbation potentials (see, e.g., the cover of this special volume).

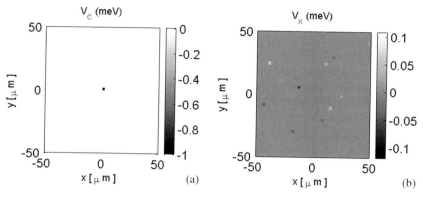

Fig. 6.11 Single particle potentials considered in the numerical calculations. Note that this is a zoom around the origin of the 400 μm × 400 μm box (with a 256 × 256 grid) used for the numerical calculations. (a) Photonic potential $V_C(x)$ (meV) (it can model an artificial or natural point defect at the origin $x = y = 0$). (b) Excitonic potential $V_X(x)$ (disordered spatial fluctuations of the exciton energy). The gray color scale is different with respect to the top panel.

For instance, the effect of an isolated excitonic defect is totally analogous to the one shown in the following for the photonic defect. As long as it remains a perturbation on the plane wave polaritonic eigenstates, a stronger disorder simply gives a higher intensity of the corresponding RRS emission. In the numerical applications here reported, we have considered a 256 × 256 spatial grid and a squared box (400 μm × 400 μm), with the photonic dot at the center of the box.

In Fig. 6.12, we show the results for the weak excitation regime, where the many-body effects produced by polariton–polariton interactions are negligible. In this linear regime, we have the conventional unperturbed dispersions of the lower and upper polariton branches. We have considered the case of resonant excitation close to the bottom of the LP branch, where the dispersion is parabolic, as one can see from Fig. 6.12(a). In Fig. 6.12(b), the intensity of the resonant Rayleigh scattering is shown, displaying the well known elastic ring. In fact, in the linear regime, the solutions of the equations $\omega_{LP}(k) = \omega_p$ are k-points on a circle, because the unperturbed polariton dispersion depends on $|k|$ only. The speckles on top of the elastic ring are due to the random nature of the excitonic potential. The width of the ring is due to the finite homogeneous broadening of the polariton modes. Note that, in order to excite the elastic ring corresponding to the wavevector k_p, the Fourier component $\tilde{V}_{C,X}(k_p)$ of the static potentials need to be finite. This condition is easily fulfilled by a typical excitonic disordered potential or by a photonic defect whose width is of the order of 1 μm.

In Fig. 6.13, we show the corresponding spatial pattern. Precisely, we have plotted the normalized quantity

$$\frac{I_C(x)}{I_C^{hom}} = \frac{|\phi_C^{ss} e^{ik_p x} + \delta\phi_C(x)|^2}{|\phi_C^{ss}|^2}, \tag{26}$$

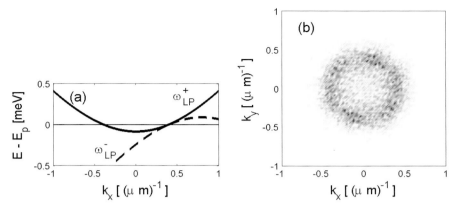

Fig. 6.12 Energy dispersion of the LP branches in the weak excitation regime. (b) Intensity (arb. units) of the photonic resonant Rayleigh scattering signal $|\delta\tilde{\phi}_c(\mathbf{k})|^2$. Note that the gray scale is inverted (white corresponds to 0). Excitation parameters: $\mathbf{k}_p = 0.4\,\mu\mathrm{m}^{-1}$ (along the x-axis), $\hbar\omega_p - \hbar\omega_{LP}(\mathbf{k}_p) = 0$ meV, $\hbar g|\Psi_X^{ss}|^2 = 0.0001$ meV. The photonic and exciton potentials are those shown in Fig. 6.11.

i.e., the total photon field intensity (homogeneous solution + potential-induced perturbation) normalized to the intensity of the homogeneous solution without the potential. For the considered potentials in Fig. 6.11, the dominant feature is due to the photonic point defect. The polariton plane wave driven by the pump is coherently scattered by the photonic defect (located at the position $x = y = 0$), producing a peculiar interference pattern, characterized by parabolic wavefronts. In fact, the polariton field scattered by the point defect is a cylindrical wave. Hence, if we consider only one defect and $\mathbf{k}_p$ is along the x-direction, the total field has the form $f(\mathbf{x}) = e^{ik_p x} + \beta\, e^{ik_p\sqrt{x^2+y^2}}$. The constant phase curves are given by the condition $k_p x + k_p \sqrt{x^2 + y^2} = 2\pi n$, whose solutions describe parabolic wavefronts with a

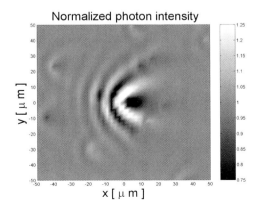

Fig. 6.13 Spatial profile of the normalized cavity photon density, i.e., $I_c(\mathbf{x})/I_c^{hom}$. Excitation parameters and potentials as in Fig. 6.12. The coherent diffusion pattern induced by the point defect at the origin is the main feature on top of the random landscape produced by the exciton disorder.

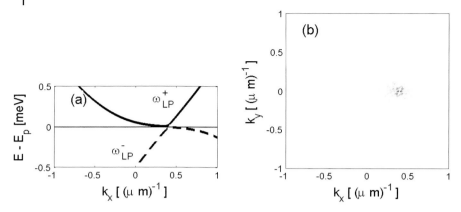

Fig. 6.14 Superfluid regime. Same parameters as in Fig. 6.12, but with $\hbar\omega_p - \hbar\omega_{LP}(k_p) = 0.467\,\text{meV}$, $\hbar g|\Psi_x^{ss}|^2 = 1\,\text{meV}$. In this superfluid regime, the RRS elastic ring has collapsed.

symmetry axis oriented along the x-direction, as nicely depicted by the exact solution in Fig. 6.13. Due to the presence of the exciton potential, additional disordered features are superimposed on the main interference pattern produced by the photonic point defect.

6.6.2
Superfluid Regime

In presence of interactions, we have seen that the spectrum of polariton Bogoliubov excitations is dramatically different from the unperturbed case. This manifestation of polariton many-body physics can be probed in a sensitive way by the resonant Rayleigh scattering emission. In Fig. 6.14, we start by considering the most spectacular regime of polariton superfluidity. This regime can be achieved when the pump is resonant with the interaction-renormalized polariton dispersion at the pump wavevector ($\Delta_p = 0$) and when the sound velocity c_s [see Eq. (21)] of the interacting polariton fluid is larger than the flow velocity $v_p = \hbar k_p/m_{LP}$ imprinted by the pump beam. This situation is more favorable to obtain for excitation close to the bottom of the LP dispersion, implying smaller pump flow velocity v_p and smaller excitation density necessary to have $c_s > v_p$. As depicted by Fig. 6.14(a), in this case, the equation $\Re[\omega_{LP}^{\pm}(\mathbf{k})] - \omega_p = 0$ has no solutions for $\mathbf{k} \neq \mathbf{k}_p$, meaning that no final states are available for the elastic scattering induced by the static potential. As a dramatic consequence, the elastic ring in Fig. 6.12(b) collapses. As shown in Fig. 6.14(b), only a weak emission around the pump wavevector $\mathbf{k}_p$ is left, due to non-resonant processes, which are allowed by the finite broadening of the polariton modes.

The real space pattern is shown in Fig. 6.15, showing that the perturbation induced by each defect remains localized around the defect positions. Hence, the polaritonic propagation is *superfluid*. In analogous way to liquid Helium and atomic

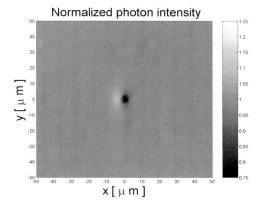

Fig. 6.15 Spatial profile of the normalized cavity photon density in the superfluid regime. Excitation parameters as in Fig. 6.14. To compare with the normal (weak) excitation regime, see Fig. 6.13.

condensates [1, 2], we can state that the polariton fluid has a superfluid behaviour according to the Landau criterion, if and only if both following conditions are satisfied:

(a) $\omega_{LP,UP}^+(k) > \omega_p$ for every $k \neq k_p$;
(b) $\omega_p > \omega_{LP}(0)$, i.e. there is an elastic ring in the weak excitation regime.

We point out that the condition (b) is necessary to have a meaningful definition of polariton superfluidity. In fact, if $\omega_p < \omega_{LP}(0)$, already in the weak excitation regime there are no real states at the pump energy and there is no resonant Rayleigh scattering elastic ring. Note that the conditions (a) and (b) are achieved not only in the resonant case $\Delta_p = 0$. Within the parabolic approximation in Eq. (18), conditions (a) and (b) are satisfied when $\Delta_p \leq 0$ and

$$-g_{LP}|\psi_{LP}^{ss}|^2 - \frac{m_{LP}v_p^2}{2\hbar} < \Delta_p < g_{LP}|\psi_{LP}^{ss}|^2 - \frac{m_{LP}v_p^2}{\hbar}. \qquad (27)$$

The calculations here reported show the robustness of the superfluid flow with respect to elastic pro-cesses, such as the scattering on static defects. As the Landau criterion for superfluidity involves also inelastic processes (e.g., emission of crystal phonons or heating of residual free carriers), it is important to note that whenever the superfluidity condition (a) for elastic scattering is fulfilled, then it is satisfied *a fortiori* also for the inelastic channels. In fact, the stability of the mean-field solution implies that, exception made for the incoherent parametric luminescence due to quantum fluctuations, no Bogoliubov quasi-particles are present above a stable mean-field solution. Whenever $\omega_{LP}^+(k) > \omega_p$, no final states are available for the coherent polaritons to lower their energy by transferring energy to the environment. This is here assumed to be at almost zero temperature, so that it can only absorb energy from the polariton system. Friction is therefore absent in this regime.

On the other hand, the incoherent population of the Bogoliubov modes corresponding to the luminescence gives rise to a sort of "normal" component of the polariton fluid analogous to the one which would appear at a finite temperature when the polariton fluid is heated because of its interaction with the phonon bath

6.6.3
Precursors of Parametric Instabilities and Branch Sticking

In the case $\Delta_p > 0$ (i.e., pump frequency higher than the renormalized frequency of the pumped mode), the resonant Rayleigh scattering response is completely different, as shown in Figs. 6.16 and 6.18. In this regime, the two LP branches stick together, while there is a splitting of their imaginary parts [see, e.g., the analogous situation in Figs. 6.8 and 6.9]. Such a scenario represents the precursor of a multi-mode parametric instability, which can be triggered by further increasing the

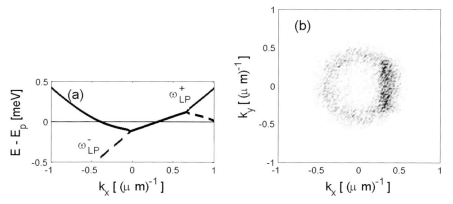

Fig. 6.16 Same parameters as in Fig. 6.12, but with $\hbar\omega_p - \hbar\omega_{LP}(\mathbf{k}_p) = 0.1\,\text{meV}$, $\hbar g |\Psi_X^{ss}|^2 = 0.07\,\text{meV}$. Branch sticking and amplified RRS are precursors of a parametric instability. Note that a portion of the image saturates the chosen gray scale.

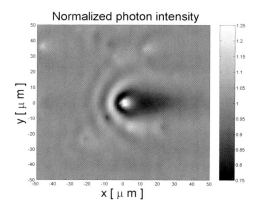

Fig. 6.17 Spatial profile of the normalized cavity photon density. Excitation parameters as in Fig. 6.16. Note that the gray scale of this plot is saturated in the region around the point defect at the origin.

6.6 Response to a Static Potential: Resonant Rayleigh Scattering | 145

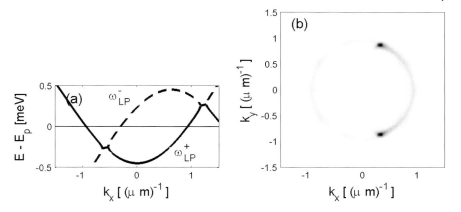

Fig. 6.18 Excitation parameters: $k_p = 0.314\,\mu\text{m}^{-1}$ (along the x-axis), $\hbar\omega_p - \hbar\omega_{LP}(k_p) = 0.47\,\text{meV}$, $\hbar g|\Psi_x^{ss}|^2 = 0.075\,\text{meV}$. Note that the gray scale of this plot is saturated around the two brightest points. Here, we consider the situation of a dominant photonic potential ($V_x = 0$).

excitation density. In contrast to the superfluid regime, a deformed RRS ring is apparent in Figs. 6.16(b) and 6.18(b). In particular, depending on the topology of the k-space region where the LP branches stick, the RRS intensity is strongly amplified either on a segment parallel to y including the point k_p [Fig. 6.16(b)], or around two points of the straight line parallel to y and passing through the point k_p [Fig. 6.18(b)].

The main consequence of this in the real-space pattern of Fig. 6.17 is an overall amplification of the density modulation induced by the defect, in stark contrast with the superfluid regime. In particular, note the long "shadow" in the downstream direction with respect to the central defect, which extends to rela- tively far distances thanks to the linewidth narrowing effect. In Fig. 6.19, the shadow of the defect is even more peculiar, showing a series of fringes parallel to the x-direction. These can be explained in terms of the interference between the pump and the two peaks in k-space shown in Fig. 6.18(b).

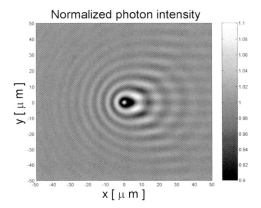

Fig. 6.19 Spatial profile of the normalized cavity photon density. Same parameters as in Fig. 6.18.

146 | 6 Quantum Fluid Effects and Parametric Instabilities in Microcavities

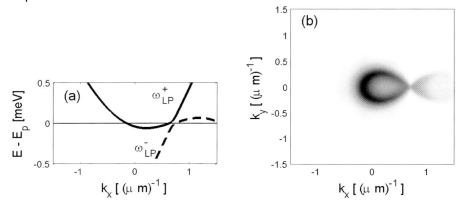

Fig. 6.20 Cherenkov regime. Parameters: $\omega_X = \omega_C(0) - 1\,\mathrm{meV}$, $k_p = 0.7\,\mu\mathrm{m}^{-1}$ (along the x-axis), $\hbar\omega_p - \hbar\omega_{LP}(k_p) = 0.599\,\mathrm{meV}$, $\hbar g |\Psi_X^{ss}|^2 = 1\,\mathrm{meV}$. Here, we consider the situation of a dominant photonic potential ($V_X = 0$).

6.6.4
Cherenkov Regime

Here, we consider the opposite case $\Delta_p \leq 0$, but with an excitation density which is not high enough to enter the superfluid regime characterized by the condition in Eq. (27). For the sake of clarity, as we have already done for Fig. 6.18 and Fig. 6.19, we take here $V_X = 0$, so to concentrate on the effect of a single defect. This situation can be realistically realized, e.g., when there is a single photonic defect (natural or artificial) which is dominant over the background excitonic disorder. In Fig. 6.20(a), we consider the resonant case ($\Delta_p = 0$), with a polariton sound speed $c_s < v_p$. As shown in Fig. 6.20(b) the weak excitation elastic RRS ring is replaced by an asymmetric pattern, which is strongly deformed and shows a singularity at the pump wavevector. The aperture angle 2θ of the singularity at k_p satisfies the

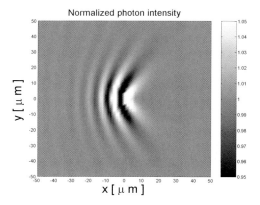

Fig. 6.21 Spatial profile of the normalized cavity photon density. Same parameters as in Fig. 6.20. The photonic point defect produces Cherenkov-like wavefronts. Note that, in order to show the peculiar shape of the wavefronts, the gray scale of this plot is saturated in the region around the point defect.

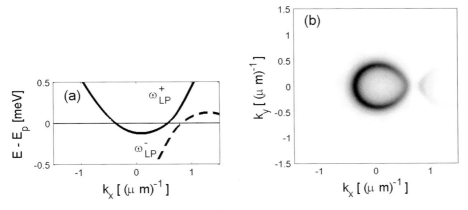

Fig. 6.22 Parameters: $\omega_x = \omega_c(0)$, $k_p = 0.7\,\mu m^{-1}$ (along the x-axis), $\hbar\omega_p - \hbar\omega_{LP}(k_p) = 0.3\,\text{meV}$, $\hbar g|\Psi_x^{ss}|^2 = 0.6\,\text{meV}$.

simple condition $\cos\theta = c_s/v_p$. In this $c_s < v_p$ regime where the polariton fluid is moving at a supersonic speed, the defect produces a peculiar real-space pattern (Fig. 6.21) showing linear Cherenkov-like wavefronts [39, 40]. The aperture 2ϕ of the Cherenkov angle has the usual value $\sin\phi = c_s/v_p$. This behavior is easily understood from a physical standpoint as follows: a moving fluid propagating along the positive x-direction in the presence of a static defect is equivalent, under a Galilean transformation, to a defect moving in the negative x-direction in a fluid at rest. This situation is a familiar one, as it corresponds to the wavefronts cre-ated by a moving duck on the surface of a lake. The rounded region on the left-hand side of the $\mathbf{k}$-space pattern in Fig. 6.20(b) was not present in the standard theory of Cherenkov emission in non-dispersive media [39] and it is due to the fact that the Bogoliubov dispersion is linear only in the neighborhood of $\mathbf{k}_p$ and then it bends upwards. A remarkable consequence of this property is the oscillatory perturbation shown by the real-space pattern upstream with respect to the defect, as apparent in Fig. 6.21.

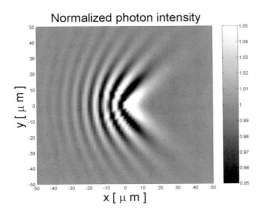

Fig. 6.23 Spatial profile of the normalized cavity photon density. Same parameters as in Fig. 6.22. The gray scale of this plot is saturated in the region around the point defect.

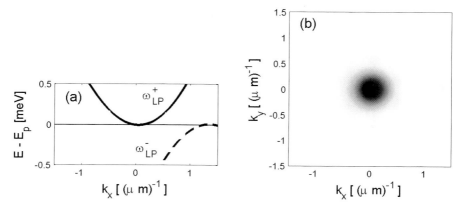

Fig. 6.24 Parameters: $\omega_x = \omega_c(0)$, $k_p = 0.7\,\mu m^{-1}$ (along the x-axis), $\hbar\omega_p - \hbar\omega_{LP}(k_p) = 0.2\,\text{meV}$, $\hbar g|\Psi_x^{ss}|^2 = 0.6\,\text{meV}$.

In the case $\Delta_p < 0$, the branches ω_{LP}^+ and ω_{LP}^- are disconnected, as well as the corresponding RRS emission pattern, as depicted in Fig. 6.22. The real-space wave fronts shown in Fig. 6.23 are still Cherenkov-like, since the separation between the two branches is relatively small.

The situation of Figs. 6.24 and 6.25 is instead different, with a full gap opened between the two LP branches. The k-space emission is then concentrated around the point k_a where the Bogoliubov dis-persion touches the line $\omega = \omega_p$. Correspondingly, the near-field pattern shows a localized perturbation around the defect, with a peculiar stripe pattern. This pattern is due to the interference of the Rayleigh scattering at k_a and the pump beam at k_p, so that the wavevector corresponding to the fringes is equal to $k_p - k_a$.

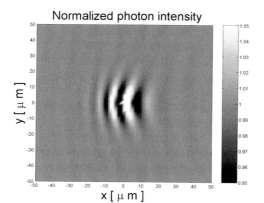

Fig. 6.25 Spatial profile of the normalized cavity photon density. Same parameters as in Fig. 6.24. The gray scale of this plot is saturated in the region around the point defect.

6.7
Conclusions

In conclusion, in this Festschrift paper, we have presented a comprehensive analysis of the exotic collective excitations of a moving polariton fluid driven by a continuous-wave optical pump, which were recently predicted in a short Letter [27].

We have analyzed in detail the interplay between the non-trivial dispersions of the collective excitations in the quantum fluid of microcavity polaritons and the onset of single-mode (Kerr) or multi-mode (parametric) instabilities. These Bogoliubov excitation spectra can be dramatically changed by tuning the pump excitation parameters, namely its frequency, incidence angle and intensity. As the present system is a non-equilibrium one, qualitatively novel features appear in the Bogoliubov spectra, which were not observed in systems close to the thermal equilibrium, such as superfluid helium or the ultracold atomic condensates.

We have then studied the propagation of the polariton fluid in the presence of a static perturbation potential acting both on the photonic and exciton component of the polariton field. In particular, we have shown the strict connection between the dispersion of the Bogoliubov excitations and the intensity and shape of the resonant Rayleigh scattering on defects. We have pointed out some experimentally accessible consequences of polaritonic superfluidity for realistic microcavity parameters. Depending on the pump excitation parameters, a very rich phenomenology has been predicted for the far- and near-field images of the resonant Rayleigh scattering emission, a phenomenology which generalizes the well-known physics of the Cherenkov effect.

Acknowledgements

LPA-ENS is a "Unité de Recherche de l'Ecole Normale Supérieure et des Universités Paris 6 et 7, associées au CNRS". CC wishes to thank all the authors contributing to the present Festschrift volume for many stimulating discussions on the physics of semiconductor microcavities. IC acknowledges stimulating discussions with C. Tozzo and F. Dalfovo on the subject of modulational instabilities in quantum fluids.

References

[1] D. Pines and P. Nozieres, The Theory of Quantum Liquids, Vols. 1 and 2 (Addison-Wesley, Redwood City, 1966).
[2] L. Pitaevskii and S. Stringari, Bose–Einstein Condensation (Oxford University Press, Oxford, 2003).
[3] A. J. Leggett, Rev. Mod. Phys. **71**, S318–S323 (1999).
[4] C. Weisbuch, M. Nishioka, A. Ishikawa, and Y. Arakawa, Phys. Rev. Lett. **69**, 3314 (1992).
[5] For a recent review, see: Special Issue on Microcavities, Guest Editors J. Baumberg and L. Viña, Semicond. Sci. Technol. **18**, No. 10 (2003).

[6] P. G. Savvidis, J. J. Baumberg, R. M. Stevenson, M. S. Skolnick, D. M. Whittaker, and J. S. Roberts, Phys. Rev. Lett. **84**, 1547 (2000).
[7] C. Ciuti, P. Schwendimann, B. Deveaud, and A. Quattropani, Phys. Rev. B **62**, R4825 (2000).
[8] M. Saba, C. Ciuti, J. Bloch, V. Thierry-Mieg, R. André, Le Si Dang, S. Kundermann, A. Mura, G. Bongiovanni, J. L. Staehli, and B. Deveaud, Nature (London) **414**, 731 (2001).
[9] P. G. Savvidis, C. Ciuti, J. J. Baumberg, D. M. Whittaker, M. S. Skolnick, and J. S. Roberts, Phys. Rev. B **64**, 075311 (2001).
[10] A. Huynh, J. Tignon, O. Larsson, Ph. Roussignol, C. Delalande, R. André, R. Romestain, and Le Si Dang, Phys. Rev. Lett. **90**, 106401 (2003).
[11] S. Kundermann, M. Saba, C. Ciuti, T. Guillet, U. Oesterle, J. L. Staehli, and B. Deveaud, Phys. Rev. Lett. **91**, 107402 (2003).
[12] J. J. Baumberg, P. G. Savvidis, R. M. Stevenson, A. I. Tartakovskii, M. S. Skolnick, D. M. Whittaker, and J. S. Roberts, Phys. Rev. B **62**, R16247 (2000).
[13] R. M. Stevenson, V. N. Astratov, M. S. Skolnick, D. M. Whittaker, M. Emam-Ismail, A. I. Tartakovskii, P. G. Savvidis, J. J. Baumberg, and J. S. Roberts, Phys. Rev. Lett. **85**, 3680 (2000).
[14] R. Houdré, C. Weisbuch, R. P. Stanley, U. Oesterle, and M. Ilegems, Phys. Rev. Lett. **85**, 2793 (2000).
[15] D. M. Whittaker, Phys. Rev. B **63**, 193305 (2001).
[16] G. Messin, J. Ph. Karr, A. Baas, G. Khitrova, R. Houdré, R. P. Stanley, U. Oesterle, and E. Giacobino, Phys. Rev. Lett. **87**, 127403 (2001).
[17] G. Dasbach, M. Schwab, M. Bayer, and A. Forchel, Phys. Rev. B **64**, 201309 (2001).
[18] G. Dasbach, M. Schwab, M. Bayer, D. N. Krizhanovskii, and A. Forchel, Phys. Rev. B **66**, 201201 (2002).
[19] C. Ciuti, P. Schwendimann, and A. Quattropani, Phys. Rev. B **63**, 041303 (2001).
[20] P. Schwendimann, C. Ciuti, and A. Quattropani, Phys. Rev. B **68**, 165324 (2003).
[21] J. Ph. Karr, A. Baas, R. Houdré, and E. Giacobino, Phys. Rev. A **69**, 031802 (2004).
[22] J. Ph. Karr, A. Baas, and E. Giacobino, Phys. Rev. A **69**, 063807 (2004).
[23] C. Ciuti, Phys. Rev. B **64**, 245304 (2004).
[24] W. Langbein, Proc. 26th Int. Conf. on the Physics of Semiconductors, ICPS 26 (Edinburgh, UK, 2002).
[25] W. Langbein, Phys. Rev. B **70**, 205301 (2004).
[26] S. Savasta, O. Di Stefano, V. Savona, and W. Langbein, cond-mat/0411314.
[27] I. Carusotto and C. Ciuti, Phys. Rev. Lett. **93**, 166401 (2004).
[28] H. Stolz, D. Schwarze, W. von der Osten, and G. Weimann, Phys. Rev. B **47**, 9669 (1993).
[29] R. Houdré, C. Weisbuch, R. P. Stanley, U. Oesterle, and M. Ilegems, Phys. Rev. B **61**, R13333 (2000).
[30] W. Langbein and J. M. Hvam, Phys. Rev. Lett. **88**, 047401 (2002), and references therein.
[31] C. Ciuti, P. Schwendimann, and A. Quattropani, Semicond. Sci. Technol. **18**, S279–S293 (2003).
[32] S. Savasta, O. Di Stefano, and R. Girlanda, Phys. Rev. Lett. **90**, 096403 (2003).
[33] B. Wu and Q. Niu, Phys. Rev. A **64**, 061603(R) (2001).
[34] A. Baas, J. Ph. Karr, H. Eleuch, and E. Giacobino, Phys. Rev. A **69**, 023809 (2004).
[35] N. A. Gippius, S. G. Tikhodeev, V. D. Kulakovskii, D. N. Krizhanovskii, and A. I. Tartakovskii, Europhys. Lett. **67**, 997 (2004).
[36] A. Baas, J. Ph. Karr, M. Romanelli, A. Bramati, and E. Giacobino, Phys. Rev. B **70**, 161307(R) (2004).
[37] R. W. Boyd, Nonlinear Optics (Academic Press, London, 1992).
[38] C. Ciuti, P. Schwendimann, and A. Quattropani, Phys. Rev. B **63**, 041303(R) (2001). D. M. Whittaker, Phys. Rev. B **63**, 193305 (2001).
[39] J. V. Jelley, Cerenkov Radiation and Its Applications (Pergamon Press, London/Oxford, 1958).
[40] I. Carusotto, M. Artoni, G. C. La Rocca, and F. Bassani, Phys. Rev. Lett. **87**, 064801 (2001).

Chapter 7
Non-Linear Dynamical Effects in Semiconductor Microcavities

Jean-Louis Staehli, Stefan Kundermann, Michele Saba, Cristiano Ciuti, Augustin Baas, Thierry Guillet, and Benoit Deveaud

7.1
Introduction

In 1992, the strong coupling regime between quantum well (QW) excitons (Xs) and microcavity (µC) photons has been observed [1]. Since then, Fabry–Perot semiconductor µCs have been intensively investigated [2]. The mixed X-photon states, the polaritons, can be excited optically by external photons sent to the planar structure. Initially, the µCs were strongly affected by disorder in the mirror and QW structures. The latter leads to inhomogeneous X broadening, which of course strongly influences the polariton spectra and in extreme cases even destroys the strong coupling regime [3–5]. Moreover, this disorder causes also Rayleigh scattering to occur in the µC, whereby the part due to the QWs is resonant [6, 7]. The spread of the polaritons, after resonant excitation by light with oblique incidence, along a ring in the wavevector plane parallel to the QW, is one of the most spectacular features of resonant Rayleigh scattering [8, 9]. It is directly related to the strong coupling regime and the particularities of the polariton dispersion that ensue. Thorough experimental and theoretical investigations were needed to understand the influence of strong coupling, disorder effects, and the combination of both [10, 11] on the optical properties and polariton dynamics in µCs.

It was not straightforward to understand even the linear optical properties of µCs [12, 13]. The observation of the linear temporal dynamics of the transmitted intensity of a very short laser pulse is an important tool to characterise a µC sample. The measured exponential temporal decay permits to confirm that the lower polariton features a Lorentzian broadening, even when the QW X is strongly affected by inhomogeneous broadening [3, 13]. Figure 7.1 also illustrates the spectacular progress which has been achieved in sample preparation, making it some times difficult to characterise a µC just by continuous waves optical spectroscopy. Time integrated (TI) four-wave mixing (FWM) spectroscopy, in the degenerate two-beam configuration, is a non-linear optical tool that is relatively easy to implement: two pulsed laser beams derived from the same laser, with wave vectors k_1 and k_2, are

Physics of Semiconductor Microcavities: From Fundamentals to Nanoscale Devices
Edited by Benoit Deveaud
Copyright © 2007 WILEY-VCH Verlag GmbH & Co. KGaA, Weinheim
ISBN: 978-3-527-40561-9

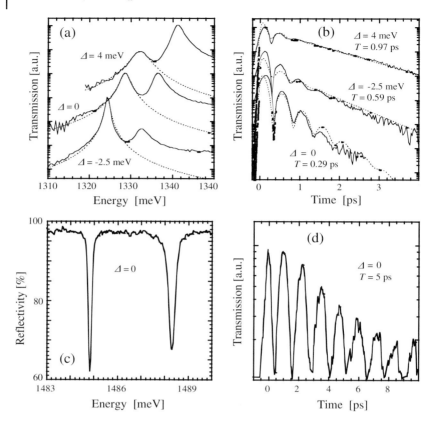

Fig. 7.1 Linear characterisation of two µC samples in the low excitation limit. (a) and (c) Transmission and reflectivity spectra, the dotted lines are fits to Lorentzians [(c) courtesy of RP Stanley]. (b) and (d) Time resolved (TR) transmission of a 100 fs long pulse through the cavity; the dotted curves are numerical solutions of Maxwell's equations. (a) and (b) The data were obtained on a $3\lambda/2$ AlGaAs/GaAs µC containing six 7.5 nm wide $In_{0.13}Ga_{0.87}As$ QWs [13]. (c) and (d) The sample is a λ µC containing a single 7.5 nm wide $In_{0.04}Ga_{0.96}As$ QW (see Fig. 7.3) [14]. Δ is the detuning between the lowest cavity mode and the lowest exciton state, and T is the exponential decay time.

focused on the sample at nearly normal incidence. The intensity of the diffracted beam, propagating along $k_4 \equiv 2k_2 - k_1$, can be dispersed in a spectrometer and measured as a function of the delay τ between the two incoming laser pulses. It is directly linked to the temporal evolution of the coherence of the optical excitations, and the temporal evolution of the FWM signals differs strongly from that of a transmitted single laser pulse (Fig. 7.2). Let us further remark that pump–probe (PP) measurements, in a geometry where the pump and probe beams hit the sample with different incidence angles and where the probe reflected or transmitted by the sample is analysed, belong to the same family as FWM experiments, since they

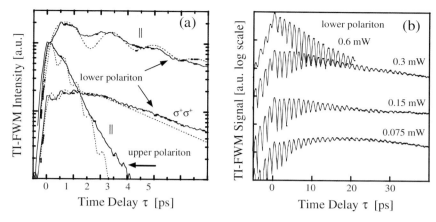

Fig. 7.2 Experimental (solid) and theoretical (dotted) TI FWM signals, emitted at the lower and upper polariton energies, versus delay time τ between the two laser pulses. The incident beams are colinearly (∥) and cocircularly ($\sigma^+\sigma^+$) polarised. Detuning $\Delta = -2$ meV for (a), $\Delta = 0$ for (b). The data were obtained using the same samples as for Fig. 7.1 [13, 14].

both permit to investigate components of the third (and/or higher) order dielectric susceptibility tensor $\chi^{(3)}(t, \tau, \omega_i, k_i, I_i)$.

Even though the interpretation of FWM measurements is not always straightforward, these experiments revealed from the beginning a characteristic property of μCs in the strong coupling regime: the long coherence times of the optical excitations, the polaritons, which can reach values close to 100 ps [14–16]. It is the particular shape of the polariton dispersion that is at the origin of the observed very long polariton dephasing times. With such long coherence times, μCs are interesting structures for future applications making use of coherence. Coherent optical excitations can be "manipulated" by ultra fast coherent control, employing one or more very short laser pulses they can be rapidly generated, switched from one state to another, and if required rapidly killed again. The temporal sequence, direction, energy and phase of the incoming pulses determine direction, energy and temporal properties of the light emitted by the cavity. Integrated optoelectronic devices containing μCs might some day be used in this way to accomplish, within a few ps or less, complex logical tasks.

The effects of strong excitation in μCs, which are featuring a very rich phenomenology, can be divided into two groups. The first one concerns excitonic non-linearities which in principle could be observed also directly in QWs outside a cavity. Since a μC acts as a narrow spectral filter, inside such a device it is possible to excite very selectively only fundamental Xs and to avoid perturbations caused by free electron–hole pairs. In this way, it has been possible eg. to observe directly the resonant Stark effect of QW Xs [17]. The phenomena of the second group are connected to the strong coupling regime, ie. polaritonic non-linearities are concerned.

They are mainly due to the peculiar properties of the dispersion of µC polaritons. The effective mass of the latter is as much as four orders of magnitude smaller than that of Xs. This property, together with the long dephasing times, makes µC polaritons good candidates for the observation of bosonic effects [18]. In fact, occupation numbers per mode of up to 10^4 can easily be reached by selective coherent excitation of a few polariton modes [19, 20]. In other words, quantum degeneracy can be reached while the polariton–polariton interactions are still quite weak. Therefore, the fermionic character of the (still strongly correlated) electrons and holes, which together with the photons are the components of the polaritons, does not perturb.

The peculiar shape of the inplane dispersion of µC polaritons gives rise to another phenomenon occurring in µCs: parametric amplification (PA). Huge parametric gain can be observed in a degenerate and coherent polariton system [20], the amplification factor depends sensitively on sample quality and in particular on the polariton line width. Besides its high scientific interest, this phenomenon appears also to be well suited for optoelectronic applications, such as high-speed switches and amplifiers, within the framework mentioned above. This paper essentially concerns the dynamics and the coherent control of PA, and the coherence of the participating polaritons. The main experimental tool used up to now for our investigations are TI PP measurements employing one or two probe pulses. They were performed as a function of pump intensity, PP delay, and the relative phase of the two probes. The dynamics in real time t of the light emitted by the µC will be discussed shortly and in a non conclusive way in the last section of this paper. All the experimental results are discussed using a theoretical model describing µC parametric polariton scattering processes, in its simplest version it just distinguishes three polarisation components, pump, signal and idler [21].

7.2
Experimental

7.2.1
The Microcavity

Most of the experimental work described here has been done on a sample of particularly high quality [22]. It is a λ microcavity (µC) with a GaAs spacer layer, the dielectric mirrors consist of 20 (top) and 26.5 (bottom) $Ga_{0.9}Al_{0.1}As/AlAs$ $\lambda/4$ layer pairs. At the centre of the structure where the electric field is strongest, there is a 7.5 nm wide $In_{0.04}Ga_{0.96}As$ quantum well (QW) (Fig. 7.3), its exciton (X) energy matches that of the lowest cavity mode (ie. the cavity is at resonance). The spacer layer is slightly tapered, which permits to vary the detuning between QW X and cavity mode by moving the laser spot on the sample (Fig. 7.4). For zero detuning, the minimum splitting amounts to 3.5 meV, the linewidths of the upper and lower polaritons (Ps) amount to about 0.3 and 0.13 meV, and the cavity lifetime is about 10 ps. Thus, the strong X-photon coupling regime is well established.

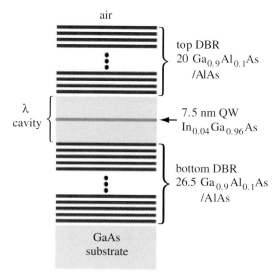

Fig. 7.3 Scheme of the microcavity. The GaAs spacer layer contains a single (In, Ga)As QW and is enclosed between (Al, Ga)As/AlAs distributed Bragg reflectors (DBR).

The exciton–photon interaction is important only for inplane wave vectors k close to zero. The dispersion of the two polariton branches is sketched in Fig. 7.5, for the case of perfect resonance between X and photon mode. The lower branch features a narrow valley with a minimum at $k = 0$ and reaches the X energy for wave vectors that are bigger than the reciprocal X radius. Ps close to $k = 0$ have a very small effective mass, some 104 to 105 times smaller than that of free Xs. Ps occupying states in the central valley interact only weakly with Ps and Xs outside the valley or with phonons, because for such pro-cesses energy and momentum conservation is difficult to achieve. These Ps are thus isolated from the other excitations of the cavity and the crystal, and have therefore very long dephasing times [15, 16, 23, 24].

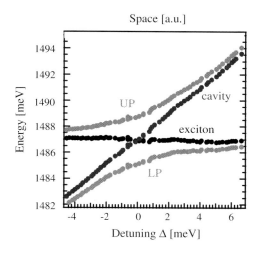

Fig. 7.4 Variation of the upper and lower polariton energies vs position on the sample, due to the wedged shape of the spacer layer. The energies of the uncoupled exciton and cavity mode have been calculated from the two measured polariton energies.

It can be said that the dispersion of the lower P branch close to $k = 0$ acts as a two dimensional trap in k space, and preserves the coherence of the optical excitations [25].

7.2.2
Pump–Probe Experiments and Parametric Amplification

Polariton–polariton scattering processes within the valley, however, are possible. The process which will be discussed in detail in this paper (sketched in Fig. 7.5) concerns Ps at $k \approx k_p$, at an energy about halfway between the bottom and the top of the trap, they can easily scatter to $k \approx 0$ and $k \approx 2k_p$ since energy and momentum are conserved for this process. In fact, this process can be stimulated, e.g. by coherently occupying the polariton states at $k \approx 0$. Or, in other words, a μC can work as an optical parametric amplifier, if it is pumped with a laser beam incident at the angle corresponding to $k \approx k_p$. In experimental configurations like ours, where one of the final P states is at $k \approx 0$, this pump angle has been called *magic*, since it is only for this angle that PA can be observed [20].

The set-up of our pump–probe experiments to investigate parametric scattering is sketched in Fig. 7.6. The sample is usually pumped at the magic angle and probed at normal incidence ($k = 0$). Both, pump and probe derive from a Ti:sapphire laser delivering 150 fs long pulses, with a repetition rate of 80 MHz. The pump beam is sent through a spectral filter of about 1 meV width, permitting to exclusively excite lower branch polaritons at $k \approx k_p$. The probe pulses remain unfiltered, having a spectral width of about 15 meV they can be used to probe polariton states anywhere in the trap. Probe and pump beam are focused onto the sample, the spot diameter is roughly 100 μm. Most of our experiments are done in transmission geometry, due to the occurrence of parametric scattering light is emitted in three different directions, with $k \approx 0$ (signal), $k \approx k_p$ (pump), and $k \approx 2k_p$ (idler). The use of

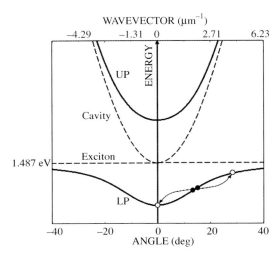

Fig. 7.5 Schematic representation of the polariton dispersion vs. angle of incidence and in-plane wavevector k, at perfect exciton–photon mode resonance. The parametric scattering process on the lower polariton (LP) branch, bringing two pump polaritons into a signal-idler pair, is sketched.

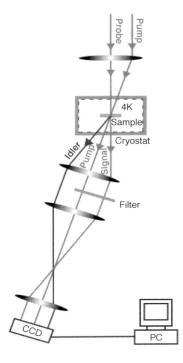

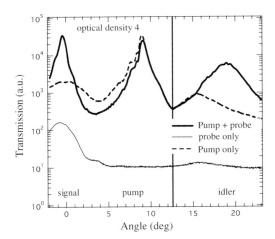

Fig. 7.6 Outline of the pump–probe experiment. While the probe hits the sample (held in a He temperature cryostat) perpendicularly, the pump impinges at the magic angle. The detection scheme, with a CCD (charge coupled device) camera, allows to achieve angle resolved measurements permitting to investigate at the same time pump, signal and idler.

Fig. 7.7 Angular distribution of the light emission of the microcavity on the backside. The emission intensity at the pump angle is comparable to that of the signal at $k = 0$, while at the idler angle, where the photon component of the polariton is small, the emission is very weak. In order to see the idler emission on the same intensity scale as pump and signal, the emissions at angles smaller than 12.5° have been attenuated by four orders of magnitude. In this measurement, the signal emission occurs at a small negative angle.

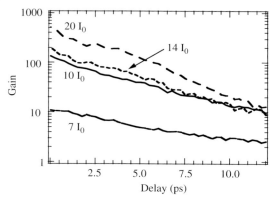

Fig. 7.8 Decay of parametric amplification with pump to probe delay. The signal gain is defined as the ratio of signal intensity measured in transmission with pump to that without pump. $I_0 \approx 4 \times 10^{10}$ photons/ (cm^2 pulse) in front of the µC is approximately the threshold intensity for PA, corresponding to a time averaged laser intensity of 0.77 W/cm^2. We estimate that roughly 3% of the pump photons penetrate into the µC.

a CCD camera permits to measure the time integrated light intensity versus emission angle. In most cases the experiments are done for different pump intensities and pump–probe delays.

The effect of stimulation on the emission of the µC pumped at the magic angle by a probe at $k \approx 0$ is shown on Fig. 7.7. When only the pump beam excites the sample at the magic angle, spontaneous parametric scattering occurs, leading to signal and idler emissions. If, contemporaneously with the pump pulses, the µC is probed at normal incidence, the polariton states at $k \approx 0$ are coherently occupied which stimulates parametric scattering of the pump polaritons, therefore the signal and idler emissions are strongly enhanced. Let us remark that the number of probe photons penetrating into the cavity is much smaller than that of the signal polaritons generated by spontaneous scattering of the pump polaritons (see Fig. 7.7), nevertheless the former have a very strong stimulation effect. We note also that the incidence angle of the pump does not exactly match the ideal (magic) value, which causes the emissions related to the spontaneous and stimulated processes to occur in slightly different directions. Moreover, the signal peak due to parametric stimulation is narrower than those due to pump and probe alone.

The decay of parametric amplification with pump–probe delay is shown in Fig. 7.8. For the lowest pump intensity of $7I_0$ the measured signal gain is relatively small. A modest increase of the pump to $10I_0$ leads to a gain that is more than ten times higher. In other words, we observe a threshold behaviour of parametric gain with respect to the pump. Parametric scattering involving two polaritons, the scattering rate is proportional to the square of pump intensity, and threshold occurs when this scattering rate overcomes pump escape and dephasing rates. Further, let us note that the decay of parametric gain with increasing delay becomes faster at high polariton densities.

7.3
A Simple Theoretical Model

A theoretical model for the µC parametric polariton amplifier has been developed by Ciuti et al. [21]. In its simplest version, the model accounts for three polarisation fields on the lower polariton branch of the µC, the pump $P_{\boldsymbol{k}_p}$ at $\boldsymbol{k} = \boldsymbol{k}_p = \boldsymbol{k}_{\mathrm{magic}}$, the signal P_0 at $\boldsymbol{k} = 0$, and the idler $P_{2\boldsymbol{k}_p}$ at $\boldsymbol{k} = 2\boldsymbol{k}_p$. The temporal behaviour of these three polarisations are described by

$$i\hbar \frac{dP_0}{dt} = \left(\tilde{E}(0) - i\gamma_0\right) P_0 + E_{\mathrm{int}} P_{2\boldsymbol{k}_p}^* P_{\boldsymbol{k}_p}^2 + F_{\mathrm{probe}}(t),\qquad(1)$$

$$i\hbar \frac{dP_{\boldsymbol{k}_p}}{dt} = \left(\tilde{E}(\boldsymbol{k}_p) - i\gamma_{\boldsymbol{k}_p}\right) P_{\boldsymbol{k}_p} + 2 E_{\mathrm{int}} P_{\boldsymbol{k}_p}^* P_0 P_{2\boldsymbol{k}_p} + F_{\mathrm{pump}}(t),\qquad(2)$$

$$i\hbar \frac{dP_{2\boldsymbol{k}_p}}{dt} = \left(\tilde{E}(2\boldsymbol{k}_p) - i\gamma_{2\boldsymbol{k}_p}\right) P_{2\boldsymbol{k}_p} + E_{\mathrm{int}} P_0^* P_{\boldsymbol{k}_p}^2.\qquad(3)$$

In the first terms of the right hand side of the above equations, the polariton energies $\tilde{E}(\boldsymbol{k})$ are renormalised with respect to the unperturbed energies $E(\boldsymbol{k})$. Due to the presence of an intense pump field $P_{\boldsymbol{k}_p}$, $\tilde{E}(\boldsymbol{k})$ is blueshifted by the repulsive polariton–polariton interactions,

$$\tilde{E}(\boldsymbol{k}) = E(\boldsymbol{k}) + 2 V_{\boldsymbol{k}\boldsymbol{k}_p 0} |P_{\boldsymbol{k}_p}|^2,\qquad(4)$$

where the effective interaction potential $V_{\boldsymbol{k}\boldsymbol{k}'\boldsymbol{p}}$ acts on two polaritons having initial wave vectors $\boldsymbol{k}$ and $\boldsymbol{k}'$ and exchanging momentum $\boldsymbol{p}$. $V_{\boldsymbol{k}\boldsymbol{k}'\boldsymbol{p}}$ is a sum of a direct contribution, a boson exchange contribution, and one due to the saturability of the exciton transition [21]. The damping rates $\gamma_{\boldsymbol{k}}$ that also appear in the first terms on the right hand side include polariton dephasing and escape from the cavity. The second terms describe parametric scattering of the pump polaritons to signal and idler polaritons, and vice versa. The scattering rate depends again on the polariton–polariton interaction potential,

$$E_{\mathrm{int}} = \tfrac{1}{2}\left(V_{\boldsymbol{k}_p \boldsymbol{k}_p \boldsymbol{k}_p} + V_{\boldsymbol{k}_p \boldsymbol{k}_p -\boldsymbol{k}_p}\right).\qquad(5)$$

The inhomogeneous terms are the pump and probe driving fields.

Let us mention some of the most important limits in the validity of this model.

(i) This model is inadequate for cases where spontaneous parametric scattering is important. It is in fact unrealistic to describe spontaneous parametric scattering by a three component model, since for a given pump wavevector $\boldsymbol{k}_p$ the ensemble of the possible wavevectors of signal and idler are not just the two points $\boldsymbol{k}_0$ and $2\boldsymbol{k}_p$ in $\boldsymbol{k}$ space, but form an eight [26, 27]. However, when stimulation is important spontaneous processes become negligible (Fig. 7.7), and in this case we expect that the theoretical model is close to the experimental reality.

(ii) For polaritons, the expectation value of a product of operators has been approximated by the product of the expectation values of the single operators. From more elaborate models [28] we know that stimulated parametric scattering is strongly influenced by signal-idler correlations. These correlations will be described only in an approximate way by our model where the expectation values of polarisation operators have been factorised as mentioned. In particular, our model might reproduce the temporal dynamics of the polarisations in a qualitative way only.

(iii) At the highest pump intensities employed in our experiments, higher order scattering processes have been observed [29] which are not described by our simple model including only terms up to the third order in the polarisations. Thus at high excitation intensities, we have to expect deviations between the predictions of the model and the experimental results.

The three coupled Eqs. (1) to (3) have been integrated numerically. The calculations discussed here have been done for single probe–pump experiments, and for pump intensities close to threshold. The following numerical parameters have been assumed: $\gamma_0 = \gamma_{k_p} = \gamma_{2k_p} = 0.15$ meV (which means that polariton damping is essentially caused by polariton escape from the μC); a polariton splitting of 3.5 meV (as for our μC containing one $In_{0.04}Ga_{0.96}As$ quantum well); exciton and cavity-mode energies 1.480 eV (at $k = 0$); duration of the probe pulses 150 fs; duration of the pump pulses 4 ps; zero delay between pump and probe pulse. Figure 7.9 shows the temporal dependence of the signal intensity. The probe to pump intensity ratio is of the order of 10^{-4}. The sharp rise close to zero is caused by the penetration of the short probe pulse into the cavity. Without pump, the probe polarisation just decays

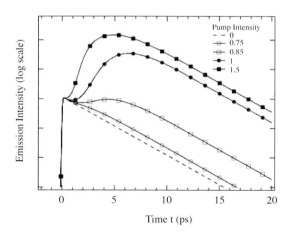

Fig. 7.9 Calculated emission intensity of the signal ($k = 0$) as a function of real time t, in a pump-single probe experiment. Pump and probe temporally coincide, and the results are shown for different relative pump intensities that are close to threshold. Pump intensity 0.9 corresponds roughly to the threshold intensity $|P_{k_p}^{thr}|^2 = \sqrt{\gamma_0 \gamma_{2k_p}}/E_{int}$ under continuous waves excitation [21]. The chosen numerical parameters are given in the text.

exponentially. There is a delay of about one or two ps before stimulated parametric sets in and the signal intensity rises. This delay tends to shorten and the signal rises more rapidly with increasing pump intensity. The delay can be attributed to the fact that signal-idler correlation has to build up before stimulated parametric scattering can set in [28]; in our simple three component model this means that idler polaritons, absent at time zero, have to be generated before stimulated scattering starts (Eqs. (1) and (3), third terms). Moreover, there is a clearly visible threshold behaviour for increasing pump intensity, since $-\gamma_0 P_0$ competes with $E_{int} P^*_{2k_p} P^2_{k_p}$ (see Eq. (1)). Also a saturation effect is visible, due to an increase of the pump intensity by 18% from 0.85 to 1.0 the signal intensity shoots up by about 30 times, while the 50% increase of the pump from 1.0 to 1.5 enhances the signal roughly by a factor of 5.

7.4
Coherent Control

Pump–probe experiments can also be done with two probe pulses, and if not only the delay but also the relative phase of the two probes can be varied, such an experiment permits to coherently control the parametric amplification process. Phase

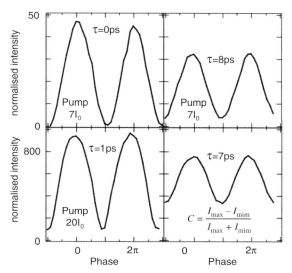

Fig. 7.10 Coherent control: Time integrated signal intensity as a function of the relative phase of the two probe pulses, for different delays τ between the two probes, and for different pump intensities ($I_0 = 4 \times 10^{10}$ photons/(cm² pulse), see Fig. 7.8). The first probe pulse coincides with the pump. The intensity has been integrated over the whole spectral width of the lower polariton emission. The contrast C of the oscillations is discussed in the text.

controlled pulse pairs can be created using a Michelson interferometer; while the delay can be varied by moving one of the mirrors with a micrometre screw, the phase is changed and actively stabilised using a piezoelectric element controlled by an optoelectronic feedback loop [30, 31]. The first probe pulse creates some polarisation near $k \approx 0$ in the μC, stimulating parametric scattering. The polarisation created by the second probe interferes with that which is already there. Depending on whether the interference is constructive or destructive, parametric scattering accelerates or slows down. Figure 7.10 shows the time integrated signal intensities collected from the back side of the sample, for different pump intensities and delays τ between the two probes. For τ close to zero of course the contrast C is practically full (close to zero intensity for a phase difference of π), but for τ as long as 8 ps the observed contrast can still be almost 80%. This means that the phases of an important part of the polaritons are conserved as long as stimulated scattering goes on. Let us remark that at higher pump densities the oscillations in the signal intensities become asymmetric, and that the contrast diminishes more rapidly with increasing delay.

Plotting the full angular dependence of the light emitted by the μC vs phase difference for a series of coherent control experiments, graphs of the kind shown in Fig. 7.11 are obtained. The oscillations of signal and idler are in phase, while those of the pump feature lower contrast and are shifted by π with respect to the other ones. This means that the pump is depleted when both probes are in phase, and stimu-

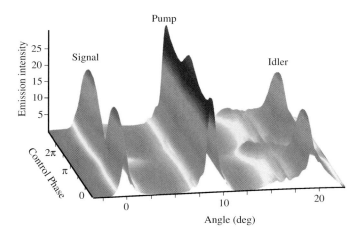

Fig. 7.11 Coherent control of the light emitted in the signal, pump and idler directions from the back of the microcavity. The surface plot shows the emission intensity as a function of the emission angle and the relative phase of the two probe pulses. The delay τ between the two phase controlled pulses is roughly 2 ps. Signal and idler intensities oscillate in phase, while the oscillations of the pump are in antiphase with respect to those of the probe. The signal and pump emissions (up to an angle of 14°) were attenuated by four orders of magnitude. In order to evidence the effects of the coherent control on the emitted pump intensity, the probe density employed here was relatively high (5×10^8 polaritons/(cm^2 pulse), and the pump density was relatively low ($\approx 10^{11}$ polaritons/(cm^2 pulse)).

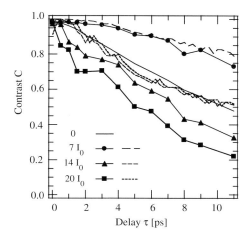

Fig. 7.12 Contrast C of the coherent control oscillations versus delay τ between the two probe pulses; for different pump intensities, $I_0 = 4 \times 10^{10}$ photons/(cm^2 pulse) is the threshold density above which parametric scattering can be stimulated (see Fig. 8). The first probe pulse is synchronous with the pump. Solid lines: measured; dashed lines: calculated using the data of Fig. 7.8, as explained in the text.

lated scattering occurs. Thus signal, idler, and to a certain extent also the pump, are controlled at the same time, as expected for a parametric amplifier. Further, when the two probe pulses are out of phase, spontaneous parametric scattering occurs in a similar way as already observed in Fig. 7.7; this leads to signal and idler emissions having a wider angular spread than in the case of stimulated scattering.

The decay of the contrast of the oscillations with increasing delay between the two probes has been carefully analysed (Fig. 7.12). Without pump the contrast decays at a rate corresponding to the total polariton lifetime (determined mainly by the emission of photons through the mirrors, and by dephasing) of $T \approx 10$ ps [31, 32]. For a pump intensity of $7I_0$, slightly above threshold, the contrast decays more slowly, evidencing a prolonged polariton coherence time. Increasing the pump intensity to 14 and $20I_0$, the decay becomes faster than without pump. Each probe pulse creates some signal polarisation in the cavity, by directly creating seeding polaritons, and indirectly by stimulating parametric scattering. At low pump powers when stimulation is weak, the effects of the two probe pulses do not influence each other. In this case the contrast can be calculated using the data of Fig. 7.8,

$$C(\tau) = \frac{|A(0) + A(\tau)|^2 - |A(0) - A(\tau)|^2}{|A(0) + A(\tau)|^2 + |A(0) - A(\tau)|^2}, \tag{6}$$

where $A(\tau)$ is the mean polarisation amplitude at $k = 0$ created by a single probe pulse at delay τ, we evaluate it by taking the square root of the time integrated signal intensity measured in the single probe experiments of Fig. 7.8. For zero pump intensity, the contrast $C(\tau)$ calculated in this way indeed coincides with the measured one. Also at a pump intensity of $7I_0$ the measured and calculated contrasts agree, i.e. the slow-down of the decay occurs in the regime where the effects of the two probes are independent; further this agreement proves that coherence of the involved polaritons is fully conserved. At higher pump intensities, the calculated contrast is close to that obtained for zero pump, while the measured one features a rapid initial decay and is lower. This shows that far above threshold the effects of

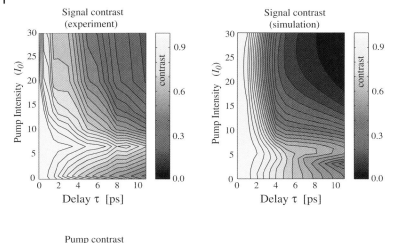

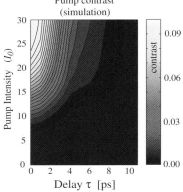

Fig. 7.13 Coherent control: Contour plots, vs. pump intensity and delay between the two probe pulses, of the contrast of the measured signal oscillations (left panel), of the calculated signal oscillations (centre panel) as explained in the text, and of the calculated pump oscillations (right panel).

the two probes are not independent any more. This can be interpreted in terms of excitation induced dephasing [33] and/or in terms of a depletion of the pump reservoir by the first probe.

The results of our detailed investigations of the contrast C vs. delay and pump intensity are shown in Fig. 7.13. The experimental data (left panel) for the signal are characterised by the following features: (i) Turning on the pump slightly increases the decay rate of C. (ii) At low pump intensities, below and slightly above threshold, there is a slow-down in the decay of the contrast with increasing pump intensity, at $I_{pump} = 7I_0$ the contrast of the interference fringes is maximum, as already mentioned. (iii) At pump intensities above $7I_0$, the contrast, at any delay, diminishes with increasing I_{pump}. Using the three component model presented in Section 7.3, the contrast has been calculated (Fig. 7.13, centre panel). Qualitatively, the model reproduces the features observed experimentally: with increasing pump intensity there is an initial increase of the decay rate, then the decay slows down and the contrast goes through a maximum. Further, the theoretical simulations permit to explain the observed asymmetry in the oscil-lations of Fig. 7.10. In Fig. 7.13, right panel, there are practically no oscillations in the number of pump polaritons, except at highest pump intensities and short delays where these oscillations reveal a (weak) depletion of the pump causing saturation of the parametric amplification and thus the appearance of asymmetric interference fringes in the signal.

7.5
Measurements Resolved in Real Time

The dynamics resolved in real time of stimulated parametric scattering has also been investigated. Figure 7.14 shows the experimental set-up employing upconversion. In a non-linear crystal the signal coming from the µC is combined with a 100 fs long reference (or gate) pulse coming directly from the laser, if the two coincide spatially and temporally light at the sum frequency (violet) is generated which can be detected with a (slow) detector. Scanning the delay of the gate permits to recover the temporal shape of the light emitted by the µC during a pump–probe experiment. A disadvantage of this technique is the relatively low sensitivity, therefore the measurements can be done only with pump intensities substantially above threshold.

Experimental results of the emitted signal intensities vs time t are shown in Fig. 7.15. Even though the probe hits the sample at the same time as the pump, initially the signal decays as if the pump were absent. It is only after a delay of 6 to 8 ps that the signal intensity increases sharply and reaches a first maximum at about 10 to 12 ps. The signal does not decay monotonously, there are some relatively slow oscillations leading to the appearance of a second maximum at about 19 ps. It has been found that the initial delay as well as the amplitude and period of the oscillations sensibly depend on the experimental conditions, e.g. how well pump and probe are in resonance with the cavity and how well the incidence angles correspond to the phase matching conditions.

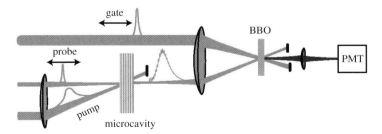

Fig. 7.14 Scheme of the experimental upconversion set-up used for PP measurements with real time resolution. If the light emitted by the µC reaches the non-linear BBO (beta barium borate) crystal at the same time as the gate, sum frequency light can be detected by the photomultiplier tube (PMT).

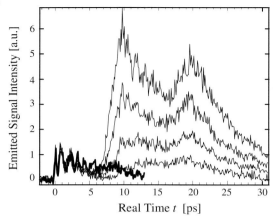

Fig. 7.15 Signal intensity resolved in real time, emitted during a PP experiment with pump–probe delay zero, for different pump intensities. Averaged external pump intensities, from bottom to top: 0 (bold), 22, 27, 34 and 42 W/cm^2; probe intensity: 0.48 W/cm^2. The delay seems to depend only weakly on the excitation density.

Theoretical calculations corresponding to these experimental data are shown in Fig. 7.16. (The wiggles in the curves are due to beatings between the emissions of the upper and lower polariton branches, the probe being spectrally wide enough to excite both; for the purpose of this discussion they can be ignored.) As for the calculations shown above in Fig. 7.9, at low excitation densities there is only a short

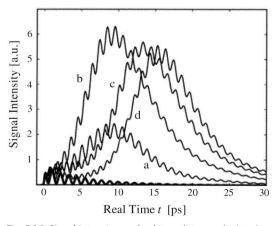

Fig. 7.16 Signal intensity resolved in real time, calculated as explained in the text for a PP experiment with pump–probe delay zero, for different pump intensities. The bold curve at the bottom is the signal intensity without pump. Averaged external pump intensities (outside the μC) are: (a) 15, (b) 30, (c) 60, and (d) 90 W/cm^2. At higher pump intensities, PA becomes saturated and it sets in with a delay.

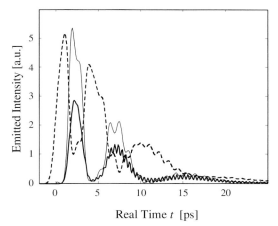

Fig. 7.17 Hypothetical signal, pump and idler intensities, resolved in real time, calculated for a PP experiment with no pump–probe delay, assuming that the exciton–exciton interactions do not cause any blueshift of the polaritons. Strong oscillations would appear, parametric scattering occurring alternatively in forward and backward directions. Bold continuous line: idler; thin continuous line: signal; dashed line: pump. The idler intensity has been enhanced by a factor of 10.

delay of about 1 ps before the signal is rising due to stimulated parametric scattering, ie. this theoreti-cally calculated delay is quite shorter than what has been observed experimentally. However, if in the calculations the pump intensity is increased to levels quite above those used experimentally (Fig. 7.16, curves c and d), the maximum in the signal emission moves to bigger delays, at 80 W/cm^2 the calculated signal intensity decreases before the rise due to parametric amplification sets in at roughly 5 ps. This is similar to what has been observed experimentally, but for quite lower pump intensities. However, the simulations predict that this signal is rising during 5 to 10 ps, while experimentally 3 to 4 ps are observed. Let us remark here that at least on an intuitive level an increase of the delay of parametric amplification with increasing pump density is unexpected. This delay cannot only be due to the build-up of signal-idler correlations as mentioned in Section 7.3, it might also be linked to the blueshift of the polariton energies which is changing with density during the experiment and which is caused by exciton–exciton interactions and saturation effects within the quite dense polariton system.

Finally, the calculations predict a monotonous signal decay, in contrast to the experiments where an oscillation is observed and a second maximum appears. If one artificially turns off the polariton blueshift due to exciton–exciton interactions which occurs at high excitation densities, we obtain what is shown in Fig. 7.17. In this hypothetical case, the signal intensity would be strongly oscillating with a period of several ps, in phase with the idler and out of phase with the pump. At the end of each period signal and idler intensities would practically come back to zero, i.e. these oscillations could be called parametric Rabi oscillations. Thus, when signal and idler intensities are decreasing parametric scattering goes backwards from signal and idler to the pump. Further, after the initial excitation signal and idler would rise with a short delay of the order of one ps, a delay corresponding to the time needed to build up signal–idler correlation as mentioned in Section 7.3.

7.6 Conclusions

Semiconductor microcavities that are in the strong coupling regime can indeed work as an efficient and rapidly responding parametric amplifier. From our results it becomes obvious that probe pulses which are much weaker than the amplified light leaving the sample can control PA in a µC. There is an optimum pump intensity, just above threshold, at which coherent control is most efficient; ie. at this optimum the polaritons at $k = 0$ remain coherent up to long delays between the two probe pulses, and thus stimulated parametric scattering can still go on. At higher excitation densities, there is a significant depletion of the pump leading to a rapid decay of the interference contrast. Our relatively simple theoretical model [21], based on just three polarisation components (pump, signal, and idler) which are coupled in third order via exciton–exciton interaction, permits to understand the major features of the dynamics of stimulated parametric scattering. However, the exact origin of the delay that occurs before stimulated PA sets in still needs to be clarified, requiring further experimental and theoretical investigations.

Acknowledgements

We thank U. Oesterle of Prof. Ilegems' group for providing the high quality µC samples used in the experimental investigations. We acknowledge stimulating discussions with G. Bongiovanni, F. Quochi, M. A. Dupertuis, A. Quattropani, P. Schwendimann, V. Savona, R. Houdré, and R. P. Stanley. This work has been supported in part by the *Fonds National Suisse de la Recherche Scientifique.*

References

[1] C. Weisbuch et al., Phys. Rev. Lett. **69**, 3314 (1992).
[2] For a review, see e.g.: V. Savona et al., Phase Transit. 68, 169 (1999); or: Special Issue on Microcavities, edited by J. J. Baumberg and L. Viña, Semicond. Sci. Technol. **18**, No. 10 (2003).
[3] R. Houdré et al., Phys. Rev. A **53**, 2711 (1996).
[4] A. V. Kavokin, Phys. Rev. B **57**, 3757 (1998).
[5] V. Savona, J. Cryst. Growth **184/185**, 737 (1998).
[6] R. Zimmermann, Nuovo Cimento D **17**, 1801 (1995).
[7] G. R. Hayes et al., Phys. Rev. B **58**, R10175 (1998).
[8] T. Freixanet et al., Phys. Rev. B **60**, R8509 (1999).
[9] R. Houdré et al., Phys. Rev. B **61**, R13333 (2000).

[10] T. Shchergov et al., Phys. Rev. Lett. **84**, 3478 (2000).
[11] W. Langbein and J. M. Hvam, Phys. Rev. Lett. **88**, 47401 (2002).
[12] R. Houdré et al., Phys. Rev. Lett. **73**, 2043 (1994).
[13] G. Bongiovanni et al., Phys. Rev. B **55**, 7084 (1997).
[14] M. Saba et al., phys. stat. sol. (a) **178**, 149 (2000).
[15] A. L. Bradley et al., in: Proc. 3rd Internat. Conf. Excitonic Processes in Condensed Matter (EXCON'98), Boston, Nov. 1–5 1998 (The Electrochemical Society, Pennington, PA, 1998), pp. 10–19.
[16] L. A. Dunbar et al., Phys. Rev. B **66**, 195307 (2002).
[17] F. Quochi et al., Phys. Rev. Lett. **80**, 4733 (1998).
[18] F. Tassone and Y. Yamamoto, Phys. Rev. B **59**, 10830 (1999).
[19] P. G. Savvidis et al., Phys. Rev. Lett. **84**, 1547 (2000).
[20] M. Saba et al., Nature **414**, 731 (2001).
[21] C. Ciuti et al., Phys. Rev. B **62**, R4825 (2000).
[22] U. Oesterle et al., J. Cryst. Growth **150**, 1313 (1995).
[23] V. Savona and C. Piermarocchi, phys. stat. sol. (a) **164**, 45 (1997).
[24] G. Cassabois et al., Phys. Rev. B **61**, 1696 (2000).
[25] J. J. Baumberg et al., Phys. Rev. Lett. **81**, 661 (1998).
[26] C. Ciuti, P. Schwendimann, and A. Quattropani, Phys. Rev. B **63**, 041303(R) (2001).
[27] W. Langbein, Phys. Rev. B **70**, 205301 (2004).
[28] P. Schwendimann, C. Ciuti, and A. Quattropani, Phys. Rev. B **68**, 165324 (2003).
[29] P. G. Savvidis et al., Phys. Rev. B **64**, 75311 (2001).
[30] M. U. Wehner, M. H. Ulm, and M. Wegener, Opt. Lett. **22**, 1455 (1997).
[31] S. Kundermann et al., Phys. Rev. Lett. **91**, 107402 (2003).
[32] S. Kundermann et al., phys. stat. sol. (a) **201**, 381 (2004).
[33] D. S. Chemla and J. Shah, Nature **411**, 549 (2001).

Chapter 8
Polariton Correlation in Microcavities Produced by Parametric Scattering

Wolfgang Langbein

8.1
Introduction

The strong coupling regime between the photon state of a Fabry–Perot cavity and the excitonic resonance of a semiconductor quantum well (QW) creates a new type of quasi-particle of mixed photon–exciton character, the planar microcavity polariton [1, 2]. These polaritons have a defined in-plane momentum k, and are typically contained in a micrometer-sized, monolithic, epitaxially-grown structure using Bragg-stacks as highly reflective mirrors. Their photon content leads to a large energy dispersion, and to the possibility of manipulating them via an external light field. The exciton content, conversely, results in a strong mutual interaction of polaritons, not present for photon modes. These peculiar properties have enabled the observation of parametric amplification of polaritons [3, 4], which was interpreted as bosonic stimulation [3, 5, 6], owing to the fact that the polaritons are, to a good approximation, bosons in the relevant exciton density regime.

In the absence of an additional injected polariton density to be amplified, parametric scattering results in a spontaneous emission of polariton pairs (signal and idler) according to momentum and energy conservation criteria. For non-resonant optical excitation, the polaritons are created by phonon-assisted relaxation of the initially excited electron–hole pairs, and accumulate due to the relaxation bottleneck at a rather large k, with a circular symmetric distribution around $k = 0$. The resulting emission is ring-shaped [7], and can show the effect of final state stimulation [8]. For resonant optical excitation, polaritons are created directly at a defined k, and the momentum and energy conservation for the scattering of a pair of these polaritons holds on an 8-shaped line of final states in k-space [9]. This mechanism has been observed under continuous wave excitation for pump, signal and idler wave-vectors lying on the symmetry axis of the 8-shape [10, 11]. Also here, a stimulation is found, that leads to a macroscopic population of the final states. The possibility of creating correlated polariton pairs was discussed [12, 13]. The 8-shaped momentum space range was observed in my previous work [14], where a quantitative comparison with the theory of parametric luminescence was presented. I report

Physics of Semiconductor Microcavities: From Fundamentals to Nanoscale Devices
Edited by Benoit Deveaud
Copyright © 2007 WILEY-VCH Verlag GmbH & Co. KGaA, Weinheim
ISBN: 978-3-527-40561-9

here an extension of these measurements of the time, spectrally, and directionally resolved emission from a semiconductor microcavity (MC) after pulsed resonant optical excitation of lower branch polaritons (LP) in a regime where the parametric polariton–polariton scattering dominates the emission. A polarization-dependent renormalization of the polariton dispersion in presence of a linearly polarized polariton density is found, corresponding to a strong resonant birefringence. The case of two pump directions is investigated. The k shapes of the observed mixed parametric processes are in agreement with the calculations using energy and momentum conservation when including the polariton renormalization. Using two pump directions, a signal polariton is corresponding to two idler polaritons. Detecting the interference of the two idler polaritons, which are both incoherent to the pump polaritons, I demonstrate the pair-correlation of the emitted signal-idler polaritons in a "which-way" experiment.

8.2
Investigated Sample and Experimental Details

The investigated sample [15] was grown by molecular beam epitaxy and consists of a 25 nm GaAs/Al$_{0.3}$Ga$_{0.7}$As single quantum well placed in the center of a λ–cavity with AlAs/Al$_{0.15}$Ga$_{0.85}$As Bragg reflectors of 25 (16) periods at the bottom (top). The sample was characterized by reflection measurements reported in Ref. [16]. The use of a wide GaAs QW leads to a negligible inhomogeneous broadening of the QW exciton due to the small effect of interface roughness and the absence of alloy disorder. The exciton–photon coupling in the MC creates three polariton resonances from the heavy-hole exciton, light-hole exciton, and cavity mode. The mixing is well described by a three-coupled-oscillator model [15]. From the measured polariton linewidths a cavity linewidth $\hbar\gamma_c$ of 0.13 meV and an excitonic linewidth of $\hbar\gamma_{hh} = \hbar\gamma_{lh} = 0.06$ meV were inferred.

The sample was held in a helium cryostat at a temperature of $T = 5$ K. The pump polaritons were excited by Fourier-limited optical pulses from a mode-locked Ti:sapphire laser of about 1 ps pulse length at 76 MHz repetition rate. The pulses were focused on the sample by a 0.5 NA aspheric lens to a diffraction-limited spot of 100 (50) µm diameter and a k-width of 0.03 (0.06) µm^{-1} for the single (dual) pump experiments. The excitation intensity I_0 corresponds to a photon flux of 21 µm^{-2} per pulse at the sample surface. The excitation was linearly polarized along $\hat{x}$, and the cross-linearly polarized emission (i.e. $\hat{y}$) was detected. Spectrally resolved emission intensities were acquired using a spectrometer and a nitrogen cooled CCD camera with a resolution of 20 µeV. Time-resolved data were taken using a synchroscan streak-camera with a time resolution of 2 ps. Directionally resolved, time and spectrally integrated intensities were taken by a video CCD camera.

8.3
Parametric Scattering for a Single Pump Direction

The case of the excitation by a single pump direction k_p has been discussed extensively in Ref. [14]. Here I show additional results concerning the density dependence and the spectral imaging. The MC dispersion measured in the photoluminescence is described using the Rabi-energies of the heavy-hole exciton of $\hbar\Omega_{hh} = 1.82$ meV, of the light hole exciton of $\hbar\Omega_{lh} = 1.1$ meV, and the effective index of the cavity mode of 3.5. The energy of the heavy-hole exciton was $E_{hh} = 1.52168$ meV, of the light-hole exciton $E_{lh} = 1.52420$ meV, and of the cavity mode $E_{cav} = 1.51930$ eV at $k = 0$, corresponding to a detuning of -2.4 meV to the heavy-hole exciton. The pump direction was chosen to be $k_p = (1.73, 0)\,\mu\text{m}^{-1}$. The final states of the parametric scattering of two pump polaritons at k_p on the lower polariton branch respecting energy and momentum conservation ($2E_{LP}(k_p) = E_{LP}(k_s) + E_{LP}(k_i)$ & $2k_p = k_s + k_i$) are shown in Fig. 8.1 together with the LP dispersion

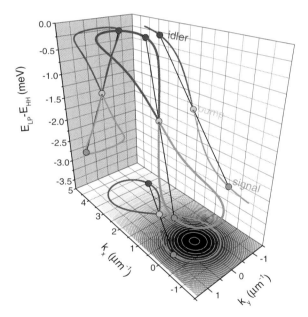

Fig. 8.1 Energy-momentum sketch of the parametric scattering process for a single pump at k_p. The signal (red) and idler (blue) polariton states for parametric scattering of two pump polaritons (green) under energy and momentum conservation are shown in a three-dimensional graph of $(k, E_{LP}(k))$ for $k_p = (1.73, 0)\,\mu\text{m}^{-1}$ with the polariton dispersion $E_{LP}(k)$ as described in the text. For illustration, a specific signal-idler combination with $k_s = (0, 0.89)\,\mu\text{m}^{-1}$ is high-lighted by circles and connected by a line. The projections on the two-dimensional planes are given. On the lower plane, $E_{LP}(k) - E_{hh}$ is shown on a grey-scale contour from -3.5 meV (black, innermost contour) to 0 meV (white), with 0.1 meV spacing.

$E_{LP}(\mathbf{k})$. They form an figure-8 shaped line S in $\mathbf{k}$-space. This shape is due to the peculiar LP dispersion, having a negative effective mass region for $|\mathbf{k}| > 1.45$ µm^{-1}. We measure the wavevector and time-resolved emission intensity $I(\mathbf{k},t)$ after excitation at $t=0$. $I(\mathbf{k},t)$ is proportional to the polariton density $N(\mathbf{k},t)$ times the escape rate to the detected photons. The escape rate is given by the cavity content $c(\mathbf{k})$ multiplied with the photon decay rate of the cavity $2\gamma_c$, so that $I(\mathbf{k},t) \propto N(\mathbf{k},t)\gamma_c c(\mathbf{k})$. The resulting measured density $N(\mathbf{k},t)$ is shown in Fig. 8.2a. For an excitation intensity of $40I_0$, the time-evolution of $N(\mathbf{k},t)$ is given in the left column. A fast buildup of the 8-shaped region of final-state density within the first 10 ps, and its decay within the next 10 ps is observed. Afterwards, only the density due to the emission of bound exciton states remains, which decays on a 100 ps timescale. The final state density shows a strong narrowing of the $\mathbf{k}$ width of the 8-shape within the first 10 ps. This is predicted by theory [14], and is a signature of the memory in the scattering process described by the anomolous signal-idler correlation [9]. The white dotted lines indicate the calculated S, which is in agreement with the measured density distribution apart from a slight size mismatch. The origin of this slight deviation can be elucidated looking at the excitation density dependence of the density distribution, which is shown in the right column. At the smallest displayed excitation intensity of $10I_0$ no significant deviation is observable. With increasing excitation intensity, the deviation is increasing. The observed behaviour can be reproduced assuming a renormalization of the polariton dispersion relative to the pump energy.

An example of the resulting S for +0.1 meV renormalization of $E_{LP}(\mathbf{k})$ is given in the graphs of $40\,I_0$ and $70\,I_0$. Such a renormalization is expected [9], but since we use a pulsed excitation, also the energy of the pump-polaritons is renormalized and the net effect is expected to vanish. However, as the detected emission is cross-linearly polarized to the pump, this result indicates a renormalization of the polaritons depending on their relative linear polarization to the pump polariton density. This can be explained considering the biexciton bound state of ≈ 1 meV binding energy [18]. This state can be created by two co-linearly polarized excitons, but not by cross-linearly polarized excitons. It is therefore to be expected that the pump-polaritons experience a negative renormalization by this state, which is not present for the cross-linearly polarized signal and idler polaritons. A similar behaviour was measured for co- and cross-circularly polarized polaritons in a pump–probe experiment[1] on the same sample [19], and has been used in the theoretical analysis of polarization dependent polariton amplification measurements [20]. The appearance of such a renormalization is even more evident from the results of the dual pump geometries discussed in the next section.

The influence of the renormalization on the polariton scattering can be directly seen in the spectrally selected polariton density $N(\mathbf{k},\omega) \propto I(\mathbf{k},\omega)/(\gamma_c c(\mathbf{k}))$ shown in Fig. 8.2b. $I(\mathbf{k},\omega)$ was measured time-integrated with a spectral resolution

---

1) Unpublished co and cross-linearly polarized pump–probe experiments confirm the presence of such a renormalization.

8.3 Parametric Scattering for a Single Pump Direction

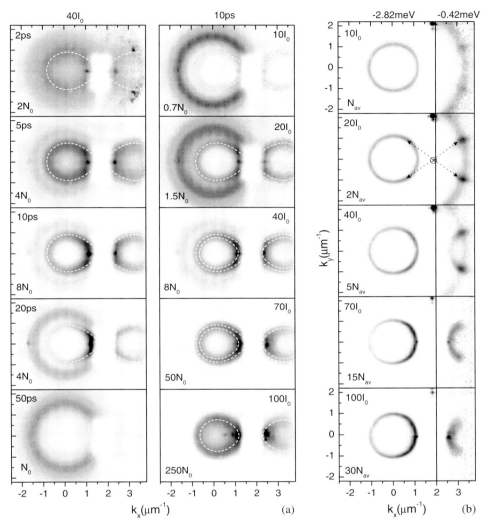

Fig. 8.2 Measured k-resolved polariton density for resonant pulsed excitation at $k_p = (1.73, 0)$ μm^{-1}. a) Time-resolved data. Left column: Different times after excitation as indicated for an excitation intensity of $40\,I_0$. Right column: Different excitation intensities as indicated 10 ps after excitation. The grey scale has a linear mapping from zero (white) to a multiple of N_0 (black), as indicated. No absolute value for the used fixed experimental population unit N_0 is available since the absolute intensity calibration was not performed to a sufficient accuracy. b) Spectrally resolved data. Linear grey scale from zero to a multiple of N_{av}, as indicated. For $k_x < 2$ μm^{-1} (signal) taken from the measured intensity at $\hbar\omega = E_{hh} - 2.63$ meV, for $k_x > 2$ μm^{-1} (idler) at $\hbar\omega = E_{hh} - 0.42$ meV. These energies are conserving the energy of the resonant pump polaritons, that was determined to 1.52015 eV by their resonant Rayleigh scattering [17] at $10\,I_0$. For $20\,I_0$, the parametric processes are indicated.

of 20 µeV. At low pump intensity $10I_0$, the idler polariton emission at $\hbar\omega = E_{hh} - 0.42$ meV is resonant to the low-intensity polariton dispersion ring that is visible due to the long-lived photoluminescence. With increasing intensity, the idler forms a ring-shaped extrusion to smaller k, which is dominating above $50 I_0$. The direction to smaller k is explained by a positive renormalization of the idler dispersion. The effect of the renormalization on the k distribution of the signal at $\hbar\omega = E_{hh} - 2.63$ meV is not as pronounced since the exciton content is smaller and the dispersion is much steeper.

8.4
Parametric Scattering for Two Pump Directions

Using two pump directions k_{p1}, k_{p2} allows, additionally to the parametric processes present for the individual pump directions, for mixed parametric processes involving one polariton of each pump. These mixed processes feature a variety of new shapes for S, some of which are displayed in Fig. 8.3. For an azimuthal separation

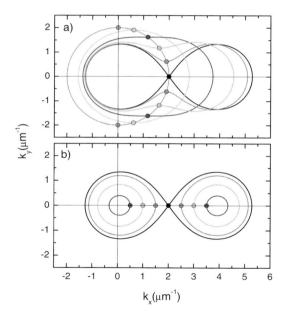

Fig. 8.3 Calculated k-space curves of energy and momentum conservation S for the mixed parametric process involving two pump directions: $k_{p1} + k_{p2} \rightarrow k_s + k_i$. The dispersion parameters used are the ones of the dual pump experiments (see text). The pump directions are indicated as circles, and are mirror symmetric around the x-axis. (a) Equal absolute value $|k_{p1}| = |k_{p2}| = 2$ µm^{-1}, for angles between k_{p1} and k_{p2} of $0, 0.2\pi, 0.4\pi, 0.6\pi, 0.8\pi, \pi$. (b) Equal directions of k_{p1} and k_{p2}, with different absolute values with $|k_{p1} + k_{p2}| = 4$ µm^{-1}, and $|k_{p1} - k_{p2}| \in \{0, 0.5, 1.0, 1.5\}$ µm^{-1}.

of the pumps, the 8-shape deforms into a peanut, then an oval and finally to a circular shape. For an axial separation instead, the 8-shape splits into two separate loops, that approach circular shape. These distinct behaviors are reflecting the different detuning of the effective pump energy $(E_{LP}(\mathbf{k}_{p1}) + E_{LP}(\mathbf{k}_{p2}))/2$ compared to the equivalent single pump case of $E_{LP}((\mathbf{k}_{p1} + \mathbf{k}_{p2})/2)$. Due to the shape of the polariton dispersion, the effective pump energy is increased for an azimuthal separation, and (initially) decreased for a radial separation. The effect of the radial separation on S is thus similar to the effect of the polarization selective dispersion renormalization observed in Fig. 8.2.

To realize the dual pump configuration, we have expanded the original experimental setup used for the single pump case by a Mach–Zender interferometer on the excitation side (see ① in Fig. 8.4). It splits the excitation into two directions and thus $\mathbf{k}$ values, which are individually adjustable by the tilt of the mirrors M1 and M2. The mirror surface is imaged onto the sample surface by the lenses L and A. Gimble mirror mounts are used for M1 and M2 to ensure that the excitation area on the sample and the temporal coincidence of the excitation pulses is independent of the mirror tilt. The MC emission is collected by the same lens combination, and imaged onto a pinhole (PH) to reject scattered light from surfaces in the setup other than the sample. The pump pulses directly reflected off the sample can be rejected

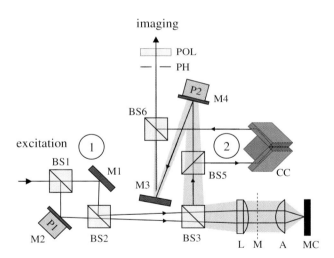

Fig. 8.4 Scheme of the optical setup used for the dual pump experiment. The two phase coherent pump pulses are created in a Mach–Zender interferometer ① consisting of the mirrors M1, M2 and the beam-splitters BS1, BS2. Their relative phase is adjustable by a piezoelectric element (P1). The pump pulses are imaged by the aspheric lens A of 0.5 numerical aperture from the mirrors onto the MC. The emission from the MC is collected by the same lens, and directed by BS3 into a second Mach–Zender ② which, by using a double mirror (M3, M4) in one arm and a corner cube mirror (CC) in the other one, produces interference between fields at $\mathbf{k} = (k_x, k_y)$ and $\mathbf{k'} = (k_x, -k_y)$. The detection phase can be adjusted by P2. All beam-splitters are non-polarizing.

by a mask (M) in the Fourier plane of lens A to avoid their scattering in the subsequent optical path. The detected emission is analyzed using a Glan–Thompson prism (POL), that transmits the emission cross-linearly polarized to the excitation pulses only. The Mach–Zender interferometer on the detection side ② is used to create an emission interference as discussed later. For the moment it is deactivated by blocking one of its arms.

The following experiments were taken at a slightly different cavity detuning compared to the single pump case presented in the previous section and in [14]. The LP dispersion was measured by spectrally and k resolved emission, and was fitted by the three-oscillator model [21] using $\hbar\Omega_{hh} = 1.68$ meV, $\hbar\Omega_{lh} = 1.1$ meV, $E_{hh} = 1.52187$ meV, $E_{lh} = 1.5246$ meV, $E_{cav} = 1.5212$ eV at $k = 0$, corresponding to a detuning of -0.7 meV to the heavy-hole exciton.

In Fig. 8.5 the measured time-integrated polariton densities $N(k)$ for the dual pump case are shown for different k_{p1}, k_{p2}. $N(k)$ is deduced from the time-integrated cross-linearly polarized emission intensity $I(k)$ measured by the video CCD using $N(k) \propto I(k)/(\gamma_c c(k))$. In a) the situation $k_{p1} = -k_{p2} = (0, 2.2)$ μm^{-1} is shown, corresponding to the angle π in the calculations of Fig. 8.3a. The k positions of the two pump beams are visible as the strongest intensity spots due to the residual transmission of the analyzer POL. The created polariton density shows a double ring structure. The inner ring issue to bound exciton emission [14], which is at smaller k as compared to Fig. 8.2 due to the different detuning. The second ring is the mixed parametric process, which shows the predicted circular shape (Fig. 8.3a). However, its radius is not equal to $|k_p|$, as predicted by the calculation (outer dotted curve), but significantly smaller. As in the single pump experiments, this gives evidence for a polarization-dependent renormalization by the pump polariton density. Using a fixed blue shift of 0.2 meV of the observed x polarized dispersion compared to the y polarized dispersion relevant to the excitation pulses, the observed circle is reproduced by the calculated S (inner dotted curve). Additionally, vertically displaced to k_{p1} and k_{p2}, a strongly enhanced emission on the parametric ring is observed. This feature is related to the cross-hatched disorder in the sample along the [110] and [1$\overline{1}$0] crystal directions [17, 21], which coincide with the x- and y-directions used in this work. The disorder allows the modes on the ring that are displaced to the pumps only in x- or y-direction to be directly excited by the pump. The resulting initial occupation of the modes leads to a stimulation of the parametric scattering, which is significantly increasing their occupation well beyond the purely additive effect. Such an interpretation is verified by the measurements for all other pump configurations b)–i) shown in Fig. 8.5. In these configurations, $|k_{p1}| \approx |k_{p2}| \approx 2.2$ μm^{-1} is retained, while the angle between them is reduced from π in a) to 0.085π in i). The observed shape of the parametric scattering can be compared to the calculations of S shown in Fig. 8.3a. An evolution from a circular (a) to an oval (b, c, d) and further to a peanut-shaped (e, f) parametric scattering region is found. In all cases, the 0.2 meV renormalized calculation of S (black dotted lines) is in good agreement with the observation. The parametric processes created by the individual pumps are given as white dotted lines in d) to i), using the same renormalization. These processes are generally weaker than the

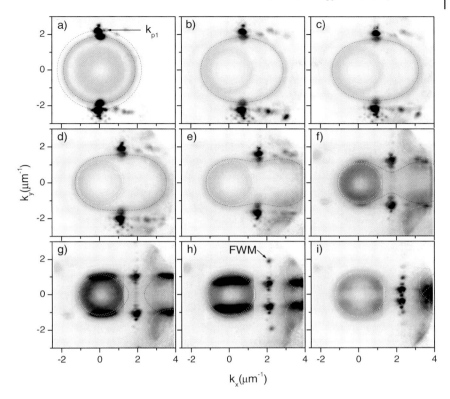

Fig. 8.5 Measured *k*-resolved time-integrated polariton density for resonant pulsed excitation with two coincident excitation pulses at k_{p1}, k_{p2} for an excitation intensity of $\approx 40\, I_0$ per pulse. Fixed linear grey scale from white (zero density) to black. The excitation wave-vectors k_{p1}, k_{p2} were adjusted mirror-symmetric around the *x*-axis, and are visible by the residual transmission of the reflected pulses through POL. The parts (a) to (i) refer to different choices of k_{p1}, k_{p2}. The dotted black curves are calculated *S* for the mixed process, while the dotted white curves are calculated *S* for the individual pumps. Some spurious signals are present due to residual reflections of the optical elements in the setup.

mixed ones, which we attribute to the larger mismatch of the excitonic contents of the involved polariton states (compare Eq. (8) in Ref. [14]). In the cases f) to i), the renormalization changes the shape of S to a two loop structure, similar to the calculations of Fig. 8.3b. Here, the strong enhancement of the parametric emission for the polaritons displaced in x-direction relative to the pumps due to the initial population mediated by the disorder is obvious. In h), a coherent four-wave mixing process (FWM) between the two pumps is observed. Higher-order mixing up to the ninth order (χ^9) is found in i) and confirmed by interference measurements similar to the ones discussed below.

In summary do the experiments show a pronounced effect of the mixed parametric scattering, and the k regions of the final states are consistent with the calculated curves of energy and momentum conservation S when taking into account a polarization-selective renormalization of the polariton dispersion.

8.5
Polariton Quantum Complementarity by Parametric Scattering

The dual pump geometry allows one to probe the quantum complementarity of polaritons, which is a result of their quantum nature [22]. Here, we test a manifestation of this property that makes quantum particles appear to be either particles or waves under different experimental conditions. The underlying idea in our experiment is the same as in Young's double-slit experiment: due to its wave-like nature, a quantum particle can travel along a quantum superposition of two different pathways, resulting in an interference pattern. If however, it is possible in principle to determine which way the particle has taken, the interference pattern is no longer observed.

To discuss the theoretical background, we follow the description given in Ref. [23]. The effective Hamiltonian describing the parametric polariton process assuming classical pump-fields is

$$\hat{H} = \sum_{k} E_{\mathrm{LP}}(k)\, \hat{p}_k^\dagger \hat{p}_k + \sum_{\substack{k,k' \\ k_s,k_i}} \left[G(k,k')\, \hat{p}_{k_s}^\dagger \hat{p}_{k_i}^\dagger + \mathrm{h.c.} \right] \delta_{k_s + k_i, k + k'}, \quad (1)$$

where the Bose operators $\hat{p}_k^\dagger$ are the polariton creation operators, and $G(k, k')$ contains details of the pump fields and the polariton interaction. The experimental scheme employs two mutually coherent pump modes of momenta k_{p1} and k_{p2} with the classical field amplitudes $P_{k_{p1}}$ and $P_{k_{p2}}$, for which $G(k,k') = g P_k P_{k'}(\delta_{k,k_{p1}} + \delta_{k,k_{p2}})(\delta_{k',k_{p1}} + \delta_{k',k_{p2}})$. The constant g is the polariton–polariton interaction amplitude, accounting for both the Coulomb interaction and the Pauli exclusion principle [9]. Two of the four products of δ's represent parametric processes driven by a single pump mode (black and red curve in Fig. 8.6). The two other terms are mixed-pump processes involving one polariton from each pump mode, whose energy-momentum conservation defines the magenta curve. We first consider a pair produced by one pump only ($P_{k_{p2}} = 0$). In the limit of low excitation intensity ($\tau = g|P_{k_{p1}}|^2 t \ll 1$), the time evolution operator applied on the polariton vacuum state $|v\rangle$ yields up to first order in τ the entangled polariton state $|\Psi\rangle = |v\rangle_{s,i} + \tau |1\rangle_s |1\rangle_i$, where s and i label a pair of signal and idler modes. This quantum state shows that the parametric process produces correlated signal-idler pairs. Using two mutually coherent pump polariton fields of equal amplitudes $P_{k_{p2}} = P_{k_{p1}} e^{i\phi}$, pairs of parametric processes sharing the signal mode are allowed (see Fig. 8.6). Such a pair of processes involves two idler modes i1 and i2 and one common signal mode s. In the limit of low excitation intensity, the time evolution operator applied on the polariton vacuum state yields in this case up to first order in τ:

$$|\Psi\rangle = |v\rangle_{s,i1,i2} + \tau |1\rangle_s \left(|1\rangle_{i1} |0\rangle_{i2} + e^{-2i\phi} |0\rangle_{i1} |1\rangle_{i2} \right). \quad (2)$$

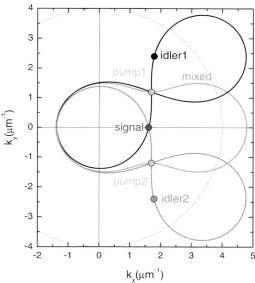

Fig. 8.6 Calculated *k*-space curves of energy and momentum conservation S involving two pump directions $\boldsymbol{k}_{p1,p2} = (1.7, \pm 1.2)$ μm^{-1}. The curves for the individual pump directions $\boldsymbol{k}_{p1}(\boldsymbol{k}_{p2})$ are black (red). The mixed parametric process of both pump directions is magenta. The dispersion parameters used are the ones of the dual pump experiments (see text). Two parametric processes sharing the signal $\boldsymbol{k}_s$ (blue dot), giving rise to mutual coherence of the idlers $\boldsymbol{k}_{i1}, \boldsymbol{k}_{i2}$ (black, red dots) are indicated. The dotted line is the border of the *k* range detected in the experiment.

The resulting polariton population at the signal mode $\langle \Psi | \hat{p}^\dagger_{\boldsymbol{k}_s} \hat{p}_{\boldsymbol{k}_s} | \Psi \rangle$ is independent of ϕ. Interference is absent due to the orthogonality of the idler states *i*1 and *i*2, that contain the information from which of the two pumps the signal originated. Interference is also absent from either idler-polariton density, since only one path is present in them. However, when observing the sum of the two idler polariton fields, again two pathways are involved. The resulting particle population is $\langle \Psi | (\hat{p}^\dagger_{\boldsymbol{k}_{i1}} + \hat{p}^\dagger_{\boldsymbol{k}_{i2}})(\hat{p}_{\boldsymbol{k}_{i1}} + \hat{p}_{\boldsymbol{k}_{i2}}) | \Psi \rangle = 2 \langle \Psi | \hat{p}^\dagger_{\boldsymbol{k}_{i1}} \hat{p}_{\boldsymbol{k}_{i1}} | \Psi \rangle (1 + \cos(2\phi))$. Interference is present since the two idler modes are pair-correlated with the same signal mode, which therefore does not contain any "which-way" information for the mixed idler mode.

The polaritons at a given $\boldsymbol{k}$ are observed by detecting the photons emitted at the same in-plane momentum [24], which carry the polariton amplitude and phase information. The interference of the two emitted idler fields was measured as a function of the relative pump phase ϕ. This is accomplished by detecting two superimposed $\boldsymbol{k}$-resolved images of the polariton emission, one of which is preliminarily inverted along the k_y axis. The two mutually coherent pump pulses are created by splitting the exciting laser pulses (see ① in Fig. 8.4), and are synchronously impinging on the microcavity with $\boldsymbol{k}_{p1} = (k_{px}, k_{py})$ and $\boldsymbol{k}_{p2} = (k_{px}, -k_{py})$. The relative phase ϕ

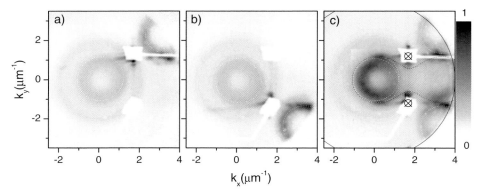

Fig. 8.7 Measured k-resolved time-integrated polariton density for resonant pulsed excitation with two coincident excitation pulses at $\boldsymbol{k}_{p1}$, $\boldsymbol{k}_{p2} \approx (1.7, \pm 1.2)$ µm^{-1} for an excitation intensity of $\approx 40\, I_0$ per pulse. Fixed linear grey scale from white (zero density) to black. The white shadows are due to a mask M in the Fourier plane blocking the directly reflected pump pulses. (a) Only $\boldsymbol{k}_{p1}$, (b) only $\boldsymbol{k}_{p2}$, (c) both pumps. The dotted lines are the calculated S for the individual (white) and mixed (black) parametric processes for 0.2 meV renormalization.

of the pulses can be adjusted using a fine adjustment of the M2 position by a piezoelectric transducer P1. The emitted photon field $\mathcal{E}(\boldsymbol{k})$ of the MC is directed through a Mach-Zender interferometer (② in Fig. 8.4) creating the superposition $\mathcal{E}_{\boldsymbol{k}} + e^{i\varphi} \mathcal{E}_{\boldsymbol{k}'}$, where $\boldsymbol{k} = (k_x, k_y)$ and $\boldsymbol{k}' = (k_x, -k_y)$. The phase φ can be adjusted using the piezoelectric transducer P2. Blocking one of the detection interferometer arms, the intensity $I(\boldsymbol{k}) = |\mathcal{E}(\boldsymbol{k})|^2$ is measured, which is proportional to the polariton number $N(\boldsymbol{k}) = \langle \Psi | \hat{p}_{\boldsymbol{k}}^\dagger \hat{p}_{\boldsymbol{k}} | \Psi \rangle$. The measured $N(\boldsymbol{k})$ is shown in Fig. 8.7 for pump 1 only, pump 2 only, and for both pumps. The observed parametric processes are in agreement with the calculated S including 0.2 meV renormalization (dotted curves).

Using both arms of the detection interferometer, the detected intensity is $\tilde{I}(\boldsymbol{k}) = |\mathcal{E}(\boldsymbol{k}) + e^{i\varphi} \mathcal{E}(\boldsymbol{k}')|^2$. Due to abberations of the collection optics, the phase φ can show some systematic variations over $\boldsymbol{k}$ space. $\tilde{I}(\boldsymbol{k})$ is proportional to the polariton number $\tilde{N}_{\boldsymbol{k}} = \langle \Psi | (\hat{p}_{\boldsymbol{k}}^\dagger + e^{i\varphi} \hat{p}_{\boldsymbol{k}'}^\dagger)(\hat{p}_{\boldsymbol{k}} + e^{-i\varphi} \hat{p}_{\boldsymbol{k}'}) | \Psi \rangle$, which depends in general on the relative phase ϕ of the pump pulses. The measured $\tilde{N}_{\boldsymbol{k}}$ are shown in Fig. 8.8 for $\varphi = \pi$ and various ϕ as indicated. To understand at which k_y the idler polaritons share the same signal mode, and thus interference is expected, we consider parametric processes driven by either one of the two pump modes. The pump at $\boldsymbol{k}_{p1}$ produces the emission of polariton pairs at $\boldsymbol{k}$ and $2\boldsymbol{k}_{p1} - \boldsymbol{k}$ respectively. Equivalently, the pump at $\boldsymbol{k}_{p2}$ produces pairs of $\boldsymbol{k}'$ and $2\boldsymbol{k}_{p2} - \boldsymbol{k}'$. By construction, the modes at $2\boldsymbol{k}_{p1} - \boldsymbol{k}$ and $2\boldsymbol{k}_{p2} - \boldsymbol{k}'$ have the same x component but opposite y components $\pm(2k_{py} - k_y)$. Consequently, idler modes at $\boldsymbol{k}$ and $\boldsymbol{k}'$ share the same signal mode if and only if $k_y = 2k_{py}$ (see dotted lines in Fig. 8.8). Interference is actually observed in the experimental data at this k_y for k_x values at which the parametric emission dominates, i.e. $k_x \approx 2.4$ µm^{-1}.

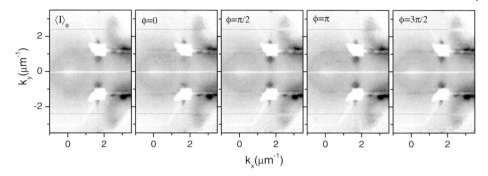

Fig. 8.8 Measured k-resolved time-integrated polariton interference density $\tilde{N}(k)$ for resonant pulsed excitation as in Fig. 8.7c. Fixed linear grey scale from white (zero density) to black. The white shadows are due to a mask M in the Fourier plane blocking the directly reflected pump pulses. Different excitation phase differences ϕ as labeled. The leftmost frame is an average over $\phi = 0 \ldots 2\pi$. The dotted lines indicates the k_y at which the idler modes share the same signal. The white region around $k_y = 0$ is due to the destructive self-interference present for the detection phase $\varphi = \pi$.

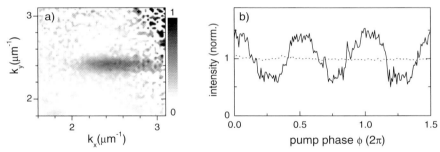

Fig. 8.9 Measured interference between the parametric emission at k and k'. (a) k resolved idler interference visibility (grey scale 0 (white) to 1 (black)). (b) $\tilde{I}(k)$ at $k \approx (2.4, 2.4)$ µm^{-1} (full curve) and at $k \approx (1.0, 0.0)$ µm^{-1} (dotted curve) versus pump phase ϕ.

For a quantitative analysis of the interference, we determine the visibility from the measurements using $V_k = \sqrt{2[\langle(\tilde{I}_k)^2\rangle_\phi / \langle \tilde{I}_k \rangle_\phi^2 - 1]}$, where $\langle \cdots \rangle_\phi$ denotes the average over ϕ. For a sinusoidal modulation, this definition is equal to the standard definition $V = (I_{max} - I_{min})/(I_{max} + I_{min})$. The measured visibility is given in Fig. 8.9a, showing that interference is actually observed at and only at[2] the expected region in k-space, with a visibility of about 0.5. The visibility is below unity due to contributions of photoluminescence and higher-order scattering processes to the emission. The measured interference shows a $\cos(2\phi)$ dependence (see Fig. 8.9b), as pre-

2) Other interferences with a different ϕ dependence, related to the mixed parametric processes and to four-wave mixing are also observed, outside of the region shown in the visibility data.

dicted by Eq. (2). Additionally, in agreement with the above analysis, the signal at $k_x = 0$ (dotted) displays no interference, even though it is created by a superposition of contributions from both pumps.

I point out that Eq. (2) is valid only in the low intensity regime, for which stimulation is insignificant. The present measurement was performed at the onset of the self-stimulated regime, corresponding to an excitation intensity of $40 I_0$ according to Ref. [14]. In the self-stimulated regime, the conclusions about the interference drawn from Eq. (2) are still valid, however, the interference could be also accounted for by a classical parametric model including random-noise driving terms [25]. In order to decide if the quantum nature of the particles is important for the observed interference, we have therefore measured the time-resolved visibility of the interference, and compared it with a quantum Langevin model [23]. From the comparison it results that only the quantum origin can explain the observed high interference visibility within the first 5 ps after excitation (not shown here).

8.6
Conclusions

In conclusion, we experimentally observed the final-state shape of the parametric polariton–polariton scattering in a semiconductor microcavity pumped by two pump directions, showing the possibility to manipulate this shape to a peanut or oval for the mixed parametric processes involving one polariton of each pump. The excitation of linearly polarized polaritons was found to result in a polarization-dependent renormalization of the polariton dispersion due to the polarization selection rules of the biexcitonic state, that can be exploited as a strong resonant birefringence. Using the dual pump scheme, pair-correlations of the parametrically created polaritons were demonstrated using a quantum mechanical "which-way" experiment. This result opens the possibility of producing many-particle entangled states of light-matter waves in a semiconductor.

Acknowledgements

The author thanks U. Woggon for continued support. The sample was grown by J. Riis Jensen at the Research Center COM and the Niels Bohr Institute, Copenhagen University. The quantum complementarity experiments were proposed and theoretically described by S. Savasta, O. Di Stefano and V. Savona.

References

[1] M. S. Skolnick, T. A. Fisher, and D. M. Whittaker, Semicond. Sci. Technol. **13**, 645 (1998).
[2] G. Khitrova, H. M. Gibbs, F. Jahnke, M. Kira, and S. W. Koch, Rev. Mod. Phys. **71**, 1591 (1999).
[3] P. G. Savvidis, J. J. Baumberg, R. M. Stevenson, M. S. Skolnick, D. M. Whittaker, and J. S. Roberts, Phys. Rev. Lett. **84**, 1547 (2000).
[4] M. Saba, C. Ciuti, J. Bloch, V. Thierry-Mieg, R. André, Le Si Dang, S. Kundermann, A. Mura, G. Bongiovanni, J. L. Staehli, and B. Deveaud, Nature **414**, 731 (2001).
[5] P. Senellart, J. Bloch, B. Sermage, and J. Y. Marzin, Phys. Rev. B **62**, R16263 (2000).
[6] J. Erland, V. Mizeikis, W. Langbein, J. R. Jensen, and J. M. Hvam, Phys. Rev. Lett. **86**, 5791 (2001).
[7] P. G. Savvidis, J. J. Baumberg, D. Porras, D. M. Whittaker, M. S. Skolnick, and J. S. Roberts, Phys. Rev. B **65**, 073309 (2002).
[8] Le Si Dang, D. Heger, R. André, F. Bœuf, and R. Romestain, Phys. Rev. Lett. **81**, 3920 (1998).
[9] C. Ciuti, P. Schwendimann, and A. Quattropani, Phys. Rev. B **63**, 041303 (2001).
[10] R. M. Stevenson, V. N. Astratov, M. S. Skolnick, D. M. Whittaker, M. Emam-Ismail, A. I. Tartakovskii, P. G. Savvidis, J. J. Baumberg, and J. S. Roberts, Phys. Rev. Lett. **85**, 3680 (2000).
[11] R. Houdré, C. Weisbuch, R. P. Stanley, U. Oesterle, and M. Ilegems, Phys. Rev. Lett. **85**, 2793 (2000).
[12] J. Ph. Karr, A. Baas, and E. Giacobino, Phys. Rev. A **69**, 063807 (2004).
[13] C. Ciuti, Phys. Rev. B **69**, 245304 (2004).
[14] W. Langbein, Phys. Rev. B **70**, 205301 (2004).
[15] J. R. Jensen, P. Borri, W. Langbein, and J. M. Hvam, Appl. Phys. Lett. **76**, 3262 (2000).
[16] P. Borri, J. R. Jensen, W. Langbein, and J. M. Hvam, Phys. Rev. B **61**, 13377 (2000).
[17] W. Langbein and J. M. Hvam, Phys. Rev. Lett. **88**, 047401 (2002).
[18] W. Langbein and J. M. Hvam, Phys. Rev. B **61**, 1692 (2000).
[19] P. Borri, W. Langbein, U. Woggon, A. Esser, J. R. Jensen, and J. M. Hvam, Semicond. Sci. Technol. **18**, S351 (2003).
[20] A. Kavokin, P. G. Lagoudakis, G. Malpuech, and J. J. Baumberg, Phys. Rev. B **67**, 195321 (2003).
[21] W. Langbein, J. Phys.: Condens. Matter **16**, S3645 (2004).
[22] M. O. Scully, B.-G. Englert, and H. Walther, Nature **351**, 111 (1991).
[23] S. Savasta, O. Di Stefano, V. Savona, and W. Langbein, Phys. Rev. Lett. **94**, 246401 (2005).
[24] V. Savona, F. Tassone, C. Piermarocchi, A. Quattropani, and P. Schwendimann, Phys. Rev. B **53**, 13051 (1996).
[25] M. O. Scully and M. S. Zubairy, Quantum Optics (Cambridge University Press, Cambridge, 1997).

Chapter 9
Spin Dynamics of Exciton Polaritons in Microcavities

Ivan A. Shelykh, Alexei V. Kavokin, and Guillaume Malpuech

9.1
Introduction

An exciton is formed by an electron and hole, i.e. by two fermions having projections of angular momenta on a given axis equal to $J_z^e = S_z^e = \pm\frac{1}{2}$ for an electron in the conduction band with S-symmetry and $J_z^h = S_z^h + M_z^h = \pm\frac{1}{2}, \pm\frac{3}{2}$ for a hole in the valence band with P-symmetry. The states with $J_z^h = \pm\frac{1}{2}$ are formed if the spin projection of the hole S_z^h is antiparallel to the projection of its mechanical momentum M_z^h. These states are called light holes. If the spin and mechanical momentum are parallel, the heavy holes with $J_z^h = \pm\frac{3}{2}$ are formed.

In the bulk samples, at $k = 0$ the light and heavy hole states are degenerate. However, in quantum wells the quantum confinement in the direction of the structure growth axis lifts this degeneracy so that energy levels of the heavy holes lie typically closer to the bottom of the well than the light-hole levels. The ground state exciton is thus formed by an electron and a heavy-hole. The total exciton angular momentum J (in the following it will be referred to as the exciton spin) has the following projections on the structure axis: ± 1 and ± 2. Bearing in mind that the photon spin is ± 1 and that the spin is conserved in the processes of photoabsorbtion, the excitons with spin projections equal ± 2 cannot be optically excited. These are so-called *spin-forbidden* or *dark* states. In quantum microcavities they are not coupled with the photonic mode so we shall neglect them in the following consideration. We note, however, that in some cases the dark states come into play: they can be mixed with the bright states by an in-plane magnetic field or be populated due to the polariton–polariton scattering.

The conservation of spin in the photoabsorbtion allows for spin-orientation of excitons by polarized light beams, the effect which manifests itself in the polarization of photoluminescence. σ^+ and σ^- circularly-polarized light excites +1 and −1 excitons, respectively. Linearly-polarized light excites a linear combination of +1 and −1 exciton-states, so that the total exciton spin projection on the structure axis is zero in this case. Optical orientation of excitons in bulk semiconductors and in quantum wells has been extensively studied in 1980s by several groups. For good reviews we address the reader to the famous volume edited by Zakharchenia and Meier [1].

The polarization of exciting light cannot be kept infinitely long by excitons. Sooner or later they lose the polarization due to inevitable spin and dipole moment relaxation. As excitons are composed by electrons and holes, let us first remind the main mechanisms of spin-relaxation for free carriers in semiconductors. Besides the interaction with structure imperfections and impurities the most important are the following three mechanisms:

1. Elliott–Yaffet mechanism [2] involving the mixing of the different spin wave functions with $k \neq 0$ as a result of the kp interaction with other bands. In quantum wells this effect plays a major role in the spin relaxation of the holes and thus can lead to the transitions between the optically active and dark exciton states, $|+1\rangle \rightleftarrows |-2\rangle$, $|-1\rangle \rightleftarrows |+2\rangle$.

2. Dyakonov–Perel' mechanism [3] caused by the spin–orbit interaction induced spin splitting of the conduction band in the off-centrosymmetric crystals and asymmetric quantum wells (Dresselhaus and Rashba terms respectively) at $k \neq 0$. This mechanism is predominant for the electrons and also leads to the transitions between the optically active and dark exciton states, $|+1\rangle \rightleftarrows |+2\rangle$, $|-1\rangle \rightleftarrows |-2\rangle$.

3. The Bir–Aronov–Pikus (BAP) mechanism [4] involving the spin-flip exchange interaction of electrons and holes. In excitons the efficiency of this mechanism is sufficiently enhanced, as the electron and the hole form the bound state. The exchange interaction consists of two parts, so called long range and short range part of the Coloumb interaction [4, 5]. The short range part leads to the coupling between heavy hole (*hhe*) and light hole excitons (*lhe*) and thus is suppressed in the quantum wells where the degeneracy of lhe and hhe is lifted. The long range part leads to the transitions within the optically active exciton doublet $|+1\rangle \rightleftarrows |-1\rangle$ and thus can lead to the inversion of the circular polarization in the time-esolved polarization measurements [6].

Maialle, de Andrada e Silva, and Sham have shown in a seminal contribution [5] that the third mechanism is predominant for quantum confined excitons. The long-range electron–hole interaction leads to the longitudinal-transverse splitting of exciton states (i.e. energy splitting between excitons having a dipole moment parallel and perpendicular to the wave-vector). This splitting is responsible for rapid spin-relaxation of excitons in quantum wells. There is no need to repeat here the arguments that led to this conclusion. It has a very important consequence: for the description of exciton polaritons in microcavities the dark states can be neglected, which allows to consider an exciton as a two level system and use the well-developed pseudospin formalism for its description. From the formal point of view the exciton can be thus described by 2×2 spin density matrix which is completely analogical to the spin density matrix for the electrons.

One should not forget, however, that the statistics of the excitons (and exciton polaritons) are closer to the bosonic than to the fermionic (excitons and exciton polaritons become bosons in the small concentration limit), and the spin dynamics of excitons and exciton polaritons can be thus very different from the dynamics of the electrons.

All the mechanisms discussed above play role in the so-called linear regime (e.g. at low optical pumping, when the polarization of photoluminescence is independent on the intensity of pumping). In the non-linear regime, at higher pumping, the exciton–exciton interactions come into play. Evidently, the pseudospin can be changed during exciton–exciton collisions. In this regime new mechanisms of spin-relaxation appear which are:

- The transformation of a pair of two bright excitons into a pair of dark ones and vice versa $|-1\rangle|+1\rangle \rightleftarrows |-2\rangle|+2\rangle$ [7].

- The self-induced Larmor precession of the in-plane pseudospin vector of the elliptically polarized polariton condensate [8].

- The 90° rotation of the polarization plane as a result of polariton–polariton collisions [9]. These effects will be discussed latter in this chapter.

The difference between spin-relaxation dynamics of exciton polaritons and spin-relaxation of pure (*mechanical*) excitons is expected to come from the different shape of dispersion curves and, consequently, different energy relaxation dynamics. The Maialle mechanism [5] of spin relaxation is strongly enhanced because of an additional splitting between the TE and TM polarized photonic modes in the cavities. It is also essential that the final state bosonic stimulation is much more efficient for polaritons than for pure excitons, which makes collective effects in their spin dynamics extremely important.

9.2
Experimental Results

The spin dynamics of cavity polaritons [10] has been experimentally studied by measurement of time-resolved polarization from quantum microcavities in the strong coupling regime. Between 2000 and 2005 a series of experimental works appeared in this field, which reported unexpected results. Let us briefly summarise the most important of them.

In the experimental work by D. Martin et al. [11] the dynamics of the circular polarization degree $\wp_c$ of the photoluminescence from the ground state of the CdTe-based microcavity has been measured. $\wp_c$ was determined as

$$\wp_c = \frac{I^+ - I^-}{I^+ + I^-} = \frac{N^+_{k=0}(t) - N^-_{k=0}(t)}{N_{k=0}(t)} \tag{1}$$

where $I^\pm$ denote the circularly-polarized intensities that were measured, $N^\pm_{k=0}$ represent the populations of polaritons with spin projections equal to ±1 in the state

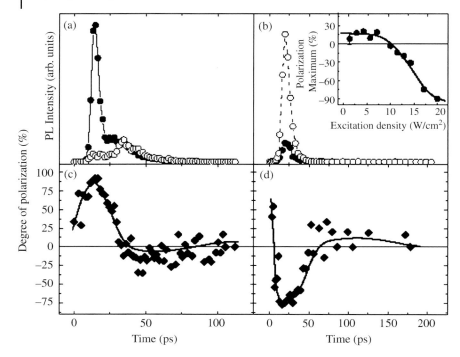

Fig. 9.1 Experimentally measured temporal evolution of the photoluminescence of a CdTe-based microcavity excited by circularly-polarized light at the positive detuning, upper polariton branch (a, $\delta = 10$ meV) and negative detuning, lower polariton branch (b, $\delta = -10$ meV). The filled circles/solid line (open circles/dashed line) denote the σ^+ (σ^-) emission. The educed time evolution of the circular polarization degree for positive and negative detunings is shown in (c) and (d), respectively. The inset shows the maximum value of the polarization degree at 20 ps in the negative detuning case. From [11].

$k = 0$, $N_{k=0}(t) = N^+_{k=0}(t) + N^-_{k=0}(t)$. The pump impulse was circularly polarized and nonresonant (i.e. the exciting laser energy was above the stop-band of the Bragg mirrors composing the microcavity). In the linear regime, when the stimulated scattering of the exciton–polaritons did not play any role, an exponential decay of the circular polarization degree of photoemission was observed. However, above the stimulation threshold $\wp_c$ exhibited a non-monotonic temporal dependence. At positive detuning, the initial polarization of ~30% first increased up to ~90% and then showed damped oscillations (Fig. 9.1). For negative detuning, the polarization degree started from a positive value ~50%, fast decreased down to strongly negative values, and then increased showing attenuated oscillations with a period of about 50 ps. As we will show in the next section this effect is connected with the appearence of the effective in-plane magnetic field created by the long-rang exchange interaction between an electron and a hole.

In the experiments carried out by the Yamamoto group [12], the microcavity has been pumped resonantly at some oblique angle (~50°). The dependence of the

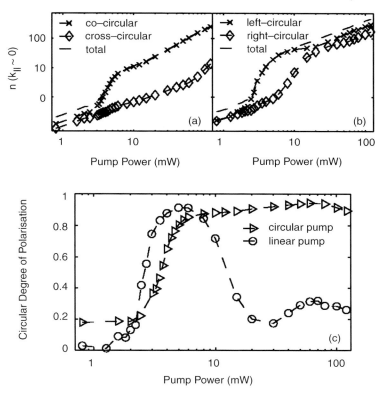

Fig. 9.2 (a) Emission from $k_\| \sim 0$ polaritons vs. pump power under the circular pump. The two circular-polarization components of the emission and their total intensity are plotted. (b) Same as (a) but for the linear pump. (c) Circular degree of polarization $\wp$ vs. pump power with circular (triangles) and linear (circles) pumps. From [12].

intensity and of the polarization of the light emitted by the ground state versus pumping polarization and power has been studied. All experiments were carried out in the continuous pumping regime. The pumping intensity corresponding to the stimulation threshold appeared to be almost twice higher in case of circular pumping than in case of linear pumping as shown on Fig. 9.2 (a, b). Figure 9.2 (c) shows the dependence of the circular polarization degree of the emitted light on the pumping intensity for different pumping polarizations. For both polarizations, $\wp_c \sim 0$ is observed far below the threshold and $\wp_c \sim 0.9$ near the threshold. This indicates that the spin relaxation is complete at low excitation density, and that stimulated scattering into a definite spin component (say, spin-up) of the ground state takes place at high densities. Above threshold, $\wp_c$ remains large in case of a circular pump. In case of a linear pump, however, $\wp_c$ decreases sharply at high pumping intensities. This effect was interpreted in terms of competition of ultra-fast scattering and spin–relaxation of exciton–polaritons.

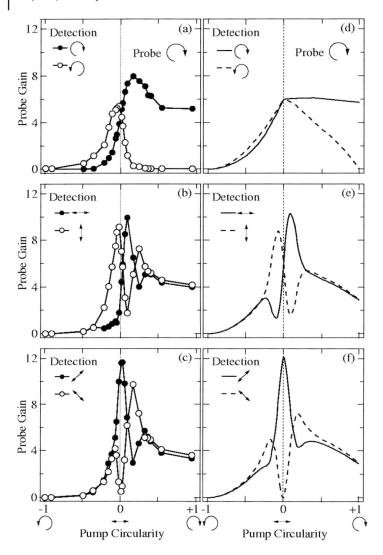

Fig. 9.3 Experimental (a–c) and theoretical (d–f) intensities of circularly (a, d) and linearly (b, c, e, f) polarized components of the light emitted by a microcavity ground state as a function of the circular polarization degree of the pumping light. From [13].

The polarization dynamics of the polariton parametric amplifiers was in focus of the experimental research carried out by Southampton, Toulouse and Sheffield groups. The parametric amplifier is realized if polaritons are created by resonant optical pumping close to the inflection point of the lower dispersion branch at the so-called *magic angle*. In this configuration the resonant scattering of two polaritons excited by the pump pulse toward the signal ($k = 0$) and the idler state is the dominant relaxation process. The scattering can be stimulated by an additional probe pulse used to create the seed of polaritons in the signal state, or it can be strong enough to become self-stimulated.

In the experiments carried out by Lagoudakis et al. [13] the polarization of the probe pulse was kept right-circular, whereas the pump polarization was changed from right- to left-circular passing through elliptical and linear polarization. Consequently, the spin-up and spin-down populations of the pump-injected polaritons were varied while the pump intensity was kept constant. They have detected two circular components of polarization of light emitted by the ground state and four in-plane linear components (vertical, horizontal and the two diagonals ones) as a function of the circular polarization degree of the excitation. These measurements are shown in Fig. 9.3(a–c). To briefly summarise these results, in the case of a linearly-polarized pump pulse and circularly-polarized probe pulse the observed signal was linearly polarized, but with a plane of polarization rotated by 45 degrees with respect to the pump polarization. In the case of elliptically-polarized pump pulses the signal also became elliptical, while the direction of the main axis of the ellipse rotated as a function of the circularity of the pump. In the case of a purely circular pump, the polarization of the signal was also circular, but its intensity was half that found for a linear pump. The polarization of the idler emission emerging at roughly twice the magic angle showed a similar behaviour, although in the case of a linearly-polarized pump the idler polarization was rotated by 90 degrees with respect to the pump polarization. To our mind, the rotation of the polarization plane in this experiment is connected with imbalance of populations of spin-up and spin-down polariton states, which provoked the spin-splitting of the polariton eigen-states. This imbalance has been introduced by the elliptically polarized pump. The whole effect, referred to as "self induced Larmor precession" will be discussed in the next section.

The polarization dynamics in a parametric oscillator without probe has been considered in the experimental work [9]. Quite surprisingly, it has been observed that for the linear pump the polarization of the signal is also linear, but turned by 90°. This effect is connected with the anisotropy of the polariton–polariton scattering as we are going to show later.

9.3
Pseudospin Formalism and Pseudospin Rotation

As we have mentioned in the introduction, the main mechanism for spin relaxation of exciton polaritons in the linear regime is the long-range electron–hole exchange interaction leading to the transitions within the optically active doublet. Consequently, the dark states can be neglected in most cases and exciton polaritons with a given in-plane wavevector k can be treated as a two level system. It can be described by the 2×2 density matrix ρ_k, which is completely analogous to the spin density matrix of the electrons.

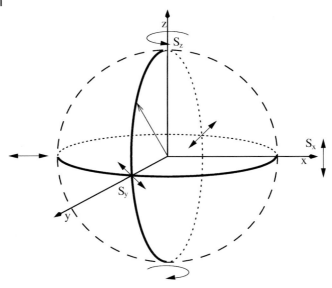

Fig. 9.4 A Poincaré sphere with a pseudospin. The equator of the sphere corresponds to different linear polarizations, while the poles correspond to two circular polarizations.

It is convinient to decompose the polariton spin density matrix as:

$$\rho_k = \frac{N_k}{2} I + S_k \cdot \sigma, \qquad (2)$$

where I is the identity matrix, σ is the Pauli-matrix vector, S_k is the pseudospin of the polariton state characterized by the wave vector k. It corresponds to the Poincaré vector of partially polarized light and describes both the exciton spin state and its dipole moment orientation (see Fig. 9.4). The common convention is to use the basis of circularly polarized states, i.e. to associate states having definite S_z with the polariton radiative states with the spin projection to the structure axis equal to ±1. These states emit circularly polarized light. Their linear combinations correspond to eigenstates of S_x and S_y yielding linearly polarized emission. The pseudospin parallel to x-axis corresponds to x-polarized light (exciton), the pseudospin antiparallel to x-axis corresponds to y-polarized light, the pseudospin oriented along y-axis describes diagonal linear polarizations.

For noninteracting polaritons the temporal evolution of the density matrix (2) is governed by the Liouville–von Neumann equation

$$i\hbar \frac{d\rho_k}{dt} = [H_k; \rho_k], \qquad (3)$$

where the Hamiltonian H_k in terms of pseudospin reads:

$$H_k = E_n(k) - g_s \mu_b \Omega_{\text{eff},k} \cdot S_k. \qquad (4)$$

Here $E_n(\mathbf{k})$ is the energy of n-th polariton branch, g_s is the effective polariton g-factor (we take $g_s = 2$ in the following), μ_B is the Bohr magneton and $\mathbf{\Omega}_{\text{eff},\mathbf{k}}$ is an effective magnetic field. Unlike real magnetic field the effective field only applies to the radiatively active doublet and does not mix optically active and dark states. We do not consider here effects of real magnetic field on the polariton spin dynamics.

The x- and y-components of $\mathbf{\Omega}_{\text{eff},\mathbf{k}}$ are non-zero if the exciton states having dipole moments in, say, x- and y-directions have different energies. This always happens for excitons having non-zero in-plane wave-vectors. The splitting of exciton states with dipole moments parallel and perpendicular to the exciton in-plane wave-vector is called *longitudinal–transverse splitting*. The longitudinal–transverse splitting of excitons in quantum wells is a result of the long-range exchange interaction. It can be described by the following reduced spin Hamiltonian [4, 5]

$$H_{ex} = \frac{3}{16} \frac{|\phi_{1s}(0)|^2}{|\phi_{3D}(0)|^2} \Delta E_{LT} \frac{f(k)}{k} \begin{pmatrix} k^2 & (k_x - ik_y)^2 \\ (k_x + ik_y)^2 & k^2 \end{pmatrix} \quad (5)$$

where ΔE_{LT} is a charachteristic energy of longitudinal–transverse exciton splitting, ϕ_{3D} and ϕ_{1s} are three and two-dimensional exciton envelope wavefunctions, the form-factor $f(k)$ is given by

$$f(k) = \int dz \int dz' \, \xi_c(z) \, \xi_{hh}(z) \, e^{-k|z-z'|} \, \xi_c(z') \, \xi_{hh}(z'), \quad (6)$$

$\xi_c(z)$, $\xi_{hh}(z)$ being the electron and heavy hole envelope functions within the quantum well. Non- diagonal terms lead to polariton spin flips and thus create an effective in-plane magnetic field $\mathbf{\Omega}_{LT,\mathbf{k}}$, further referred to the Maialle field. As off-diagonal elements of the Hamiltonian are proportional to $(k_x - ik_y)^2$, $\mathbf{\Omega}_{LT,\mathbf{k}}$ is in general not parallel to $\mathbf{k}$, but makes with the x-axis twice the same angle as $\mathbf{k}$. The effective magnetic field is zero for $\mathbf{k} = 0$ and increases as a function of $\mathbf{k}$, following a square root law at large k. For more details the reader can refer to Tassone [14].

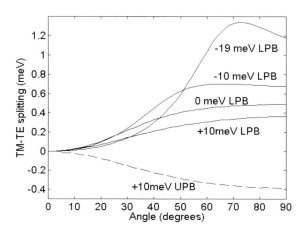

Fig. 9.5 Longitudinal–transverse polariton splitting calculated for detunings: +10 meV, 0 meV, –10 meV, –19 meV. Solid lines: lower polariton branch, dashed lines: upper polariton branch. From [6].

In microcavities, splitting between longitudinal and transverse polariton states is amplified due to the exciton coupling with the cavity mode. Note that the cavity mode frequency is also split in TE- and TM-light polarizations [15]. The resulting polariton splitting strongly depends on the detuning between the cavity mode and the exciton resonance and, in general, increases with k. Figure 9.5 shows the longitudinal-transverse polariton splitting Δ_{LT} calculated for a CdTe-based microcavity sample from Ref. [11] for different detunings. For these calculations, polariton eigenfrequencies in two linear polarizations have been found numerically by the transfer matrix method. One can see that the splitting is very sensitive to the detuning, and may achieve 1 meV, which exceeds by an order of magnitude the bare exciton longitudinal-transverse splitting.

Using the pseudospin representation, one can rewrite the Liouville–von Neumann equation for the density matrix as

$$\frac{d\mathbf{S}_k}{dt} = \frac{g_s \mu_b}{\hbar} [\mathbf{S}_k \times \mathbf{\Omega}_{\text{eff},k}]. \tag{7}$$

The Maialle field thus induces precession of the pseudospin of the ensemble of circularly polarized polaritons and can cause oscillations of the circular polarization degree of the emitted light in time as it was observed experimentally [11].

The precession of the polariton pseudospin about the Maialle field resembles the precession of electron spins about the Rashba effective magnetic field [16]. The physical origin of these two fields is, however, different. The Rashba field is created by the spin–orbit interaction in asymmetric quantum wells, whereas the Maialle field appears because of the long-range electron–hole interaction. The orientation of these two fields is also different. The Rashba field is perpendicular to the wavevector, whereas the Maialle field is neither perpendicular nor parallel to it, in general.

The z-component of $\mathbf{\Omega}_{\text{eff},k}$ splits $J = +1$ and $J = -1$ exciton states, and is null if the polariton–polariton interactions are neglected. However, in the nonlinear regime it can arise due to the difference of concentrations of spin-up and spin-down polaritons [8]. To show this, let us first consider the connection of the pseudospin formalism with the second quantization representation.

If the polariton concentration is less than the saturation density, they can be treated as good bosons, and thus a pair of bosonic destruction operators $a_{k\uparrow}, a_{k\downarrow}$ can be introduced to describe the polariton doublet of wave vector $\mathbf{k}$. The occupation numbers of spin up, spin down polaritons and z-component of the pseudospin can be found as

$$\begin{aligned}
N_{k\uparrow} &= \text{Tr}\,[\rho_k a_{k\uparrow}^+ a_{k\uparrow}] \equiv \langle a_{k\uparrow}^+ a_{k\uparrow}\rangle, \\
N_{k\downarrow} &= \text{Tr}\,[\rho_k a_{k\downarrow}^+ a_{k\downarrow}] \equiv \langle a_{k\downarrow}^+ a_{k\downarrow}\rangle, \\
S_{k,z} &= \tfrac{1}{2}[\langle a_{k\uparrow}^+ a_{k\uparrow}\rangle - \langle a_{k\downarrow}^+ a_{k\downarrow}\rangle].
\end{aligned} \tag{8}$$

For in-plane components of the pseudospin one should introduce the bosonic operators for linear polarized polaritons as follows:

$$a_{\mathbf{k},+x} = \frac{1}{\sqrt{2}}(a_{\mathbf{k},\uparrow} + a_{\mathbf{k},\downarrow}), \quad a_{\mathbf{k},-x} = \frac{1}{\sqrt{2}}(a_{\mathbf{k},\uparrow} - a_{\mathbf{k},\downarrow}),$$
$$a_{\mathbf{k},+y} = \frac{1}{\sqrt{2}}(a_{\mathbf{k},\uparrow} + ia_{\mathbf{k},\downarrow}), \quad a_{\mathbf{k},-y} = \frac{1}{\sqrt{2}}(a_{\mathbf{k},\uparrow} - ia_{\mathbf{k},\downarrow}). \tag{9}$$

These formulas allow to obtain $S_{\mathbf{k},x}$ and $S_{\mathbf{k},y}$ as:

$$S_{\mathbf{k},x} = \tfrac{1}{2}[\langle a^+_{\mathbf{k},+x} a_{\mathbf{k},+x}\rangle - \langle a^+_{\mathbf{k},-x} a_{\mathbf{k},-x}\rangle] = \operatorname{Re}\langle a^+_{\mathbf{k},\downarrow} a_{\mathbf{k},\uparrow}\rangle,$$
$$S_{\mathbf{k},y} = \tfrac{1}{2}[\langle a^+_{\mathbf{k},+y} a_{\mathbf{k},+y}\rangle - \langle a^+_{\mathbf{k},-y} a_{\mathbf{k},-y}\rangle] = \operatorname{Im}\langle a^+_{\mathbf{k},\downarrow} a_{\mathbf{k},\uparrow}\rangle. \tag{10}$$

Using this operator algebra we are now going to consider the polariton pseudospin dynamics in the non-linear regime. Let us start from the simplest case where all the polaritons are in the same quantum state, i.e. form a "condensate", so that the scattering to the other states is completely suppressed. Of course, this is an over-simplification of the real situation, but nevertheless we will analyse this toy model as it allows to introduce the concept of the self-induced Larmor precession of the in-plane polariton pseudospin which is of crucial importance in the experiments with resonant pumping. The general form of the interaction Hamiltonian reads

$$H = \varepsilon\,(a^+_\uparrow a_\uparrow + a^+_\downarrow a_\downarrow) + V_1(a^+_\uparrow a^+_\uparrow a_\uparrow a_\uparrow + a^+_\downarrow a^+_\downarrow a_\downarrow a_\downarrow) + 2V_2 a^+_\uparrow a_\uparrow a^+_\downarrow a_\downarrow, \tag{11}$$

where the index corresponding to the polariton state $\mathbf{k}$ in the reciprocal space is omitted. In (11) ε is the free polariton energy, the matrix elements V_1 and V_2 correspond to the forward scattering of the polaritons in the triplet configuration (parallel spins) and in the singlet configuration (antiparallel spins). If the polariton–polariton interactions were spin-isotropic, i.e. the Hamiltonian (11) was covariant with respect to the linear transformation of the operators (9), the matrix elements would be interdependent, so that

$$V_1 - V_2 = 0. \tag{12}$$

However, this situation is not realized in microcavities, where the major contribution to polariton–polariton interaction is given by the exchange term, so that the polariton–polariton interaction is in fact anisotropic, and $V_1 \gg V_2$. This anisotropy manifests itself experimentally in the optically induced splitting of the spin-up and spin-down polariton states [17], which has been measured as a function of the circular polarization degree of the excitation (directly linked to the imbalance of populations of spin-up and spin-down states) (see Fig. 9.6).

The Hamiltonian (11) conserves $N_\uparrow$ and $N_\downarrow$, but not the in-plane components of the pseudospin. It becomes evident, if one calculates the commutator of the Hamiltonian (11) with the operator $a^+_\downarrow a_\uparrow$ which governs the linear polarization. This commutator is zero only if the condition (12) is satisfied, which is never the case

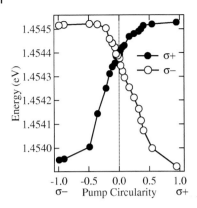

Fig. 9.6 Dependence of the spin-splitting of the exciton resonance on pump polarization degree measured experimentally for a GaAs-based microcavity. From [17].

experimentally. The equation of motion for $\langle a_\downarrow^+ a_\uparrow \rangle$ can be obtained from (3) and (11) and reads

$$\frac{d\langle a_\downarrow^+ a_\uparrow \rangle}{dt} = \frac{i}{\hbar} 2(V_1 - V_2)[\langle a_\downarrow^+ a_\downarrow a_\downarrow^+ a_\uparrow \rangle - \langle a_\uparrow^+ a_\uparrow^+ a_\downarrow a_\uparrow \rangle]. \qquad (13)$$

In the mean field approximation, the fourth-order correlators in the right side of (13) can be decoupled and Eq. (13) transforms to the equation of precession of the pseudospin in an effective magnetic field $\boldsymbol{\Omega}_{\text{int}}$ directed along the structure growth axis. The absolute value of the field is determined by the difference between the populations of spin-up and spin down polaritons,

$$g_s \mu_B \boldsymbol{\Omega}_{\text{int}} = 2(V_1 - V_2)(N_\downarrow - N_\uparrow). \qquad (14)$$

Experimentally, the effect manifests itself as a rotation of the polarization plane of photoemission if the σ^+ and σ^- populations are imbalanced.

To conclude this section, we underline that the effective magnetic field acting on the polariton pseudospin has two components, in general, $\boldsymbol{\Omega}_{\text{eff},\boldsymbol{k}} = \boldsymbol{\Omega}_{\text{LT},\boldsymbol{k}} + \boldsymbol{\Omega}_{\text{int},\boldsymbol{k}}$. The in-plane component is governed by the long range electron–hole interaction. It is concentration independent and leads to the beats between the circularly polarized components of the photoemission. The component parallel to the structure growth axis arises because of the anisotropy of the polariton–polariton interaction and depends on the imbalance between spin-up and spin-down polaritons. It leads to beats between linearly polarized components of the photoemission.

9.4
Interplay Between Spin and Energy Relaxation

In the previous section we have introduced the concept of the polariton pseudospin and analyzed the mechanisms of formation of the effective magnetic field acting on it. This has allowed us to explain qualitatively the experimentally observed oscilla-

9.4 Interplay Between Spin and Energy Relaxation

tions of the circular polarization degree and the 45° rotation of the polarization plane in the polariton parametric amplifier. However, the scattering of polaritons leading to their relaxation in the reciprocal space and their energy relaxation has not been described. Consequently, some experimental data such as the difference between the polarization dynamics below and above stimulation threshold under nonresonant pump [11] or the linear polarization inversion in a parametric oscillator [9] cannot be explained in this framework. Formally, kinetic equations for the occupation numbers and pseudospins can be decoupled only in the linear regime. Thus, a theoretical description of non-linear processes require the self consistent accounting of both energy and spin relaxation processes as shown below.

Our starting point is the Hamiltonian of the system in the interaction representation. Polariton interaction with acoustic phonons and polariton–polariton scattering will be taken into account. Only the lower polariton branch will be considered, coupling with the upper branch and dark exciton states is neglected.

$$H = H_{\text{shift}} + H_{\text{scatt}} . \tag{15}$$

Here the "shift" term describes interaction of exciton polaritons having the same momentum but possibly different spins:

$$H_{\text{shift}} = \sum_{k,\sigma=\uparrow,\downarrow} \left(g_B \mu_B \Omega_{\text{LT},k} a^+_{\sigma,k} a_{-\sigma,k} + V^{(1)}_{k,k,0}(a^+_{\sigma,k} a_{\sigma,k})^2 + V^{(2)}_{k,k,0} a^+_{\sigma,k} a^+_{-\sigma,k} a_{\sigma,k} a_{-\sigma,k} \right)$$
$$+ \sum_{k,k'\neq k,\sigma=\uparrow,\downarrow} \left(V^{(1)}_{k,k',0} a^+_{\sigma,k} a^+_{\sigma,k'} a_{\sigma,k} a_{\sigma,k'} + V^{(2)}_{k,k',0} a^+_{\sigma,k} a^+_{-\sigma,k'} a_{\sigma,k} a_{-\sigma,k'} \right). \tag{16}$$

Here $a_{\uparrow,k}, a_{\downarrow,k}$ are annihilation operators of the spin-up and spin-down polaritons, $\Omega_{\text{LT},k} = \Omega_{\text{LT},k,x} + i\Omega_{\text{LT},k,y}$ ($\Omega_{\text{LT},k,x}$ and $\Omega_{\text{LT},k,y}$ are the x and y projections of the effective magnetic field). The "scattering term" describes scattering between states with different momenta:

$$H_{\text{scatt}} = \frac{1}{4} \sum_{k,k',q\neq 0, \sigma=\uparrow,\downarrow} e^{\frac{it}{\hbar}(E(k)+E(k')-E(k+q)-E(k'-q))}$$
$$\times \left\{ V^{(1)}_{k,k',q} a^+_{\sigma,k+q} a^+_{\sigma,k'-q} a_{\sigma,k} a_{\sigma,k'} + 2V^{(2)}_{k,k',q} a^+_{\sigma,k+q} a^+_{-\sigma,k'-q} a_{\sigma,k} a_{-\sigma,k'} \right\}$$
$$+ \text{H.c.} + \frac{1}{2} \sum_{k,q\neq 0,\sigma=\uparrow,\downarrow} e^{\frac{it}{\hbar}(E(k)+\hbar\omega_q - E(k+q))} U_{k,q} a^+_{\sigma,k+q} a_{\sigma,k} b_q + \text{h.c.} \tag{17}$$

In (17) b_q is an acoustic phonon operator, $U_{k,q}$ is the polariton-phonon coupling constant, $E(k)$ is the dispersion of the low polariton branch. The matrix elements $V^{(1)}_{k,k',q}$ and $V^{(2)}_{k,k',q}$ describe scattering of two polaritons in the triplet and the singlet configurations, respectively. As it has been discussed in the previous section, in real microcavities the polariton–polariton interaction is strongly anisotropic: the triplet scattering is usually much stronger than the singlet one [18]. Moreover, the matrix elements $V^{(1)}_{k,k',q}$ and $V^{(2)}_{k,k',q}$ can have opposite signs [9]. Indeed, the interaction between two polaritons with parallel spins is always repulsive, while polaritons with opposite spins are characterised by an attractive interaction and can even form a bound state (bipolariton) [19].

We use the Hamiltonian (1) to write the Liouville equation for the total density matrix of the system ρ:

$$i\hbar \frac{d\rho}{dt} = [H(t), \rho] = [H_{\text{shift}}(t) + H_{\text{scatt}}(t), \rho]. \tag{18}$$

We shall use the Born–Markov approximation familiar in quantum optics [20]. The Markov approximation means physically that the system is assumed to have no phase memory. It is, in general, not true for the coherent processes described by the Hamiltonian H_{shift} but is a reasonable approximation for the scattering processes involving a momentum transfer [21]. We apply the Markov approximation to the scattering part of Eq. (18) which therefore rewrites:

$$\frac{d\rho}{dt} = \frac{1}{i\hbar}[H_{\text{shift}}(t), \rho] - \frac{1}{\hbar^2} \int_{-\infty}^{t} [H_{\text{scatt}}(t), [H_{\text{scatt}}(\tau), \rho(\tau)]] d\tau. \tag{19}$$

The next step is to perform the Born approximation which is to factorize the density matrix of the whole system as a product of the phonon density matrix and polariton density matrix corresponding to the different states in the reciprocal space: $\rho = \rho_{\text{ph}} \otimes \prod_k \rho_k$. The phonons are then "traced out", i.e. in the calculation their occupation numbers are treated as fixed parameters determined by the temperature. The density matrices ρ_k are given by Eq. (2). They contain information about both occupation numbers and pseudospin components of all states in the reciprocal space.

Equations (15) to (19) together with the formulas (8) to (10) for occupation numbers and pseudospins are sufficient to derive a closed set of dynamics equations for $N_{\uparrow,k}, N_{\downarrow,k}$ and $S_{k,\perp} = e_x S_{k,x} + e_y S_{k,y}$ as shown in [22]. Maialle spin–relaxation because of the TE-TM splitting and self-induced Larmor precession are reduced in this model to the precession of the polariton pseudospin about an effective magnetic field $\Omega_{\text{eff},k}$ that arises from the Hamiltonian term H_{shift}. Once polariton populations and pseudospins are known, the intensities of the circular and linear components of photoemission are given by

$$\begin{aligned}
I_k^+ &= N_{\uparrow,k}, \\
I_k^- &= N_{\downarrow,k}, \\
I_k^x &= \frac{N_{\uparrow,k} + N_{\downarrow,k}}{2} + S_{x,k}, \\
I_k^y &= \frac{N_{\uparrow,k} + N_{\downarrow,k}}{2} - S_{x,k}.
\end{aligned} \tag{20}$$

Note that light polarizations parallel to x- and y-axes correspond to the pseudospin parallel and antiparallel to x-axis, respectively.

The general form of the kinetic Eqs. (15) to (19) is complicated and their solution requires heavy numerical calculations. However, some particular cases can be considered here.

If the pumping intensity is weak, the polariton–polariton scattering is dominated by the scattering with acoustic phonons. The polariton–polariton interaction terms can be thus neglected and the system of kinetic equations becomes quite simple and can be easily solved numerically. For occupation numbers and pseudospin we have:

$$\frac{dN_k}{dt} = -\frac{1}{\tau_k}N_k + \sum_{k'}\left[(W_{k'k} - W_{kk'})\left(\frac{1}{2}N_k N_{k'} + 2(S_k \cdot S_{k'})\right) + (W_{kk'}N_{k'} - W_{kk'}N_k)\right], \quad (21)$$

$$\frac{dS_k}{dt} = -\frac{1}{\tau_{sk}}S_k + \sum_{k'}\frac{1}{2}\left[(W_{k'k} - W_{kk'})(N_k S_{k'} + N_{k'} S_k) + (W_{k'k}S_{k'} - W_{kk'}S_k)\right]$$
$$+ \frac{g_s \mu_B}{\hbar}\left[S_k \times \Omega_{LT,k}\right], \quad (22)$$

where the transition rates are

$$W_{kk'} = \begin{cases} \frac{2\pi}{\hbar}|U_{k,k'-k}|^2 n_{ph,k'-k}\delta(E(k') - E(k) - \hbar\omega_{k'-k}), & |k'| > |k| \\ \frac{2\pi}{\hbar}|U_{k,k'-k}|^2 (n_{ph,k'-k} + 1)\delta(E(k') - E(k) - \hbar\omega_{k'-k}), & |k'| < |k|. \end{cases} \quad (23)$$

The Dirac delta functions account for the energy conservation during the scattering acts. Mathematically, they appear from the integration of the time-dependent exponents in the second term of Eq. (19). Writing energetic delta functions in (23) we assume that the polariton longitudinal transverse splitting does not modify strongly the polariton dispersion curve. Of course, in numerical calculations the delta functions should be replaced by resonant functions having a finite amplitude which can be estimated as an inwerse energy broadening of the polariton state. The polariton lifetime τ_k has been introduced in Eqs. (21) and (22) to take into account the radiative decay of polaritons. The pseudospin life-time can be less then τ_k and should be estimated as $\tau_{sk}^{-1} = \tau_k^{-1} + \tau_{sl}^{-1}$ where τ_{sl} is the characteristic time of the spin–lattice relaxation (by this we mean all the processes of spin relaxation within the polariton doublet apart those provided by the longitudinal transverse splitting).

The last term in (22) is the same as in Eq. (7). It describes the pseudospin precession about an effective in-plane magnetic field given by the polariton longitudinal-transverse splitting. This term is responsible for oscillations of the circular polarization degree of the emitted light.

Equations (21) and (22) were first obtained in Ref. [6] using the method of unitary transformation of the spin density matrix. They simplify significantly if the problem has a cylindrical symmetry. This case was analyzed numerically in [6], where the reader can find the details of the calculation. In brief, we have considered the microcavity of Ref. [11] at two different pumping powers, below and above stimulation threshold, and with a pump circularly polarized. Below the threshold, the spin system is in the collision dominated regime, i.e. relaxation of polaritons down to the ground state goes through a huge number of random passes each corresponding to the scattering act with an acoustic phonon. The polarization degree displays a monotonic decay in this case. It is easy to understand, as in each scattering event the direction of the effective magnetic field changes randomly, so that in average no oscillation can be observed. The situation is completely analogical to those for the electrons undergoing spin relaxation in an effective Rashba field. Spin relaxation of spin-up and spin-down polaritons go with the same rate, in general.

The situation changes dramatically above the stimulation threshold. In this case, the relaxation rates for spin-up and spin-down polaritons become different, in general. Once the ground state is populated preferentially by polaritons having a given spin orientation, the relaxation rate of polaritons having this spin orientation is enhanced. This stimulated scattering process makes increase the circular polarization degree of the emission in time, as it was experimentally observed. Also the polarization degree is found to oscillate with a period sensitive to the pumping power and the detuning. The detuning (difference between bare exciton and cavity

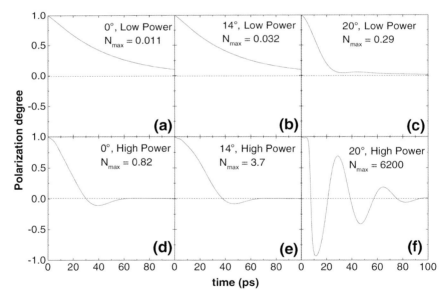

Fig. 9.7 Polarization degree of the photoluminescence at different angles (0°, 14°, and 20°) from the model CdTe microcavity at the detuning −10 meV, calculated using Eqs. (15) to (17) for a pulsed non-resonant excitation with the excitation power of 7 μJ per pulse (a–c) and excitation power of 700 μJ per pulse (d–f). The maximum value polariton occupation numbers achieved at the given angles are indicated on the figures. From [6].

9.4 Interplay Between Spin and Energy Relaxation

mode energies) has an important effect on the polariton spin relaxation. First of all, the longitudinal transverse splitting versus wave vector strongly depends on detuning as is shown in Fig. 9.5. Also the energy relaxation is extremely sensitive to detuning. We have found numerically that for zero detuning, the polarization degree of the ground state only weakly oscillates. On the contrary, for −10 meV detuning a strong bottleneck effect arises and the polarization degree at the bottleneck is found to influence the polarization degree of all lower states. Indeed, in contrast to the zero detuning case, the polarization at the ground state and at 14° look similar, both experiencing damped oscillation (Fig. 9.7). Physically, the appearence of the oscillations in the stimulation regime means that the scattering is no more random, but is due to the effect of the bosonic stimulation is "directed" from the pump region to the bottleneck and then to the ground state.

The equation set (21) to 23) has also been used in to interprete experimental results of the Yamamoto group [12]. In this case, the inhomogeneous term describing the time-independent pump should be added into the right hand side of (21) and (22). Also, in numerical simulations the anisotropy of the polariton distribution in the reciprocal space induced by the pump should be taken into account.

The results of the corresponding numerical modeling are shown in Fig. 9.8. Figure 9.8(a) and (b) show the σ^+- and σ^--polarized populations of the ground state versus the pumping density. Figure 9.8(a) is computed for circular pumping and Fig. 9.8(b) for linear pumping. The resulting polarization of the polariton ground state is shown in Fig. 9.8(c). Due to the effective spin-relaxation, the emission is found unpolarized below the stimulation threshold for both polarizations of excitation. This situation evolves dramatically above the threshold. The circular polarization degree of the emitted light increases up to more than 60% for the linear pump case and it increases up to 90% for the circular pump case.

The increase of the circular polarization degree versus the pump intensity for the circular excitation is easy to understand qualitatively. In fact, the increase of the polariton concentration leads to the switch on of bosonic stimulated scattering, which accelerates the scattering toward the ground state. Polaritons have no more enough time to rotate their spin and the polarization of the signal coincides with the polarization of the pump. The photoluminescence circular polarization degree can be roughly estimated as

$$\wp_c \sim \frac{1}{1+\left\langle \mathbf{\Omega}_{LT}^2 \right\rangle \tau_{rel}^2} \tag{24}$$

where τ_{rel} is the relaxation time from the pump state toward the ground state which decreases versus the pump intensity.

The case of the linear pump is not so easy to understand. If the average value of the vector product of the polaritons pseudospin and the in-plane effective magnetic field is nonzero (which requires a strong anisotropy of the distribution of polaritons in the reciprocal space), the polariton pseudo-spin acquires a z-component (circular polarization). This process is also controlled by the polariton redistribution over the

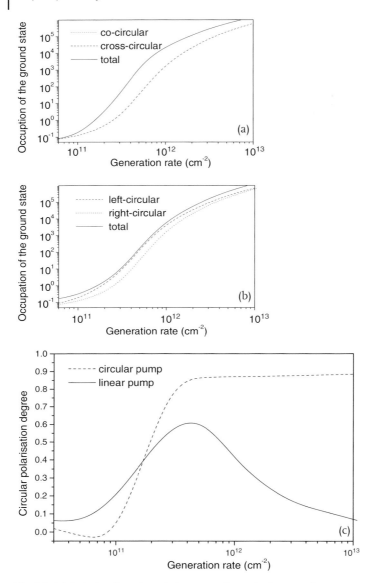

Fig. 9.8 (a) Emission from the polariton ground state in the case of a circular pumping. Total emission (solid line), emission co-circular (dotted line) and cross-circular (dashed line) are shown as functions of the pumping intensity. The co-circular and total emission lines are not distinguishable. (b) The same as a) but for the linear pumping. (c) Circular polarization degree of light emitted by the polariton ground state as function of the pumping intensity for the linear pumping (solid line) and circular pumping (dotted line). From [12].

elastic circle in the k-space. As the direction of the effective in-plane magnetic field depends on the orientation of k, this redistribution leads to a random changing of the pseudospin rotation direction and to a suppression of the z-component of the pseudospin. For low polariton concentration, where the stimulated bosonic scattering can be neglected, the diffusion along the elastic circle is much faster than the relaxation towards the ground state. As a result, before attaining the ground state the polaritons are redistributed almost homogeneously on the elastic circle and no perpendicular component of the pseudospin appears. On the other hand, above the stimulation threshold the processes of bosonic stimulation scattering toward the ground state play an important role, and dominate over the scattering along the elastic circle. The distribution of polaritons in reciprocal space remains strongly anisotropic and a preferable axis for the pseudospin rotation appears. This leads to the appearance of a strong z-component of the pseudospin. However, further increase of the pump power has an opposite effect: the polaritons scatter to the ground state very quickly and have no time to rotate their pseudospin. The competition between these two mechanisms leads to the appearance of a maximum in the polarization degree as a function of the pump power. For the linear pump $\wp$ can be roughly estimated as

$$\wp \sim \frac{\langle \boldsymbol{\Omega}_{\mathrm{LT}} \rangle \tau_{\mathrm{rel}} \tau_{\mathrm{ec}}}{\tau_{\mathrm{rel}} + \tau_{\mathrm{ec}}(1 + \langle \boldsymbol{\Omega}_{\mathrm{LT}} \rangle^2 \tau_{\mathrm{rel}}^2)} \tag{25}$$

where τ_{ec} is a scattering time on the elastic circle independent on the pump intensity. The polarization degree so determined has a maximum if $\tau_{\mathrm{rel}} = \langle \boldsymbol{\Omega}_{\mathrm{LT}}^{-1} \rangle$.

The numerical simulations also predict a slightly lower threshold at linear pumping than at circular pumping in qualitative agreement with the experiment. This effect comes in our model from different rates of stimulated spin–scattering in the two cases. It might be enhanced in the experiment by a stronger absorption of light in case of linear pumping. Actually, if the TE–TM splitting is comparable with the spectral width of the exciting light (which is the case in experiment), simultaneous excitation of TE- and TM-polarized modes by a circular polarized light is less probable than the resonant excitation of either TE- or TM-mode by a linearly polarized light.

The deviation between experiment and numerical modelling is in the maximum value of circular polarization of emission at linear pumping: it is about 60 percent in theory and more than 90 percent in the experiment. This difference can result from the optically-induced spin splitting of the ground state above the stimulation threshold which favors relaxation to the lowest energy spin state.

9.5
Spin-dynamics of Polariton–Polariton Scattering

In Eqs. (21) and (22) we have described polariton relaxation assisted by acoustic phonons and their spin-relaxation because of the Maialle field. Polariton–polariton scattering has not been considered. On the other hand, in many cases polariton–polariton scattering is the dominant mechanism of the polariton redistribution in the reciprocal space. In particular, this scenario is realised in the large negative detuning situation, where a strong bottleneck effect suppresses the phonon-assisted mechanisms for the relaxation toward the ground state. It is also the case in the polariton parametric oscillator where a pair of polaritons pumped at the magic angle scatters into the signal and idler states by polariton–polariton scattering.

If the polariton–polariton interactions are the only mechanism of the redistribution in the reciprocal space, Eq. (19) allows deriving the system of kinetic equations for polariton occupation numbers $N_{k\uparrow}$ and $N_{k\downarrow}$ and in-plane pseudospins $\mathbf{S}_{\perp,k}$ as:

$$\begin{aligned}
\frac{dN_{k\uparrow}}{dt} &= \mathrm{Tr}\left(a^+_{k\uparrow}a_{k\uparrow}\frac{d\rho}{dt}\right) = -\frac{1}{\tau_k}N_{k\uparrow} + \frac{g_S\mu_B}{\hbar}\mathbf{e}_z\cdot[\mathbf{S}_{\perp,k}\times\boldsymbol{\Omega}_{LT,k}] \\
&+ \sum_{k',q}\{W^{(1)}_{k,k',q}[(N_{k\uparrow}+N_{k'\uparrow}+1)N_{k+q\uparrow}N_{k'-q\uparrow} - (N_{k+q\uparrow}+N_{k'-q\uparrow}+1)N_{k\uparrow}N_{k'\uparrow}] \\
&\quad + W^{(2)}_{k,k',q}[(N_{k\uparrow}+N_{k'\downarrow}+1)(N_{k+q\uparrow}N_{k'-q\downarrow}+N_{k+q\downarrow}N_{k'-q\uparrow}+2(\mathbf{S}_{\perp,k+q}\cdot\mathbf{S}_{\perp,k'-q})) \\
&\quad - (N_{k\uparrow}N_{k'\downarrow}+(\mathbf{S}_{\perp,k}\cdot\mathbf{S}_{\perp,k'}))(N_{k+q\uparrow}+N_{k'-q\downarrow}+N_{k+q\downarrow}+N_{k'-q\uparrow}+2)] \\
&\quad + 2W^{(12)}_{k,k',q}[N_{k\uparrow}(\mathbf{S}_{\perp,k'}\cdot\mathbf{S}_{\perp,k'-q})+N_{k'-q\uparrow}(\mathbf{S}_{\perp,k'}\cdot\mathbf{S}_{\perp,k+q})-N_{k\uparrow}\mathbf{S}_{\perp,k'}\cdot(\mathbf{S}_{\perp,k'-q}+\mathbf{S}_{\perp,k+q})] \\
&\quad + W^{(12)}_{k,k',q}[(\mathbf{S}_{\perp,k}\cdot\mathbf{S}_{\perp,k+q})(N_{k'-q\uparrow}+N_{k'-q\downarrow}-N_{k\uparrow}-N_{k'\downarrow}) \\
&\quad + (\mathbf{S}_{\perp,k}\cdot\mathbf{S}_{\perp,k'-q})(N_{k+q\uparrow}+N_{k+q\downarrow}-N_{k\uparrow}-N_{k'\downarrow})]\},
\end{aligned} \quad (26)$$

$$\begin{aligned}
\frac{d\mathbf{S}_{\perp,k}}{dt} &= -\frac{1}{\tau_{sk}}\mathbf{S}_{\perp,k} + \frac{g_S\mu_B}{\hbar}[\mathbf{S}_{\perp,k}\times\boldsymbol{\Omega}_{int,k}] + \frac{g_S\mu_B(N_{k\uparrow}-N_{k\downarrow})}{2\hbar}\boldsymbol{\Omega}_{LT,k} \\
&+ \sum_{k',q}\Bigg\{\frac{W^{(1)}_{k,k',q}}{2}\mathbf{S}_{\perp,k}[N_{k+q\uparrow}N_{k'-q\uparrow}+N_{k+q\downarrow}N_{k'-q\downarrow}-N_{k'\uparrow}(N_{k+q\uparrow}+N_{k'-q\uparrow}+1) \\
&\quad -N_{k'\downarrow}(N_{k+q\downarrow}+N_{k'-q\downarrow}+1)] \\
&\quad + W^{(1)}_{k,k',q}(\mathbf{S}_{\perp,k+q}(\mathbf{S}_{\perp,k'}\cdot\mathbf{S}_{\perp,k'-q})+\mathbf{S}_{\perp,k'-q}(\mathbf{S}_{\perp,k'}\cdot\mathbf{S}_{\perp,k+q})-\mathbf{S}_{\perp,k'}(\mathbf{S}_{\perp,k+q}\cdot\mathbf{S}_{\perp,k'-q})) \\
&\quad + \frac{W^{(2)}_{k,k',q}}{2}[2(\mathbf{S}_{\perp,k}+\mathbf{S}_{\perp,k'})(N_{k+q\uparrow}N_{k'-q\downarrow}+N_{k+q\downarrow}N_{k'-q\uparrow}+2(\mathbf{S}_{\perp,k+q}\cdot\mathbf{S}_{\perp,k'-q})) \\
&\quad -(\mathbf{S}_{\perp,k}(N_{k'\uparrow}+N_{k'\downarrow})+\mathbf{S}_{\perp,1}(N_{k\uparrow}+N_{k\downarrow})) \\
&\quad \times (N_{k+q\uparrow}+N_{k'-q\uparrow}+N_{k+q\downarrow}+N_{k'-q\downarrow}+2)] \\
&\quad -2W^{(12)}_{k,k',q}\mathbf{S}_{\perp,k}((\mathbf{S}_{\perp,k'}\cdot\mathbf{S}_{\perp,k+q})+(\mathbf{S}_{\perp,k'}\cdot\mathbf{S}_{\perp,k'-q})) + \frac{W^{(12)}_{k,k',q}}{2}\mathbf{S}_{\perp,k'-q} \\
&\quad \times [2((N_{k'\uparrow}+1)N_{k+q\uparrow}+(N_{k'\downarrow}+1)N_{k+q\downarrow})+(N_{k+q\uparrow} \\
&\quad +N_{k+q\downarrow}-N_{k'\uparrow}-N_{k'\downarrow})(N_{k\uparrow}+N_{k\downarrow})] \\
&\quad + \frac{W^{(12)}_{k,k',q}}{2}\mathbf{S}_{\perp,k+q}[2((N_{k\uparrow}+1)N_{k'-q\uparrow}+(N_{k'\downarrow}+1)N_{k'-q\downarrow}) \\
&\quad + (N_{k'-q\uparrow}+N_{k'-q\downarrow}-N_{k\uparrow}-N_{k'\downarrow})(N_{k\uparrow}+N_{k\downarrow})]\Bigg\},
\end{aligned} \quad (27)$$

where $\boldsymbol{e}_z$ is a unitary vector in the direction of the structure growth axis, $\tilde{\boldsymbol{\Omega}}_{LT,\boldsymbol{k}}$ is obtained from $\boldsymbol{\Omega}_{LT,\boldsymbol{k}}$ by a rotation by 90° about the structure growth axis, the absolute value of $\boldsymbol{\Omega}_{int,\boldsymbol{k}}$ is given by the following expression

$$\boldsymbol{\Omega}_{int,\boldsymbol{k}} = \frac{2\boldsymbol{e}_z}{g_s \mu_B} \sum_{\boldsymbol{k}'} (V^{(1)}_{\boldsymbol{k},\boldsymbol{k}',0} - V^{(2)}_{\boldsymbol{k},\boldsymbol{k}',0})(N_{\boldsymbol{k}'\uparrow} - N_{\boldsymbol{k}'\downarrow}). \tag{28}$$

The equation for spin-down occupation numbers can be obtained from (24) by changing the spin indices. The transition rates are as follows

$$\begin{aligned} W^{(1)}_{\boldsymbol{k},\boldsymbol{k}',\boldsymbol{q}} &= \frac{2\pi}{\hbar} \left|V^{(1)}_{\boldsymbol{k},\boldsymbol{k}',\boldsymbol{q}}\right|^2 \delta(E(\boldsymbol{k}) + E(\boldsymbol{k}') - E(\boldsymbol{k}+\boldsymbol{q}) - E(\boldsymbol{k}'-\boldsymbol{q})), \\ W^{(2)}_{\boldsymbol{k},\boldsymbol{k}',\boldsymbol{q}} &= \frac{2\pi}{\hbar} \left|V^{(2)}_{\boldsymbol{k},\boldsymbol{k}',\boldsymbol{q}}\right|^2 \delta(E(\boldsymbol{k}) + E(\boldsymbol{k}') - E(\boldsymbol{k}+\boldsymbol{q}) - E(\boldsymbol{k}'-\boldsymbol{q})), \\ W^{(12)}_{\boldsymbol{k},\boldsymbol{k}',\boldsymbol{q}} &= \frac{2\pi}{\hbar} \operatorname{Re}(V^{(1)}_{\boldsymbol{k},\boldsymbol{k}',\boldsymbol{q}} V^{*(2)}_{\boldsymbol{k},\boldsymbol{k}',\boldsymbol{q}}) \delta(E(\boldsymbol{k}) + E(\boldsymbol{k}') - E(\boldsymbol{k}+\boldsymbol{q}) - E(\boldsymbol{k}'-\boldsymbol{q})). \end{aligned} \tag{29}$$

As in Eq. (23) for the matrix elements of polariton-phonon interaction, the delta functions ensure the energy conservation. The signs of the $W^{(1)}_{\boldsymbol{k},\boldsymbol{k}',\boldsymbol{q}}$, $W^{(2)}_{\boldsymbol{k},\boldsymbol{k}',\boldsymbol{q}}$ and $W^{(12)}_{\boldsymbol{k},\boldsymbol{k}',\boldsymbol{q}}$ can differ. Although $W^{(1)}_{\boldsymbol{k},\boldsymbol{k}',\boldsymbol{q}}$, $W^{(2)}_{\boldsymbol{k},\boldsymbol{k}',\boldsymbol{q}}$ are always positive, $W^{(12)}_{\boldsymbol{k},\boldsymbol{k}',\boldsymbol{q}}$ can be positive or negative depending on the phase shift between the matrix elements of the singlet and triplet scattering. In particular it is negative if these matrix elements are real and have opposite signs.

Equations (24) and (25) are too complicated to allow for easy modelling in case of nonresonant pumping, where all the quantum states in the reciprocal space should be taken into account. However, the system (24) and (25) can be used for the modelling of the spin dynamics of polariton parametric oscillator, where we can restrict our consideration by only three states in the reciprocal space $\boldsymbol{p}$, $\boldsymbol{s}$, $\boldsymbol{i}$ corresponding to the pump, signal and idler [23]. Before presenting the numerical results, let us discuss a qualitatively the spin properties of parametric oscillators.

First, let us consider the polariton parametric oscillator geometry of the experiment of the group from Toulouse [9]. They pumped at the magic angle using TE-polarized light and no probe. In this case there is no initial imbalance between σ^+ and σ^- populations and no effective magnetic field in the z-direction. Moreover, the pseudospin of the pump polaritons is parallel to $\boldsymbol{\Omega}_{LT}$, so that the in-plane magnetic field does not rotate it. The spin relaxation is thus provided only by parametric scattering. Naively one would expect that the polarization of the signal would correspond to one of the pump. However, the experiment shows that the signal polarization is rotated by 90° with respect to pump polarization. Let us see how this can be explained by our formalism.

In the spontaneous scattering regime, when $N_{s\uparrow} = N_{s\downarrow} = 0$, $\boldsymbol{S}_{\perp,s} = 0$ the system of kinetic equations (24) and (25) strongly simplifies and for the signal we have:

$$\frac{dN_{s\uparrow}}{dt} = W^{(1)} N^2_{p\uparrow} + 2W^{(2)}(N_{p\uparrow} N_{p\downarrow} + (\boldsymbol{S}_{\perp,p} \cdot \boldsymbol{S}_{\perp,p})), \tag{30}$$

$$\frac{dN_{s\downarrow}}{dt} = W^{(1)}N_{p\downarrow}^2 + 2W^{(2)}(N_{p\uparrow}N_{p\downarrow} + (\mathbf{S}_{\perp,p} \cdot \mathbf{S}_{\perp,p})), \tag{31}$$

$$\frac{d\mathbf{S}_{\perp,s}}{dt} = 2W^{(12)}\mathbf{S}_{\perp,p}(N_{p\uparrow} + N_{p\downarrow}). \tag{32}$$

Equation (32) shows that the direction of the in-plane pseudospin of the signal state is determined by the sign of $W^{(12)}$. For $W^{(12)} < 0$ the pseudospin of the signal is opposite to the pseudospin of the pump. The inversion of the pseudospin corresponds to the 90° rotation of the polarization plane of photoemission (see Fig. 9.4 and discussion of this figure in Section 9.3). This allows the theoretical explanation of the experimental result of Toulouse group [9].

The linear polarization degree of the signal $\wp_{L,s}$ in the spontaneous scattering regime can be estimated as

$$\wp_{L,s} = \wp_{L,p} \frac{4W^{(12)}}{W^{(1)} + 4W^{(2)}}, \tag{33}$$

where $\wp_{L,s}$ is a linear polarization degree of the pump. As triplet scattering normally dominates, $|W^{(12)}| \ll W^{(1)}$ and thus $\wp_{L,s}$ is very small (percents in [9]). However, enhancement of the pump power can induce a considerable increase of the linear polarization degree of the photoemission. Indeed, the spontaneous scattering created in the signal state a seed with weak linear polarization perpendicular to the pump. The polarization of the seed can be then enhanced by bosonic stimulation if the intensity of the pump exceeds the stimulation threshold.

The 45° rotation in the pump–probe experiments of Lagoudakis and coauthors [13] is an effect which combines both the self-induced Larmor precession and polarization inversion due to the polariton–polariton scattering. The circular probe enhances the parametric scattering of the polaritons with a spin parallel to those of the probe spin. This introduces an imbalance between spin-up and spin-down polaritons in the pump state which gives rise to an effective magnetic field in the z-direction as described by Eq. (14). This field rotates the in-plane polariton polarization in a direction which depends on the sign of $\wp_c$ (the self-induced Larmor precession). The polaritons are then scattered toward the ground state. The in-plane polarization is in addition rotated by 90 degrees during this scattering event. The total angle of the rotation is determined by the parameters of the cavity and the intensities of the pump and probe. Numerical simulations carried out in [8, 22] have shown that for the geometry of the Lagoudakis experiment it is about 45°, which agrees with the experimental data.

To conclude this section, the Born–Markov treatment of the Liouville equation for the density matrix of the polaritonic system allowed us to obtain the set of the kinetic equation for occupation numbers and pseudospins of polariton quantum states in the reciprocal space, which takes into account all the main mechanisms of the spin and energy relaxation. The equations derived have allowed reproducing qualitatively the experimental results on the polarization dynamics of quantum microcavities obtained both at non-resonant pumping and in the parametric oscillator regime.

9.6
Perspective: Toward "Spin-optronic" Devices

In the conclusion of [6] the concept of "spin-optronic" devices has been formulated. Such devices would be based on manipulations with the polarization of light on pico-second and nano-meter scales. The expression "spin-optronic" has not been appreciated by the editors of Physical Review B who replaced it in the title of Ref. [12] by "spin dependent optoelectronic" which is a long but clear expression anyway. Nowadays optical communication technologies are based on intensity and frequency modulation. The light polarization modulation is very rarely used. On the other hand, polarization of light represents three additional degrees of freedom that could be efficiently used for encoding of information. The polarization modulated signals can hardly be used for long-range telecomunication as polarization is quickly lost in optical fibers. However short range optical information transfer, or even optical information transfer on a processor scale are excellent application areas for "spin-optronic" devices. Indeed nano-range information transfer could take advantage on the use of the polarization degree of freedom. This means that one should produce amplifiers and switches sensitive to polarization. One should also produce polarization converters, polarization modulators, and even more important stable sources of polarized light. Several functionalities could be embedded within a single compound. All these devices should have extremely low power consumption in order to be integrated on chip together with classical electronic functions. The control on the system, and therefore on some of the devices should be electronic in order to be an efficient interface with the information processing unities. In other words, one cannot envisage to control the system with a Ti:sapphire laser! Future polariton devices seem able to fulfil this program. The experiments performed by Lagoudakis [13, 17] in the resonant-pumping geometry shows that the microcavity is able to convert the polarization, or to act as a switch sensitive to the polarization of a reference beam. The main obstacle on the road is of course the need to have an electrically pumped polariton laser operating at room temperature. An utlralow stimulation threshold is required as well, but it seems to be a minor problem. If these problems are solved, polariton devices would respond to real technological needs. In this framework they actually represent a valuable solution for the next generation of optoelectronic devices.

Acknowledgements

We wish to thank K. V. Kavokin, M. M. Glasov, D. D. Solnyshkov, P. Bigenwald, P. G. Lagoudakis, J. J. Baumberg and L. Vina for valuable collaboration. This work has been supported by the Marie Curie project "Clermont 2", contract MRTN-CT-2003-503677.

References

[1] F. Meier and B. P. Zakharchenya (Eds.), Optical Orientation (North-Holland, Amsterdam, 1984).
[2] R. J. Elliot, Phys. Rev. **96**, 266 (1954).
[3] M. I. D'yakonov and V. I. Perel, Fiz. Tverd. Tela **13**, 3851 (1971) [Sov. Phys. – Solid State **13**, 3023 (1972)].
[4] G. E. Pikus and G. L. Bir, Zh. Eksp. Teor. Fiz. **60**, 195 (1971) [Sov. Phys. – JETP **33**, 108 (1971)].
[5] M. Z. Maialle, E. A. de Andrada e Silva, and L. J. Sham, Phys. Rev. B **47**, 15776 (1993).
[6] K. V. Kavokin, I. A. Shelykh, A. V. Kavokin, G. Malpuech, and P. Bigenwald, Phys. Rev. Lett. **89**, 077402 (2004).
[7] I. A. Shelykh, L. Viña, A. V. Kavokin, N. G. Galkin, G. Malpuech, and R. André, phys. stat. sol. (c) **1**, 1351 (2004).
[8] I. Shelykh, G. Malpuech, K. V. Kavokin, A. V. Kavokin, P. Bigenwald, Phys. Rev. B **70**, 115301 (2004).
[9] K. V. Kavokin, P. Renucci, T. Amand, X. Marie, P. Senellart, J. Bloch, and B. Sermage, phys. stat. sol. (c) **2**, 763 (2005).
[10] A. Kavokin and G. Malpuech, Cavity Polaritons, Vol. 32 of the series "Thin solid films and Nanostructures", edited by V. M. Agranovich (Elsevier, North Holland, Amsterdam, 2003).
[11] M. D. Martin, G. Aichmayr, L. Viña, and R. André, Phys. Rev. Lett. **89**, 77402 (2002).
[12] I. Shelykh, K. V. Kavokin, A. V. Kavokin, G. Malpuech, P. Bigenwald, H. Deng, G. Weihs, and Y. Yamamoto, Phys. Rev. B **70**, 035320 (2004).
[13] P. G. Lagoudakis, P. G. Savvidis, J. J. Baumberg, D. M. Whittaker, P. R. Eastham, M. S. Skolnick, and J. S. Roberts, Phys. Rev. B **65**, 161310 (2002).
[14] F. Tassone, F. Bassani, and L. C. Andreani, Phys. Rev. B **45**, 6023 (1992).
[15] G. Panzarini, L. C. Andreani, A. Armitage, D. Baxter, M. S. Skolnick, V. N. Astratov, J. S. Roberts, A. V. Kavokin, M. R. Vladimirova, and M. A. Kaliteevski, Phys. Rev. B **59**, 5082 (1999).
[16] Y. A. Bychkov and E. I. Rashba, JETP Lett. **39**, 78 (1984).
[17] A. Kavokin, P. G. Lagoudakis, G. Malpuech, and J. J. Baumberg, Phys. Rev. B **67**, 195321 (2003).
[18] C. Ciuti, V. Savona, C. Piermarocchi, A. Quattropani, and P. Schwendimann, Phys. Rev. B **58**, 7926 (1998).
[19] A. L. Ivanov, P. Borri, W. Langbein, and U. Woggon, Phys. Rev. B **69**, 075312 (2004).
[20] H. J. Carmichael, Statistical methods in quantum optics 1: Master equations and Fokker–Planck equations (Springer-Verlag, Berlin, Heidelberg, 1999).
[21] C. W. Gardiner and P. Zoller, Phys. Rev. A **55**, 2902 (1997).
[22] M. M. Glazov, I. A. Shelykh, G. Malpuech, K. V. Kavokin, and A. V. Kavokin, Solid State Commun. **134**, 117 (2005).
[23] C. Ciuti, P. Schwendimann, B. Deveaud, and A. Quattropani, Phys. Rev. B **62**, R4825 (2000).

Chapter 10
Bose–Einstein Condensation of Microcavity Polaritons

Vincenzo Savona and Davide Sarchi

10.1
Introduction

It was back in 1995, sitting right in Marc Ilegems office, that Claude Weisbuch asked: "Do you think they might condense?". The question was about exciton polaritons in semiconductor planar microcavities (MCs) and had tickled Claude's scientific curiosity since his first observation of strongly coupled polaritons in microcavities, a few years before [1]. The question, at that time, had been stimulated by the remark that MC polaritons at low density behave as Bose quasiparticles with an exceptionally light effective mass, five orders of magnitude smaller than the free electron mass. As the critical temperature for Bose–Einstein condensation (BEC) of noninteracting particles is inversely proportional to their mass, MC polaritons were a good system to look at when dreaming of high-T_c BEC. And yet, although this idea has been storming around within the polariton community until present, it was clear from the very beginning that, in one way or another, it was too easy to be true.

This paper is a non exhaustive overview of how ideas about MC polariton BEC and related phenomena were developed in the last decade. In what follows, we briefly summarize the basic concepts about BEC. Then, we consider MC polaritons and review the efforts that have been spent to investigate nonlinear coherent phenomena. Several aspects of the problem are discussed, including the role of interactions, the energy relaxation mechanisms, or the polariton radiative lifetime. The basic literature is reviewed. We conclude by making some considerations about what steps are in our opinion still required – in theory as well as in experiments – in order to give a satisfactory answer to Claude's question.

This work is far from being a review work. Many important aspects of BEC in semiconductors, such as the fascinating saga of excitonic BEC [2], will purposely not be covered. Other related aspects of the problem, such as polariton resonant parametric processes or the polariton laser are discussed in deeper detail by other authors in this issue.

10.2
Bose–Einstein Condensation: Basic Facts

The phenomenon of Bose–Einstein condensation has been the object of renewed interest since the recent experimental realisation of a condensate of diluted ultracold alcali atoms in 1995 [3]. Since then, a number of thorough reviews spanning the various aspects of this interesting problem have been written [4–7]. Here, we sketch the basic facts about BEC which are relevant to the problem of polaritonic BEC and refer to these reviews for a more complete account of the experimental and theoretical facts behind BEC.

A gas of N free noninteracting Bose particles in thermal equilibrium at temperature T is distributed in momentum space according to the Bose statistics

$$n_{k} = \frac{1}{e^{\beta(\varepsilon_{k}-\mu)} - 1}, \qquad (1)$$

where $\varepsilon_{k} = \hbar^{2}k^{2}/(2m)$ is the energy-momentum dispersion of free particles, and $\beta = (k_{B}T)^{-1}$. The chemical potential μ is always smaller than the lowest energy eigenvalue ε_{0} and is fixed by the condition

$$\sum_{k} n_{k} = N. \qquad (2)$$

We see that when μ approaches the value ε_{0}, the occupation of the lowest level n_{0} becomes increasingly large. On the other hand, under the same limit, it is easy to show that the total occupation of the excited states

$$N_{E} = \sum_{k \neq 0} n_{k}, \qquad (3)$$

remains finite for a three-dimensional system, even when $\mu = \varepsilon_{0}$, and is an extensive quantity proportional to the system volume V. This *macro-occupation* n_{0} of the lowest energy level is at the basis of BEC. For a fixed temperature T, two situations are possible. If the total number of particles N can be accommodated in the excited levels for $\mu < \varepsilon_{0}$, then the occupation of the ground state n_{0} is of the order of unity and the system is not condensed. If on the other hand N is larger than the critical number of particles, defined as $N_{c} = N_{E}(T, \mu = \varepsilon_{0})$, then a large number of particles have to occupy the ground state in order to fulfill the normalization condition (2) and BEC takes place. It is important to point out that, for a finite volume V and a finite number of particles N, the chemical potential stays strictly smaller than ε_{0} even when the system is condensed. It is only in the thermodynamic limit, $N, V \to \infty$ with $N/V =$ constant, that $\mu = \varepsilon_{0}$. In this case, each n_{k} is finite whereas n_{0} diverges in such a way that n_{0}/N approaches a finite value, smaller than one, which is called the condensate fraction. For a fixed particle number, on the other hand, $N_{c}(T)$ is a continuous function of the temperature which becomes smaller than N

for a critical temperature T_c, well defined in the thermodynamic limit, below which BEC occurs. For an ideal, noninteracting Bose gas, the critical temperature is given by

$$k_B T_c = \frac{2\pi\hbar^2}{m}\left(\frac{n}{2.612}\right)^{2/3}. \tag{4}$$

For a system of finite size, the above considerations seem to suggest that BEC is just characterized by the quantitative criterion that n_0 be significantly larger than $n_{k\neq 0}$. This naturally leads to the question whether BEC makes any sense for a finite size system. It turns out that there is actually a more stringent criterion for BEC, which emerges if we rewrite the distribution (1) in the continuum limit $\sum_k \to V/(2\pi)^3 \int d\mathbf{k}$, valid for large volume V. Above T_c the distribution is then given by

$$n(\mathbf{k}) = \frac{V}{(2\pi)^3} \frac{1}{e^{\beta(\varepsilon(\mathbf{k})-\mu)} - 1}. \tag{5}$$

Let us define the field operator $\hat{\psi}(\mathbf{r})$, describing the quantum field of the many-body Bose system. A quantity of high relevance for BEC is the one-body density matrix $n(\mathbf{r},\mathbf{r}') = \langle \hat{\psi}^\dagger(\mathbf{r}) \hat{\psi}(\mathbf{r}') \rangle$. The average $\langle \cdots \rangle$ is taken over the equilibrium state of the system. This quantity describes the correlation between the particle quantum field at two different positions in space. It can easily be shown that the the one-body density matrix is equal to the Fourier transform of the momentum distribution $n(\mathbf{k})$ of the particle density. Above T_c we can therefore use expression (5) to show that the one-body density matrix decays with the distance $s = |\mathbf{r}-\mathbf{r}'|$ according to an exponential law, with a characteristic length given by the thermal De Broglie wavelength $\lambda_T = \sqrt{2\pi\hbar^2/(mk_B T)}$. This means that for a normal gas, quantum correlations survive only on the scale λ_T, whereas at larger distance the system is uncorrelated as expected for a classical gas. Things are quite different below T_c, where the momentum distribution is

$$n(\mathbf{k}) = n_0 \delta(\mathbf{k}) + \frac{V}{(2\pi)^3} \frac{1}{e^{\beta\varepsilon(\mathbf{k})} - 1}. \tag{6}$$

This expression implies that the two-point density matrix $n(\mathbf{r},\mathbf{r}')$ is nonvanishing in the limit $s \to \infty$. The nonvanishing contribution at long distance originates from the condensate term $n_0\delta(\mathbf{k})$ in (6). This expression allows to state the Penrose–Onsager criterion [7, 8], Stating that a Bose-condensed system is characterized by a finite long-range one-body correlation. This feature, often called *Off-Diagonal Long Range Order* or ODLRO, is the true peculiar property of BEC, at the origin of spectacular effects like the matter-wave interference and amplification [7].

An alternative way of understanding BEC is provided by expressing field operator in second quantization, $\hat{\psi}(\mathbf{r}) = \sum_k \hat{a}_k \phi_k(\mathbf{r})$, with Bose commutation rules

$[\hat{a}_{k}, \hat{a}_{k'}^{\dagger}] = \delta_{kk'}$. As $\langle \hat{a}_0^{\dagger} \hat{a}_0 \rangle = n_0 \gg 1$, the commutator $[\hat{a}_0, \hat{a}_0^{\dagger}] = 1$ can be neglected with respect to n_0 and the operator $\hat{a}_0$ can be approximated by a c-number $\sqrt{n_0}$. This limit, first introduced by Bogoliubov, is equivalent to treating the condensate as a classical field, so that the field operator is rewritten as $\hat{\psi}(r) = \psi_0(r) + \delta\hat{\psi}(r)$. The function $\psi_0(r)$ is the wave function of the condensate and plays the role of an order parameter for the Bose–Einstein phase transition. It is a complex quantity whose square modulus determines the condensate density whereas the phase is important in characterizing the condensate spatial coherence. We point out that the Bogoliubov ansatz implies that the expectation value $\langle \hat{\psi}(r) \rangle = \psi_0(r)$ has a well defined phase in the complex plane. The Hamiltonian of the system, however, does not depend on the phase of the order parameter. This broken gauge symmetry is analogous to that emerging from the semiclassical theory of a laser [9–12]. A contradiction then arises, as the Heisenberg uncertainty principle states that the expectation values of the phase and of the number of particles cannot be simultaneously determined. A dilute Bose gas made of a finite number of atoms cannot therefore be described by a condensate wave function in the Bogoliubov limit. It was recently pointed out that, for an interacting system, the time evolution of a state with an initially well defined ψ_0 is characterized by a phase instability [13]. As a consequence, number-conserving theories [14, 15] were developed which are essentially analogous to the quantum theory of laser including noise [9], where the system is described by a statistical mixture of all possible values of the field phase, while for the field expectation value $\langle \hat{\psi} \rangle = 0$ still holds, as the gauge symmetry requires.

Interactions between particles play an essential role in the physics of BEC (see e.g. the discussion by P. Nozières in [4]). Indeed, when we consider the dynamics of a gas undergoing condensation, particle-particle collisions are required to fill the condensate while emptying the excited states. Another essential aspect is the compressibility of the condensate, which turns out to be zero in the noninteracting case and finite in presence of interactions [7]. A finite compressibility is at the origin of most of BEC physics, in particular the Bogoliubov energy spectrum, superfluidity, and the collective nature of the excitations. Interactions in a diluted Bose gas are often described in terms of an isotropic contact potential, in the s-wave approximation. For most diluted systems, BEC physics is well accounted for by a self-consistent mean-field approximation of the interaction Hamiltonian, sometimes referred to as the Hartree–Fock–Bogoliubov (HFB) approximation [16, 17]. HFB results in a Hamiltonian which is quadratic in the field operator $\hat{\psi}(r)$, and depends self-consistently on the particle density and one-body correlations. As a consequence, the Hamiltonian can be formally diagonalized by a Bogoliubov canonical transformation from the particle creation $\hat{a}_k^{\dagger}$ and annihilation $\hat{a}_k$ operators to the collective excitation operators $\hat{b}_k^{\dagger}$ and $\hat{b}_k$. The energy dispersion of these collective excitations is given by

$$\varepsilon_k = \left[\frac{gn}{m} \hbar^2 k^2 + \left(\frac{\hbar^2 k^2}{2m} \right)^2 \right]^{1/2}, \qquad (7)$$

where g is the interaction constant appearing in the Hamiltonian and n the particle density. For small momentum this expression is linear in k. This linear spectrum is typical of phonon-like waves in a continuous compressible medium. For BEC they essentially represent many-body collective excitations and, together with the long-range correlation, characterize an interacting Bose gas undergoing condensation. For larger momenta, the second term in square brackets dominates and a quadratic dispersion typical of single-particle excitations is recovered. The Bogoliubov spectrum has been accurately measured for a diluted Bose gas [7]. It is important to remark that it occurs only in presence of interactions, as expression (7) indicates. As Landau suggested, the Bogoliubov linear energy-momentum dispersion prevents energy dissipation in the flow of the Bose gas. BEC of an interacting gas is therefore expected to give rise to superfluidity, a phenomenon well known for ^{4}He and also verified in the case of diluted gas BEC [7].

For the purpose of the present work it is essential to extend the discussion to two-dimensional systems. In two dimensions, the critical density $N_c(T)$ diverges for any finite temperature. This amounts to the fact that BEC only occurs at $T = 0$ [7, 18]. A more general result holds for interacting particles where, according to the Hohenberg–Mermin–Wagner theorem, the thermal fluctuations of the phase of the order parameter destroy the condensate at any finite temperature. As a result, the one-body density matrix does not approach a constant value at large distance, but vanishes according to a power law. This behaviour is still significantly different from that of a normal gas and can be shown to still exhibit the phenomenon of superfluidity below a finite superfluid critical temperature $\tilde{T}_c$. This superfluid-normal transition is however not allowed, as another critical phenomenon occurs at temperatures lower than $\tilde{T}_c$. In two dimensions, a compressible fluid can form vortices by spending a finite energy. This is not possible in three dimensions where a vortex-line costs a macroscopic energy proportional to its length. Vortices in two dimensions, however, can only be spontaneously created above a critical temperature T_{BKT} which characterizes the so called Berezinskii–Kosterlitz–Thouless transition [19]. Above T_{BKT} the presence of vortices results in a nonvanishing friction and the superfluid behaviour disappears. For a two dimensional system the relation $T_{BKT} < \tilde{T}_c$ is always fulfilled and the KBT mechanism is the one determining the transition from a superfluid to a normal phase.

Spatial confinement cannot be neglected when studying the properties of BEC of diluted atoms, as these systems are always characterized by a spatial magnetic trap used to increase the gas density. Spatial confinement is equally important for the present discussion. In fact polaritons, like excitons and other excitations in semiconductors, are always localized by the spatial inhomogeneity of artificial semiconductor heterostructures [20]. Furthermore, following the analogy with diluted gases, proposals have recently been made to investigate polariton BEC in spatially confined systems. The nonuniform Bose gas is theoretically described in terms of a quantum-field theory under the same assumptions as in the spatially uniform case but including the effect of a confining potential. To the lowest order in the density of noncondensed particles, this theory reduces to the nonlinear Schrödinger equation, more generally known as the Gross–Pitaevskii equation [7], which successfully describes the spatial and time evolution of the condensate wave function. The main features of BEC, such as a collective excitation spectrum and superfluidity, are also found in the case of confined Bose systems. Spatial confinement has also important consequences on the thermodynamics of BEC, modifying the critical temperature and the condensate fraction. In two dimensions, confinement has a more dramatic effect on the basic properties of BEC and its role is still poorly understood. An ideal two-dimensional Bose gas, differently from its spatially uniform counterpart, undergoes BEC at finite temperature. The Hohenberg–Mermin–Wagner theorem in fact applies only to uniform systems. Physically, a confined system has discrete energy levels and therefore the effects of phase fluctuations, responsible for the absence of BEC, are quenched at low enough temperature. In presence of interactions the physics of BEC in two dimensions is not well understood and in particular it is not clear how the spontaneous formation of vortices typical of the KBT transition is influenced by the spatial inhomogeneity of the quantum fluid.

The feature of BEC which is most relevant to the present discussion is the dynamics of condensation and, more precisely, the dynamical evolution of a Bose gas towards the equilibrium BEC. Many theoretical efforts have been spent on the subject [4, 7, 21–25]. A generally accepted result for a weakly interacting Bose gas in three dimensions is that BEC is achieved in more than one step, involving different timescales [4, 22]. The first step is nucleation of a population in the lowest energy levels. This process can be described by a standard kinetic theory, in terms of the Boltzmann equation. The scattering rate which enters this kinetic theory, in the s-wave approximation, is $W = 4ma^2/(\pi\hbar^3)$, where a is the characteristic scattering length of the interacting system. Already at this stage, the scattering rate is proportional to the mass and to the scattering length, and the formation is therefore inhibited for very light particles or for very weak interactions. After this kinetic process has taken place, however, the system is still characterized by short-range correlation and large nonequilibrium fluctuations of the order parameter phase. Only after a time much longer than the kinetic time the phase fluctuations relax and a long range order sets in, thus forming a genuine condensate according to the Penrose–Onsager criterion.

10.3
Review of Exciton and Polariton BEC

The idea of BEC of excitons in solids traces back to the pioneering works by Moskalenko [26], Blatt [27], and Keldysh and Kozlov [28]. Excitonic BEC was studied in several systems including bulk semiconductors [4, 29–31] (exciton molecules have also been considered for BEC [32, 33]), quantum wells [34–37], and on more exotic systems such as Hall bilayers [38]. The interest in excitons as possible candidates for BEC basically stems from two remarks. First, excitons have a very light effective mass, of the order of the free electron mass. According to (4), this implies a very high critical temperature for BEC at typical density, provided thermal equilibrium can be obtained. Second, due to the charge neutrality, exciton quasiparticles experience weak mutual interactions which allow describing the system as a weakly interacting Bose gas, as opposed to strongly interacting systems like superfluid ^{4}He. The difficulty with excitons, however, is related to their finite lifetime which can range from a few picoseconds to microseconds, as is the case for orthoexciton in Cu$_2$O. As we have pointed out, BEC requires a finite time for the long range correlations to build up. In particular, the kinetic time for the formation of a quasicondensate is inversely proportional to the mass and to the scattering length. The drawback for having a light mass and thus a high critical temperature is therefore a very long condensation time. This time might even be longer than the actual exciton lifetime, thus preventing BEC. Exciton equilibrium BEC is therefore only expected in systems where excitons have a very long lifetime, such as ortho- or para-excitons in Cu$_2$O [29–31] or indirect excitons in type-II quantum wells [36, 37].

Especially in bulk semiconductors, excitons cannot be described without taking into account the strong coupling to the electromagnetic field which gives rise to mixed modes, called polaritons. In this respect, BEC of bulk polaritons has been postulated by several authors [4, 39, 40]. The polariton dispersion E_k in bulk semiconductors can be characterized by the equation

$$\left(\frac{\hbar c k}{\varepsilon_B E_k}\right)^2 = 1 + \frac{4\pi\beta}{[1+(E_k/(\hbar\omega_k))]^2} \, , \tag{8}$$

where β is the exciton polarization constant and $\hbar\omega_k$ the exciton dispersion. The energy-momentum dispersion of the lowest polariton branch deviates from a quasiparticle behaviour at low momentum, where the coupled modes show an anticrossing in the dispersion curves. In particular, for $k < k_0 = \sqrt{\varepsilon_B}\omega_0/c$ the polariton is almost totally photon-like and the linear dispersion of photons holds. In this region, the group velocity of polaritons approaches the light speed and the rate of polariton escape through recombination at the system boundaries increases. A bottleneck effect thus arises, as the rate of polariton relaxation through acoustic phonon emission becomes slower than the radiative recombination rate. In steady-state, a polariton population builds up at the momentum region close to k_0, where photoluminescence is experimentally observed [41]. It has been argued by several authors [28, 39] that the population buildup at the bottleneck might give rise to BEC. At present, however, no experimental evidence of bulk polariton BEC exists and a detailed theoretical model of polariton dynamics and interactions has never been studied.

In 1996 it was first suggested by Imamoglu et al. [42] that microcavity polaritons might undergo BEC. This work was intended to explore the possibility of an excitonic laser operating without population inversion. It was suggested that, while in the limit of a photon-like polariton the usual population inversion criterion had to be fulfilled for lasing, in the opposite limit of strong exciton-like character, excitonic BEC would result in a population buildup and thus in coherent single-mode emission without population inversion. The intermediate case was called for the first time an *exciton–polariton laser*. It was pointed out that a system particularly eligible for polariton lasing would be that of microcavity polaritons. The polariton dispersion in microcavities is approximately given by the roots of the equation [43]

$$\left(E_k - \hbar\sqrt{\omega_{C,0}^2 + c^2 k^2/\varepsilon_B}\right)\left(E_k - \hbar\omega_k\right) = \frac{\hbar^2 \Omega_R^2}{4}, \tag{9}$$

where $\omega_{C,0}$ is the frequency of the resonant cavity mode at zero in-plane momentum, $\hbar\omega_k$ the in-plane exciton dispersion and $\hbar\Omega_R$ the vacuum-field Rabi splitting. In contrast to bulk polaritons, microcavity polaritons at zero exciton-cavity detuning, are characterized by a nonvanishing exciton component throughout the whole lower polariton branch, and by a quasiparticle (quadratic) dispersion with a very low effective mass (of the order of 10^{-5} free electron masses) close to $k = 0$. Though based on very simple considerations, the work by Imamoglu et al. was the first to address the system of microcavity polaritons and to stress the role played by relaxation and recombination dynamics in the context of polariton BEC. Shortly after this work, the first experimental claim of polaritonic BEC in microcavities by Pau et al. [44] followed. The polariton photoluminescence spectrum under steady-state nonresonant excitation showed a nonlinear increase of the lower polariton peak as a function of pump intensity, which was interpreted as a population buildup driven by final-state stimulation. The word "boser" was specially created to indicate this phenomenon. It is important to remark at this point that neither the suggestion by Imamoglu et al., nor the experiment by Pau et al. ever related BEC to the formation of a complex order parameter and of long range order. The claim of polaritonic BEC was solely based on the idea of population buildup. The first boser experiment was however very controversial and its interpretation was strongly criticized shortly after its publication [45]. The main criticism was that, at the measured density, the exciton binding energy and the polariton Rabi splitting would be at least partially bleached and the system would be more correctly described in terms of a plasma of unbound electron–hole pairs, obeying Fermi statistics. The nonlinearity would then be explained in terms of a standard laser action. The authors of the first boser experiment finally had to withdraw their initial claim and to revert to a more conventional interpretation [46]. At the time of the boser controversy, it was generally accepted within the polariton community that the required condition for boser action would be the simultaneous observation of the polariton energy-momentum dispersion with a finite vacuum field Rabi splitting. This would ensure that the emission actually originated from polaritons undergoing final-state stimulation. In reality, a population buildup is only the first step towards BEC which requires also the formation of an ordered phase displaying long-range correlations. Closely following the boser result, at least three other experiments provided clear evidence of polariton final-state stimulation. These experiments were performed under nonresonant high-energy excitation in the electron–hole continuum [47–49] or at the upper polariton [50]. The most recent claim of polariton BEC was made by Deng et al. [51, 52]. In this work, a nonlinear increase of the lower polariton emission as a function of pump intensity is observed as in previous works. Here, however, the polariton dispersion, the second order correlation function in time of the emitted field, the momentum and real-space profiles of the emission are simultaneously measured. Results show that above threshold the second-order correlation function at zero time delay slowly decreases from a value of 1.8 slightly above threshold to 1.5 for a pump intensity 17 times larger than threshold, while one might expect a behaviour similar to that of a laser transition [9, 53], where the second order correlation at zero delay rapidly changes from the value of 2, expected for a thermal population below threshold, to the value of 1, indicating a pure coherent state above threshold. Both the momentum and the spatial distribution of polaritons measured by Deng et al. indicate a polariton buildup at the lowest energy level and remind of the results obtained in the case of diluted atom BEC. When, however, the expected superfluid velocity is estimated from the polariton-polariton interaction constant (see Eq. (3) of Ref. [52]), it turns out that the linear Bogoliubov dispersion expected above threshold would strongly deviate from the perfectly quadratic dispersion which is actually measured up to a pump intensity 7.6 times its threshold value. This suggests that, in this strongly non equilibrium condition of the experiment, polaritons recombine before the long-range order required for BEC is actually formed, and a description in terms of a single particle picture still holds.

On the theoretical side, a true polariton quantum field theory analogous to the Bogoliubov theory of BEC is still lacking. The polariton population buildup has been described by several authors in terms of a simple Boltzmann equation including scattering via phonon absorption or emission as well as mutual polariton scattering [48, 54–59]. These results, while predicting a population buildup arising from final-state stimulation, cannot account for the actual collective behaviour of a condensate, namely the collective Bogoliubov spectrum and the long-range correlations. More recently, some authors have focussed on the analogy between the polariton nonequilibrium BEC and laser, proposing a simple polariton laser model based on the quantum theory of laser [53, 60–62]. This approach however, does not account for the important role played by polariton–polariton interaction between condensate and noncondensate, which increases the quantum correlations and produces a scattering into the condensate that gives rise to the BEC phenomenon. Some authors have proposed a quantitative analysis in terms of the thermal equilibrium Kosterlitz–Thouless equation of state [63]. Given however the strong nonequilibrium character of polariton relaxation dynamics, such analysis might be oversimplified. Other works [64–66] have treated the polariton Bose gas in the framework of BCS theory, assuming high carrier density and weak exciton binding. This situation is however not relevant to most experiments where vacuum field Rabi splitting is observed.

An analysis of polariton BEC in terms of quantum field theory should describe consistently the time and space evolution of the polariton quantum field in two dimensions $\hat{\psi}_j(\mathbf{r}, t)$ ($j = L, U$ for lower and upper polariton respectively) and of their quantum correlations. The starting polariton Hamiltonian is written as (the time-dependence is omitted from the notation from now on)

$$H = H_C + H_X + H_R + H_I, \qquad (10)$$

where

$$H_C = \hbar \int d\mathbf{r}\, \hat{\psi}_C^\dagger(\mathbf{r})\, \omega_C(-i\nabla)\, \hat{\psi}_C(\mathbf{r}) \tag{11}$$

is the single particle Hamiltonian of the cavity photon field $\hat{\psi}_C(\mathbf{r})$, having energy dispersion $\omega_C(\mathbf{k}) = \sqrt{\omega_{C,0}^2 + c^2 k^2/\varepsilon_B}$,

$$H_X = \int d\mathbf{r}\, \hat{\psi}_X^\dagger(\mathbf{r}) \left(-\frac{\hbar^2 \nabla^2}{2 m_{ex}} \right) \hat{\psi}_X(\mathbf{r}) \tag{12}$$

is the single particle Hamiltonian of the exciton field $\hat{\psi}_X(\mathbf{r})$, and

$$H_R = \hbar \Omega_R \int d\mathbf{r}\, \hat{\psi}_X^\dagger(\mathbf{r})\, \hat{\psi}_C(\mathbf{r}) + \text{h.c.} \tag{13}$$

is the exciton–photon linear coupling term, which is taken in function of the vacuum-field Rabi splitting $\hbar \Omega_R$. The interaction hamiltonian H_I describes the Coulomb interaction between excitons and the saturation of the photon-exciton coupling and reads

$$H_I = \frac{\hbar g}{4} \int d\mathbf{r}\, \hat{\psi}_X^\dagger(\mathbf{r}) \hat{\psi}_X^\dagger(\mathbf{r}) \hat{\psi}_X(\mathbf{r}) \hat{\psi}_X(\mathbf{r}) - \frac{\hbar \Omega_R}{n_{sat} A} \int d\mathbf{r}\, \hat{\psi}_C^\dagger(\mathbf{r}) \hat{\psi}_X^\dagger(\mathbf{r}) \hat{\psi}_X(\mathbf{r}) \hat{\psi}_X(\mathbf{r}) + \text{h.c.}, \tag{14}$$

where g is the Coulomb matrix element in the contact approximation, n_{sat} is the excitonic saturation density and A is the quantization area [67]. This formalism neglects photon leakage out of the cavity mirrors, responsible for the finite polariton lifetime, which might be included in the resulting field equations, e.g. via Langevin fluctuation and dissipation terms. They also approximate the exciton–exciton interaction as a short-range interaction, a common assumption when dealing with the small momenta involved in polariton dynamics. To these Hamiltonians one should then add other scattering mechanisms, such as the exciton–acoustic–phonon scattering Hamiltonian in the deformation potential approximation [68], in order to account for the dynamics of energy relaxation. One approach to the dynamics of the polariton field could be, first, the diagonalization of the single-particle Hamiltonian $H_0 = H_C + H_X + H_R$, resulting in the lower and upper polariton quantum fields $\hat{\psi}_j(\mathbf{r})$ ($j = L, U$). The various interaction terms should then be expressed in the polariton basis. Starting from there, various field-theoretical approaches are possible. For example the Hartree–Fock–Bogoliubov approach to a quantum fluid could be generalized to our polariton problem, by introducing the separation of each polariton field into

$$\hat{\psi}_j(\mathbf{r}, t) = \tilde{\psi}_j(\mathbf{r}, t) + \phi_j(\mathbf{r}, t), \tag{15}$$

where $\phi_j(\mathbf{r}, t)$ is the polariton condensate field, while $\tilde{\psi}_j(\mathbf{r}, t)$ is the zero-amplitude Bose fluctuation field. At this point, two approaches are suitable. In the standard U(1) symmetry breaking approach to BEC [16, 17], $\tilde{\psi}_j(\mathbf{r}, t)$ is a Bose quantum field while $\phi_j(\mathbf{r}, t)$ is a classical field, i.e. a complex function. In this case, coupled field equations for the condensate and noncondensate polaritons can be derived in the self-consistent mean-field approximation, including anomalous correlation terms [17]. A possible variation on this theme is the number-conserving approach [15], which better accounts for the condensate phase diffusion [13]. In this approach, $\phi_j(\mathbf{r}, t) = \Phi_j(\mathbf{r}, t) \hat{a}_j$, where $\Phi_j(\mathbf{r}, t)$ is the condensate wave function and $\hat{a}_j$ is a Bose operator. In this case, the commutation relation for the fluctuation field has to be consistently modified. The Gross–Pitaevskii equation for the condensate wave function $\Phi_j(\mathbf{r}, t)$ then arises from consistency relations. Finding a solution for the polariton fields within this formalism, preserving the space dependence of the field, is a very cumbersome task and to our knowledge has never been tried. Only very recently, Carusotto and Ciuti [69–71] have proposed the first step in this direction. In Ref. [69] they derived in the mean field approximation the analogous of the Gross–Pitaevskii equation [7] in the case of polaritons, aiming at a generalized description of the parametric photoluminescence. For a weakly interacting Bose system, this equation is the first approximation describing the dynamics of the condensate at zero temperature. For polaritons, the generalized Gross–Pitaevskii model should describe the polariton condensate in the limit where quantum as well as thermal fluctuations are negligible. Very recently, in Ref. [71] the inclusion of fluctuations allows the investigation of the spontaneous formation of coherence above the parametric emission threshold, a phenomenon that was previously predicted by a simplified three-mode approach [72]. Ref. [69] is also the first prediction of the experimentally accessible consequences of Bogoliubov-like spectrum expected for a polariton condensate under resonant pump conditions.

Before concluding this short overview, let us comment on the fact that high-energy excitation is one of the key ingredients for possibly producing polariton BEC. As we know, BEC is a process where a macroscopic quantum state exhibiting long-range correlation spontaneously forms out of a (thermal) uncorrelated gas of particles. In semiconductors, high-energy excitation of electron–hole pairs is followed by energy relaxation through interaction with a thermal bath. The initial correlation induced by the excitation laser is thus lost after interaction with the bath, and it can be reasonably assumed that if a correlation is present after population buildup in the lowest energy levels, this is due to the BEC mechanism. Population buildup can also be obtained by direct resonant laser excitation of low energy levels. Polariton formation by resonant excitation is more efficient than high-energy excitation, as relaxation through the polariton band is strongly suppressed due to the bottleneck effect [73]. The drawback of resonant excitation, however, is that any correlation observed in the lowest polariton levels might be a residual of the coherent excitation, thus making the interpretation in terms of BEC more ambiguous. Resonant excitation can also give rise to parametric polariton processes, which are strongly resonant on the energy-momentum curve and display a nonlinear threshold in the polariton emission (see Baumberg and Ciuti in this issue). Parametric scattering is a driven wave-mixing process and therefore does not directly relate to BEC in the thermodynamic sense. We should point out, however, that the special case of parametric photoluminescence bears a strong analogy with BEC. In a parametric process, two identical polaritons created by the pump resonantly scatter to a pair of *signal* and *idler* polaritons, conserving the total energy and momentum. In the case of parametric amplification, the process is stimulated by an external laser beam which resonantly probes the signal polariton, and is fully described in terms of wave-mixing of classical fields. In parametric photoluminescence, on the contrary, the process is driven by vacuum-field fluctuations of the signal and idler polaritons, and the polariton is a quantum fluctuating field with zero classical amplitude [74]. Recently [72], it was described how the signal polaritons involved in the parametric photoluminescence might undergo a symmetry-breaking transition, driven by the quantum correlations, with the formation of an order parameter – the polariton classical field amplitude – similar to BEC. As already mentioned, an analysis of this quantum state in terms of the self-consistent mean-field theory [69–71] shows that a collective behaviour spontaneously arises above the parametric emission threshold, with the polariton dispersion changing from the single-particle behaviour to a Bogoliubov-like energy spectrum.

10.4
Some Considerations on Microcavity Polariton BEC

What are the advantages and disadvantages of the microcavity polariton system in the context of BEC? The initial excitement caused by the light effective mass has finally turned into deception. If on one hand the light mass implies a very high critical temperature for BEC, the concept of thermal equilibrium itself is scarcely relevant for polaritons. It is in fact well known that the polariton radiative lifetime, of the order of 10 ps at most, is much shorter than all typical relaxation times in the system. This leads to the bottleneck effect and to an "inverted" energy distribution, with higher energy levels much more populated than the ground level in a typical situation under steady state nonresonant excitation. This large occupation of excited states can easily reach the exciton saturation density [75], thus causing the bleaching of exciton oscillator strength from which polaritons originate. Moreover, the time required for BEC to take place, given the light polariton mass, might be significantly longer than the polariton lifetime itself, thus preventing the formation of the order parameter and of long-range correlations. The formation of correlations is further limited by the presence, in a semiconductor, of other phase-destroying mechanisms like phonon emission and absorption or scattering with unwanted free carriers or with defects, which are always present in semiconductor heterostructures. It is presently not clear how these decoherence mechanisms will affect the off-diagonal spatial coherence of the polariton quantum fluid. We know that polaritons at high density are increasingly robust against dephasing, showing an increase of time-coherence at the stimulation threshold [67, 76]. The question whether the space-coherence shows a similar behaviour is only now being considered [71]. If the coherence length induced by scattering with a thermal reservoir (e.g. of phonons or free carriers) is significantly shorter than the system size, then the Penrose–Onsager criterion might no-longer be satisfied, making the concept of BEC inappropriate.

An advantage of polaritons might be the strongly suppressed polariton-polariton Coulomb interaction, essentially related to the small phase space available due to the light effective mass [77]. Therefore, the polariton system is much closer to the picture of a weakly interacting Bose gas than other exciton systems. Another advantage is represented by the optical excitation and detection, allowing a precise determination of the energy-momentum distribution or alternatively of a space- and time-resolved detection of the polariton dynamics [78, 79]. In the end, the true motivation of all investigations on polaritonic BEC is the perspective of having a quantum fluid available in an artificial nanostructure at a reasonably high temperature. This, together with the ease in fabrication and the scalability of semiconductor nanostructures, holds great promise for the implementation of devices that would take advantage of the phase coherence of BEC for quantum information technology. To this purpose, however, one remark concerning spatial confinement shoud be made. If a Bose system is confined on a scale comparable or shorter than the thermal wavelength λ_T, then the definition of a condensate in terms of the Penrose–Onsager criterion no longer makes sense. In fact, on such a small scale, the two-point density matrix does not decay significantly over a distance equal to the system size, and the situation displaying long-range correlations can no longer be distinguished from the exponentially decaying case typical of a normal gas. Spatial confinement in the case of dilute Bose gases, always occurs on a length scale at least ten times longer than the thermal wavelength λ_T which, in the typical case, is of the order of 1 µm at 100 nK for e.g. Rubidium atoms. For polaritons, this requirement might represent a further drawback of the very light polariton mass, equal to approximately $10^{-5} m_0$. In fact, the thermal wavelength at a temperature of 3 K is approximately 10 µm for polaritons. In the absence of a theoretical estimate of the two-point correlation function out of equilibrium, we can make the simplifying assumption that the spatial correlation of the polariton system does not differ significantly from the equilibrium one. Then spatial confinement must be designed on a scale much larger than 10 µm for any superfluid behaviour to be expected.

On the experimental side, all measurements carried out under nonresonant excitation bring evidence of polariton population buildup, without clear evidence of the formation of a long-range order. Different conclusions can be drawn from experiments on polariton parametric scattering. In this case, the problems related to the relaxation bottleneck are not present and all experimental observations are well explained in terms of quantum-field theory. Recent theoretical works [69–72] predict the formation of an order parameter and superfluidity. Presumably, parametric scattering represents the most promising context for the observation of polaritonic BEC.

On the theoretical side, it appears that the microcavity polariton domain is not yet benefitting from the vast theoretical knowledge that is available in the field of BEC. Almost all the theoretical approaches until present have treated the system either with the help of ad-hoc assumptions or within Boltzmann-like models. These models have indeed helped understanding many features of the polariton laser. However, we feel that an extension to the polariton problem of the full quantum field theory of BEC, including mutual interaction and accounting for quantum correlations both in space and time domains, is still lacking. Only very recently, the first effort towards such an approach has appeared in the literature [69, 71].

In conclusion, BEC of microcavity polaritons always constitutes an exciting challenge, nine years after it was first suggested [42]. The most important questions are still unanswered. In the future, a significant improvement of microcavity structures will be required in order to increase the polariton radiative lifetime, at the same time suppressing the structural disorder and the scattering mechanisms affecting the polariton coherence. More accurate theoretical models will bring us a better understanding of the interplay between BEC and the various mechanisms taking place in semiconductors. When considering the efforts taken to achieve diluted gas BEC, it turns out that polariton BEC is still a very young research field which might lead in the next future to a significant breakthrough.

10.5
Afterword

Our interest for microcavity polaritons is the direct consequence of the fruitful collaboration with Marc Ilegems and his research group. This collaboration started back in 1993 when one of us (VS) joined the Ecole Polytechnique Fédérale de Lausanne as a PhD student with a project on polaritons. Since then, this interest has constantly grown, leading to exciting physics and to a lot of fun. For this, we are very grateful to Marc and we wish him a successful and happy life.

References

[1] C. Weisbuch, M. Nishioka, A. Ishikawa, and Y. Arakawa, Phys. Rev. Lett. **69**, 3314 (1992).
[2] D. Snoke, Science **298**, 1368 (2002).
[3] W. Ketterle, Rev. Mod. Phys. **74**, 1131 (2002).
[4] A. Griffin, D. W. Snoke, and S. Stringari (Eds.), Bose–Einstein Condensation, Cambridge University Press, Cambridge, 1995.
[5] F. Dalfovo, S. Giorgini, L. P. Pitaevskii, and S. Stringari, Rev. Mod. Phys. **71**, 463–512 (1999).
[6] A. J. Leggett, Rev. Mod. Phys. **73**, 307–356 (2001).
[7] L. Pitaevskii and S. Stringari, Bose–Einstein Condensation, Clarendon Press, Oxford, 2003.
[8] O. Penrose and L. Onsager, Phys. Rev. **104**, 576 (1956).
[9] L. Mandel and E. Wolf, Optical Coherence and Quantum Optics, Cambridge University Press, 1995.
[10] V. DeGiorgio and M. O. Scully, Phys. Rev. A **2**, 1170 (1970).
[11] M. O. Scully, Phys. Rev. Lett. **82**, 3927 (1999).
[12] V. V. Kocharovsky, M. O. Scully, S.-Y. Zhu, and M. Suhail Zubairy, Phys. Rev. A **61**, 023609 (2000).

[13] M. Lewenstein and L. You, Phys. Rev. Lett. **77**, 3489 (1996).
[14] C. W. Gardiner, Phys. Rev. A **56**, 1414 (1997).
[15] Y. Castin and R. Dum, Phys. Rev. A **57**, 3008 (1998).
[16] P. C. Hohenberg and P. C. Martin, Ann. Phys. (N.Y.) **34**, 291 (1965).
[17] A. Griffin, Phys. Rev. B **53**, 9341 (1996).
[18] Y. E. Lozovik and V. I. Yudson, Physica A **93**, 493 (1978).
[19] J. M. Kosterlitz and D. J. Thouless, J. Phys. C **6**, 1181 (1973).
[20] R. Zimmermann, E. Runge, and V. Savona, in: Quantum Coherence, Correlation and Decoherence in Semiconductor Nanostructures, edited by T. Takagahara (Elsevier Science, 2003), pp. 89–165.
[21] C. W. Gardiner, P. Zoller, R. J. Ballagh, and M. J. Davis, Phys. Rev. Lett. **79**, 1793 (1997).
[22] C. W. Gardiner, M. D. Lee, R. J. Ballagh, M. J. Davis, and P. Zoller, Phys. Rev. Lett. **81**, 5266 (1998).
[23] D. Jaksch, C. W. Gardiner, and P. Zoller, Phys. Rev. A **56**, 575 (1997).
[24] O. M. Schmitt, D. B. T. Thoai, L. Bányai, P. Gartner, and H. Haug, Phys. Rev. Lett. **86**, 3839 (2001).
[25] B. Mieck and H. Haug, Phys. Rev. B **66**, 075111 (2002).
[26] S. A. Moskalenko, Fiz. Tverd. Tela **4**, 276 (1962).
[27] I. M. Blatt, K. W. Boer, and W. Brandt, Phys. Rev. **126**, 1691 (1962).
[28] L. V. Keldysh and A. N. Kozlov, Sov. Phys. JETP **27**, 521 (1968).
[29] D. W. Snoke, J. P. Wolfe, ande A. Mysyrowicz, Phys. Rev. Lett. **64**, 2543 (1990).
[30] D. W. Snoke, Jia Ling Lin, and J. P. Wolfe, Phys. Rev. B **43**, 1226 (1991).
[31] E. Fortin, S. Fafard, and A. Mysyrowicz, Phys. Rev. Lett. **70**, 3951 (1993).
[32] N. Peyghambarian, L. L. Chase, and A. Mysyrowicz, Phys. Rev. B **27**, 2325 (1983).
[33] M. Hasuo, N. Nagasawa, T. Itoh, and A. Mysyrowicz, Phys. Rev. Lett. **70**, 1303 (1993).
[34] Yu. E. Lozovik and V. I. Yudson, Sov. Phys. JETP **44**, 389 (1976).
[35] X. Zhu, P. B. Littlewood, M. S. Hybertsen, and T. M. Rice, Phys. Rev. Lett. **74**, 1633 (1995).
[36] D. Snoke, S. Denev, Y. Liu, L. Pfeiffer, and K. West, Nature **418**, 754 (2002).
[37] L. V. Butov, A. C. Gossard, and D. S. Chemla, Nature **418**, 751 (2002).
[38] J. P. Eisenstein and A. H. Macdonald, Nature **432**, 691 (2004).
[39] E. Hanamura and H. Haug, Phys. Rep. **33**, 209 (1977).
[40] C. Comte and P. Nozières, J. Physique (Paris) **43**, 1069 (1982).
[41] L. C. Andreani, in: Confined Electrons and Photons, NATO ASI Series B vol. 340, edited by E. Burstein and C. Weisbuch (Plenum Press, New York, 1994), p. 57.
[42] A. Imamoglu, R. J. Ram, S. Pau, and Y. Yamamoto, Phys. Rev. A **53**, 4250 (1996).
[43] V. Savona, L. C. Andreani, P. Schwendimann, and A. Quattropani, Solid State Commun. **93**, 733 (1995).
[44] S. Pau, H. Cao, J. Jacobson, G. Björk, Y. Yamamoto, and A. Imamoglu, Phys. Rev. A **54**, R1789 (1996).
[45] M. Kira, F. Jahnke, S. W. Koch, J. D. Berger, D. V. Wick, T. R. Nelson, Jr., G. Khitrova, and H. M. Gibbs, Phys. Rev. Lett. **79**, 5170 (1997).
[46] H. Cao, S. Pau, J. M. Jacobson, G. Björk, Y. Yamamoto, and A. Imamoglu, Phys. Rev. A **55**, 4632 (1997).
[47] L. S. Dang, D. Heger, R. André, F. Boeuf, and R. Romestain, Phys. Rev. Lett. **81**, 3920 (1998).
[48] P. Senellart and J. Bloch, Phys. Rev. Lett. **82**, 1233 (1999).
[49] F. Boeuf, R. André, R. Romestain, L. Si Dang, E. Péronne, J. F. Lampin, D. Hulin, and A. Alexandrou, Phys. Rev. B **62**, R2279 (2000).
[50] J. Bleuse, F. Kany, A. P. de Boer, P. C. M. Christianen, R. André, and H. Ulmer-Tuffigo, J. Cryst. Growth **184/185**, 750 (1998).
[51] H. Deng, G. Weihs, C. Santori, J. Bloch, and Y. Yamamoto, Science **298**, 199 (2002).
[52] H. Deng, G. Weihs, D. Snoke, J. Bloch, and Y. Yamamoto, Proc. Natl. Acad. Sci. USA **100**, 15318 (2003).

[53] F. P. Laussy, G. Malpuech, A. Kavokin, and P. Bigenwald, Phys. Rev. Lett. **93**, 016402 (2004).
[54] F. Tassone and Y. Yamamoto, Phys. Rev. B **59**, 10830 (1999).
[55] R. Huang, F. Tassone, and Y. Yamamoto, Phys. Rev. B **61**, R7854 (2000).
[56] D. Porras, C. Ciuti, J. J. Baumberg, and C. Tejedor, Phys. Rev. B **66**, 085304 (2002).
[57] J. Erland, V. Mizeikis, W. Langbein, J. R. Jensen, and J. M. Hvam, Phys. Rev. Lett. **86**, 5791 (2001).
[58] H. T. Cao, T. D. Doan, D. B. Tran Thoai, and H. Haug, Phys. Rev. B **69**, 245325 (2004).
[59] I. Shelykh, G. Malpuech, K. V. Kavokin, A. V. Kavokin, and P. Bigenwald, Phys. Rev. B **70**, 115301 (2004).
[60] Yu. G. Rubo, F. P. Laussy, G. Malpuech, A. Kavokin, and P. Bigenwald, Phys. Rev. Lett. **91**, 156403 (2003).
[61] G. Malpuech, Y. G. Rubo, F. P. Laussy, P. Bigenwald, and A. V. Kavokin, Semicond. Sci. Technol. **18**, S395 (2003).
[62] I. Shelykh, F. P. Laussy, A. V. Kavokin, and G. Malpuech, cond-mat/0503402 (2005).
[63] A. Kavokin, G. Malpuech, and F. P. Laussy, Phys. Lett. A **306**, 187 (2003).
[64] P. R. Eastham and P. B. Littlewood, Phys. Rev. B **64**, 235101 (2001).
[65] F. M. Marchetti, B. D. Simons, and P. B. Littlewood, Phys. Rev. B **70**, 155327 (2004).
[66] J. Keeling, P. R. Eastham, M. H. Szymanska, and P. B. Littlewood, Phys. Rev. Lett. **93**, 226403 (2004).
[67] C. Ciuti, P. Schwendimann, and A. Quattropani, Semicond. Sci. Technol. **18**, S279 (2003).
[68] G. Mahan, Many-Particle Physics (Plenum, New York, 1990).
[69] I. Carusotto and C. Ciuti, Phys. Rev. Lett. **93**, 166401 (2004).
[70] C. Ciuti, phys. stat. sol. (b) **242**, No. 11 (2005) (this issue).
[71] I. Carusotto and C. Ciuti, cond-mat/0504554 (2005).
[72] V. Savona, P. Schwendimann, and A. Quattropani, Phys. Rev. B **71**, 125315 (2005).
[73] F. Tassone, C. Piermarocchi, V. Savona, A. Quattropani, and P. Schwendimann, Phys. Rev. B **56**, 7554 (1997).
[74] C. Ciuti, P. Schwendimann, and A. Quattropani, Phys. Rev. B **63**, 041303R (2001).
[75] G. Khitrova, H. M. Gibbs, F. Jahnke, M. Kira, and S. W. Koch, Rev. Mod. Phys. **71**, 1591 (1999).
[76] D. Porras and C. Tejedor, Phys. Rev. B **67**, 161310 (2003).
[77] C. Ciuti, V. Savona, C. Piermarocchi, A. Quattropani, and P. Schwendimann, Phys. Rev. B **58**, R10123 (1998).
[78] W. Langbein, Phys. Rev. B **70**, 205301 (2004).
[79] W. Langbein, in: Proceedings of the 26th International Conference on the Physics of Semiconductors, Edinburgh (Institute of Physics Publishing, 2002).

Chapter 11
Polariton Squeezing in Microcavities

Antonio Quattropani and Paolo Schwendimann

11.1
Introduction

Marc Ilegems and ourselves were interested in microcavities and microcavity polaritons since 1993. Thanks to Mark Ilegems we had the opportunity to start a fruitful collaboration with his research group thus gaining an experimental basis and a strong motivation for our theoretical investigations. Polariton squeezing was one of the subjects that we discussed from the beginning of our collaboration and this paper reviews our main results on the subject.

Polaritons are mixed elementary excitations in a solid resulting from the coupling between photons and excitons or phonons [1]. In the last fifteen years it has been shown [2–4] that these excitations have remarkable statistical properties as e.g. squeezing. Squeezing has been extensively studied in Quantum Optics in the last thirty years and is characterized as follows: Any radiation field amplitude may be separated into two components that are dephased by $\pi/2$. As we show in Section 11.2, a state of the radiation field is squeezed when the fluctuations in one component of the field amplitude can be reduced below its standard quantum limit whereas the fluctuations in the second component become larger than this limit. This property has been theoretically discussed and experimentally demonstrated for several quantum optical systems including parametric amplifiers, two-level systems and semiconductor lasers. As we shall see in Section 11.3, bulk polaritons states are intrinsically squeezed. Intrinsic squeezing is also present in quantum well polaritons and in microcavity polaritons. The intrinsic squeezing is small in theses systems and is not easily detected for bulk and quantum well polaritons, as we shall discuss later. On the contrary, as we show in Section 11.4 an important squeezing is found in a quantum well embedded in a microcavity when considering non-linearly interacting polaritons leading to parametric effects. In this case, squeezing is related to the presence of anomalous correlations between the interacting polariton modes. The effect has been experimentally demonstrated in a particular configuration corresponding to the one of a degenerate parametric system [5, 6].

Physics of Semiconductor Microcavities: From Fundamentals to Nanoscale Devices
Edited by Benoit Deveaud
Copyright © 2007 WILEY-VCH Verlag GmbH & Co. KGaA, Weinheim
ISBN: 978-3-527-40561-9

11.2
Squeezed States

As it is well known, Heisenberg's uncertainty relations define a lower limit of the quantum fluctuations in the amplitudes of the electromagnetic field. Consider a field mode whose amplitude is described by the Bose operator a. This amplitude operator may be expressed in terms of the two components d_1 and d_2 defined as

$$d_{1/2} = \frac{\sqrt{\pm 1}}{2}(a^\dagger \pm a) \quad \text{with} \quad [d_1, d_2] = \frac{i}{2}, \tag{1a}$$

in analogy to what is done for a classical field amplitude. The new operators represent the two field mode components having a phase difference of $\pi/2$. The operators d_1 and d_2 are related to the canonical operators p and q describing a harmonic oscillator. The mean square deviations of the operators d_1 and d_2 satisfy the uncertainty relations

$$\langle (\Delta d_1)^2 \rangle^{1/2} \langle (\Delta d_2)^2 \rangle^{1/2} \geq 1/4. \tag{1b}$$

The equality holds when the expectation values are taken over coherent states. Moreover, for a generic coherent state it holds $\langle (\Delta d_1)^2 \rangle = \langle (\Delta d_2)^2 \rangle = 1/4$. The coherent state represents the state with the lowest possible quantum noise in both field components that may be generated without violating Heisenberg's uncertainty relations. Squeezed states are peculiar states of the electromagnetic field that allow to overcome at least partially the limits imposed by Heisenberg's relations (1b). A squeezed state has the following characteristic property

$$\langle (\Delta d_1)^2 \rangle^{1/2} < 1/2 \quad \text{and} \quad \langle (\Delta d_2)^2 \rangle^{1/2} > 1/2 \tag{2}$$

or vice versa, with

$$\langle (\Delta d_1)^2 \rangle^{1/2} \langle (\Delta d_2)^2 \rangle^{1/2} \geq 1/4.$$

Therefore, in a squeezed state the quantum noise in one of the field components may be reduced below the quantum noise of a coherent state at the expense of the noise in the second component such that (1b) is not violated. Squeezing and squeezed states have been extensively studied in Quantum Optics both theoretically and experimentally. We refer the interested reader to the existing literature [7]. In this Section we want only to introduce some concepts that will be useful in the following. Squeezed states may be generated from the vacuum state of the mode by the unitary transformation

$$S(z) = \exp[(za^{\dagger 2} - z^* a^2)/2] \quad \text{with} \quad z = re^{i\psi}. \tag{3}$$

The operator $S(z)$ allows also defining the new creation and annihilation operators

$$A^\dagger(z) \equiv S(z)\, a^\dagger S(z)^\dagger = \cosh(r)\, a + \sinh(r)\, a^\dagger \tag{4}$$

verifying the commutation rule $[A(z), A^\dagger(z)] = 1$. Using these definitions one shows that the state $S(z)|0\rangle$ is squeezed. In fact one obtains

$$\langle 0|\, S(z)\, d_1^{\,2} S(z)^\dagger\, |0\rangle = \exp(-r)/4, \tag{5a}$$

$$\langle 0|\, S(z)\, d_2^{\,2} S(z)^\dagger\, |0\rangle = \exp(r)/4. \tag{5b}$$

These equations show that the squeezing transformation $S(z)$ allows calculating squeezing, either from the d-operators averaged over the squeezed vacuum state or from the transformed d-operators averaged over the original vacuum state, leading to the same result. One may also generate a squeezed state out of a coherent state viz. $|z, \alpha\rangle = S(z)|\alpha\rangle$. This squeezed state, also called the two photon coherent state [8], has the following property: $A(z)|z, \alpha\rangle = S(z)\, a S(z)^\dagger S(z)|\alpha\rangle = \alpha |z, \alpha\rangle$. One verifies that the state $|z, \alpha\rangle$ is indeed a squeezed state by calculating the dispersion of d_1. A straightforward calculation gives

$$\langle z, \alpha|\, (\Delta d_1)^2\, |z, \alpha\rangle = \cosh(2r) - \sinh(2r) \cos(\phi - 2\psi), \tag{6}$$

where ϕ is the phase of z and ψ is the phase of the coherent state. The maximum squeezing is obtained for $\phi = 2\psi$. Squeezed states are produced in non-linear optical processes as e.g. second order parametric effects [7]. Since parametric effects play an important role in Section 11.4, we give here a very short account of them. A parametric effect is characterized as follows: an intense radiation field impinges on a crystal and generates two related radiation fields called the signal and the idler field respectively. The sum of the frequencies of these fields is equal to the frequency of the impinging field. In a particular configuration the so-called degenerate parametric amplifier the signal and the idler have the same frequency. The degenerate parametric amplifier is well described by the following Hamiltonian

$$H = \hbar\omega a^\dagger a + \chi^{(2)} E(a^\dagger a^\dagger + a a), \tag{7a}$$

whereas the parametric amplifier is described by the Hamiltonian

$$H = \hbar\omega_s a_s^\dagger a_s + \hbar\omega_i a_i^\dagger a_i + \chi^{(2)} E(a_i^\dagger a_s^\dagger + a_i a_s). \tag{7b}$$

Here $\chi^{(2)}$ is the second order susceptibility of the medium. The index s in (7b) refers to the signal and the index i to the idler, respectively. The Hamiltonian (7a) is diagonalized through the unitary transformation $S(z)$ and its ground state is a squeezed state. This system is therefore intrinsically squeezed. The same procedure may be introduced for the Hamiltonian (7b). We shall see some details of this calculation in the next section.

Finally it is useful to generalize the definition of the operators d_1 and d_2 by introducing a phase

$$d_{1/2} = \frac{\sqrt{\pm 1}}{2}\left(a^\dagger e^{i\phi} \pm a e^{-i\phi}\right). \tag{8}$$

The phase ϕ may represent the phase of the external field acting on the considered mode, as it is the case in a heterodyne measurement. We shall make extensive use of the material presented here in the next sections.

11.3
Intrinsic Squeezing of Polaritons

Squeezing effects may be found in systems other than the ones considered in Quantum Optics. In particular squeezed states may be found in solids when discussing their elementary excitations, like excitons. These squeezed states acquire a particular interest when they may be observed through the emission of a radiation field, whose statistical properties reproduce the statistics of the excitation in the solid. In fact, these states may offer new and efficient sources of squeezed radiation. A particularly interesting solid state excitation in this respect is the polariton. Polaritons are known to be mixed states of radiation and solid state excitations like excitons or phonons [1]. Because of their hybrid nature between matter and radiation they show some properties that are characteristic of a radiation field. Furthermore, the radiation component of polaritons is in principle observable outside the solid.

11.3.1
Intrinsic Squeezing of Bulk Polaritons

Polaritons have been theoretically predicted [1, 9] and observed experimentally [10, 11] in bulk semiconductors and insulators. The polariton states may be considered as the propagation states of a radiation field in a crystal. Both exciton-polaritons resulting from the interaction between excitons and radiation in a semiconductor and phonon-polaritons resulting from the interaction between phonons and radiation in an insulator have been addressed. Due to the translational symmetry in a crystal, momentum is conserved in the interaction between photons and excitations such that each radiation mode couples to only one excitation mode in k-space. Phonons are known to be bosons. Excitons are composite particles resulting from the interaction between electron and hole in a crystal. They are not really bosons, but may be considered as such under condition of low excitation intensity [1]. Therefore, the same bosonic Hamiltonian describes both exciton- and phonon-polaritons. Polariton states are the eigenstates of the Hamiltonian

$$H = \sum_k \frac{\hbar c k}{\sqrt{\varepsilon_\infty}} A_k^{1\dagger} A_k^1 + \sum_k \hbar \omega_k A_k^{2\dagger} A_k^2 + \sum_k i C_k \left(A_k^{1\dagger} + A_{-k}^1 \right)\left(A_k^2 - A_{-k}^{2\dagger} \right)$$
$$+ \sum_k i D_k \left(A_k^{1\dagger} + A_k^1 \right)\left(A_{-k}^{1\dagger} + A_{-k}^1 \right), \tag{9a}$$

$$\omega_k = \omega_0 + \gamma k^2, \tag{9b}$$

$$C_k = \hbar \omega_0 \left(\frac{\pi \chi \omega_k}{k c \sqrt{\varepsilon_\infty}} \right)^{1/2}, \tag{9c}$$

$$D_k = \left(\frac{\pi \chi \hbar \omega_0^2}{k c \sqrt{\varepsilon_\infty}} \right), \tag{9d}$$

where χ is the polarisability of the medium and

$$\left[A_k^\nu, A_{k'}^{\mu\dagger} \right] = \delta_{\nu,\mu} \delta_{k,k'}. \tag{9e}$$

Here the A_k^ν operators are the photon [$\nu=1$] and excitation (exciton or phonon) [$\nu=2$] annihilation operators respectively. The couplings (9c) and (9d) depend on the matrix elements between the ground state of the crystal and the excitation, as well as on the excitation energy for each value of the wave vector k of the excitation. This quantum Hamiltonian, which was first introduced by Hopfield [1], has been extensively studied in the literature [10, 11]. As it is well known, the Hamiltonian (9) is diagonalized by introducing new Bose operators $B_k^\nu (\nu = 1, 2)$, which are linear combinations of the exciton and photon operators [1]. The energy eigenvalues of the Hamiltonian (9) are functions of the wave vector k and consist of two branches that are presented in Fig. 11.1. The upper branch tends asymptotically to the dispersion of free photons and the polaritons whose energy lies on this branch are called upper- or photon-polariton. The lower branch tends asymptotically to the exciton dispersion and the corresponding polaritons are called lower- or exciton-polaritons.

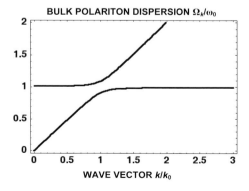

Fig. 11.1 Schematic plot of the dispersion Ω_k/ω_0 of upper- and lower polaritons in bulk materials as a function of the the normalized wave vector k/k_0. The frequency ω_0 denote exciton frequency and $\omega_0 = c k_0 / \sqrt{\varepsilon_\infty}$.

More explicitly the Hopfield Hamiltonian is diagonalized by the linear transformation

$$B_k^\nu = W_\nu A_k^1 + X_\nu A_k^2 + Y_\nu A_{-k}^{1\dagger} + Z_\nu A_{-k}^{2\dagger}, \tag{10}$$

where ν denotes the upper [$\nu = 2$] and lower [$\nu = 1$] polariton, respectively. The explicit form of the coefficients in (10) is given in [1] and has been generalized to the case of spatial dispersion in [12]. We notice that in the Hamiltonian (9) several terms appear consisting of products of two creation or two annihilation operators, respectively. These terms are also present in the Hamiltonians (7) and are responsible for intrinsic squeezing. Furthermore, the linear transformation relating the new polariton operators to the ones of the excitation and of the radiation is a generalization of (4) for a two-mode case. Therefore, we expect that some squeezing effect should appear in the polariton states. Indeed, this is the case as it is found by looking for the unitary operator leading to the transformation (10). The eigenstates of the diagonalized Hamiltonian are constructed from a new vacuum state, which is related to the original vacuum of the uncoupled excitons and photons by a unitary transformation U. Equation (10) takes the form

$$B_k^\nu = U^\dagger A_k^\nu U. \tag{11a}$$

The polariton vacuum $|0'\rangle$ is related to the exciton–photon vacuum by the transformation

$$|0'\rangle = U^\dagger |0\rangle. \tag{11b}$$

The details on the construction of the unitary transformation U may be found in [13–15]. The transformation takes the form $U = U^1 U^2$ with

$$U^1 = \prod_k \exp\left[i\mu_1(k)\left(C_k^{1\dagger}C_k^1 + C_{-k}^{1\dagger}C_{-k}^1\right)\right]\exp\left[\Xi_1(k)C_k^{1\dagger}C_{-k}^{1\dagger} - \Xi_1^* C_k^1 C_{-k}^1\right]. \tag{12a}$$

An analogous form holds for U^2. In (12) we introduce the operators

$$C_{\pm k}^1 = \frac{\left(A_{\pm k}^1 + A_{\pm k}^2\right)}{\sqrt{2}}, \tag{12b}$$

$$C_{\pm k}^1 = \frac{\left(A_{\pm k}^1 + A_{\pm k}^2\right)}{\sqrt{2}}, \tag{12c}$$

$$\Xi_1(k) = |\Xi_1(k)|\exp[i\theta_1(k)] = \tanh^{-1}\left(\frac{|Y_1 + Z_1|}{|W_1 + X_1|}\right)\exp[i\theta_1(k)], \tag{12d}$$

$$\Xi_2(\mathbf{k}) = |\Xi_1(\mathbf{k})| \exp[-i\theta_1(\mathbf{k})], \tag{12e}$$

$$\theta_1(\mathbf{k}) = -\theta_2(\mathbf{k}) = \tanh^{-1}\left(\frac{X_1 Y_1 - Z_1 W_1}{X_1 Z_1 - Y_1 W_1}\right), \tag{12f}$$

$$\mu_1(\mathbf{k}) = -\mu_2(\mathbf{k}) = \tan^{-1}\left(\frac{X_1}{iW_1}\right). \tag{12g}$$

The operators U^1 and U^2 act each as a product of a phase transformation times a two mode squeezing transformation [16], which is a generalization of the squeezing operator defined in Section 11.2. From the expression for U, it follows that the ground state of the polaritons is intrinsically squeezed with respect to the variables $C^\nu_{\pm\mathbf{k}}$. Therefore, in order to calculate the amount of squeezing we consider the variance of the following linear combination of the operators $C^\nu_{\pm\mathbf{k}}$

$$d_1 = 2^{-2/3}\left(C^1_\mathbf{k} + C^1_{-\mathbf{k}} + C^{1\dagger}_\mathbf{k} + C^{1\dagger}_{-\mathbf{k}}\right), \tag{13}$$

that generalizes the definition (3) to the two-mode case. The expression for the intrinsic squeezing in the ground state is given by

$$\langle 0'|(\Delta d_1)^2|0'\rangle = \tfrac{1}{4}[1 + 2\sinh^2|\Xi_1(\mathbf{k})| - \cos(\theta_1)\sinh|2\Xi_1(\mathbf{k})|]. \tag{14}$$

Squeezing is present when $\langle 0'|(\Delta d_1)^2|0'\rangle < \tfrac{1}{4}$. The amount of squeezing depends on the parameters of the material contained in the definitions of the coefficients in the relation (2). Squeezing can be found also for the field quantities alone in the two opposite modes combination $(A^1_{\pm\mathbf{k}} + A^2_{\pm\mathbf{k}})/\sqrt{2}$. In Table 11.1 we present the explicit values of $\sqrt{\langle 0'|(\Delta d_1)^2|0'\rangle}$ evaluated for different exciton- and phonon-polaritons and for a definite value of the normalized wave vector $\mathbf{k}$.

Table 11.1 Values of the variance defined in (14) and percentage squeezing for different materials. These values are for $k/k_0 = 0.1$ with $k_0 = \omega_0\sqrt{\varepsilon_\infty}/c$ of a table.

Material	Quadrature variance $\sqrt{\langle 0'|(\Delta d_1)^2|0'\rangle}$	Squeezing (%) $100\left(1 - 2\sqrt{\langle 0'|(\Delta d_1)^2|0'\rangle}\right)$
SiO2 (phonons)	0.4343	13.14
CdS (phonons)	0.4492	10.17
GaP (phonons)	0.4794	4.11
PbI2 (excitons)	0.4994	0.11
CuCl (excitons)	0.4996	0.07
CdS (excitons)	0.4998	0.03
ZnSe (excitons)	0.4998	0.02

A squeezing effect is present in all considered materials. However, for exciton polaritons its magnitude is very small whereas it becomes non-negligible for phonon polaritons. This characteristic is related to the fact that Wannier excitons exhibit a small longitudinal–transverse splitting. This splitting is related to the quantity $2\pi\chi$, which gives the coupling to the photons and whose magnitude determines that of the squeezing. For phonon polaritons $2\pi\chi$, is of about two orders of magnitude larger than for excitons and therefore leads to a much larger squeezing. The measurement of the amount of intrinsic squeezing of bulk polaritons is an extremely difficult task. First of all polaritons are eigenstates of the Hamiltonian which cannot be directly excited by an external field. Secondly, the radiation emitted by a bulk crystal in which polaritons are present, doesn't carry direct information on the polariton state. In fact, at the surface of the crystal the polariton modes are coupled to all the modes of the external radiation field. Therefore the contributions to the emission of the individual polariton modes are mixed and it is no more possible to reconstruct the original polariton modes outside of the crystal. Since only indirect measurements give access to the polariton in the crystal itself, the experimental evidence for intrinsic bulk polariton squeezing becomes exceedingly difficult.

11.3.2
Intrinsic Squeezing of Polaritons in Confined Systems

The difficulty in demonstrating experimentally polariton squeezing in bulk systems may be overcome by considering a two-dimensional confined system like a quantum well. In this system the translational symmetry in the growth direction is lost, whereas it holds in the plane of the quantum well. In this structure the excitons are two-dimensional objects confined in the plane of the well, but they couple to all three-dimensional field modes. Since momentum is not conserved in the growth direction an exciton couples to all field modes in this direction. As a consequence, the exciton acquires a finite spontaneous broadening and its emission can be measured. This characteristic of a confined exciton system may be exploited in order to measure the properties of polaritons. Polariton squeezing in a quantum well may be calculated [17] but not easily measured. In fact the lower-polariton exists inside the quantum well where momentum conservation in the plane still holds and thus behaves like a bulk polariton. On the contrary the upper-polariton cannot exist, because of the finite lifetime of the exciton. Therefore, squeezing of quantum well polaritons is not easier to be measured than squeezing in the bulk. Things change when the quantum well is embedded in a semicondutor microcavity. In this case, the exciton couples to the cavity modes and mixed exciton-photon modes with a finite lifetime develop. These microcavity polaritons that were first observed in 1992 [18], have the advantage of being directly observable through their photon component. Indeed, this component is transmitted into the outside world due to the finite transmission of the microcavity mirrors. Therefore, it is appropriate to look for the squeezing of microcavity polaritons. The polariton Hamiltonian for quantum well excitons embedded in a microcavity reads

$$H = \sum_{\boldsymbol{q}} \hbar\omega_q A_{\boldsymbol{q}}^{2\dagger} A_{\boldsymbol{q}}^{2} + \sum_{\boldsymbol{Q},\lambda} \hbar v |\boldsymbol{\gamma}| A_{\boldsymbol{Q},\lambda}^{1\dagger} A_{\boldsymbol{Q},\lambda}^{1} + i \sum_{\boldsymbol{q},\boldsymbol{Q},\lambda} C_{-}^{\lambda} (A_{-\boldsymbol{q}}^{2} - A_{-\boldsymbol{q}}^{2\dagger})(A_{\boldsymbol{Q},\lambda}^{1} + A_{-\boldsymbol{Q},\lambda}^{1\dagger})$$

$$+ \left(\frac{1}{\hbar\omega_q}\right) \sum_{\boldsymbol{Q}',\boldsymbol{Q},\lambda,\lambda'} C_{\boldsymbol{Q}}^{\lambda} C_{\boldsymbol{Q}'}^{\lambda'*} (A_{\boldsymbol{Q}',\lambda'}^{1} + A_{-\boldsymbol{Q}',\lambda'}^{1\dagger})(A_{\boldsymbol{Q},\lambda}^{1} + A_{-\boldsymbol{Q},\lambda}^{1\dagger}), \tag{15}$$

where $C_{\boldsymbol{Q}}^{\lambda} C_{\boldsymbol{Q}'}^{\lambda'*}$ in (15) corresponds to D_k in (9e), $A_{\boldsymbol{q}}^{2\dagger}$ and $A_{\boldsymbol{q}}^{2}$ are the creation and annihilation operators of the two-dimensional excitons, $\boldsymbol{q}$ is the component of the wave vector $\boldsymbol{Q}$ perpendicular to the z-direction and k_z is the component of $\boldsymbol{Q}$ in the z-direction. $A_{\boldsymbol{Q}}^{1\dagger}$ and $A_{\boldsymbol{Q}}^{1}$ are the field operators. The coupling constants are defined as

$$C_{\boldsymbol{Q}}^{\lambda} = C_{\boldsymbol{q},k_z}^{\lambda} = C_{-\boldsymbol{q},-k_z}^{\lambda*} = \sqrt{\frac{2\pi\hbar v}{L'|\boldsymbol{Q}|}}\, \omega_q \mu_{cv} \cdot \hat{e}_{\boldsymbol{q},k_z,\lambda}\, \frac{1}{c} F(0) \frac{1}{L} \int_{-L/2}^{L/2} dz\, \rho(z)\, e^{i k_z z}, \tag{16}$$

where $v = c/\sqrt{\varepsilon_\infty}$ is the velocity of light in the medium, $\omega_q = \omega_0 + \kappa q^2$ is the exciton frequency and $A_{\boldsymbol{Q},\lambda}^{1} = A_{\boldsymbol{q},k_z,\lambda}^{1}$, $A_{-\boldsymbol{Q},\lambda}^{1\dagger} = A_{-\boldsymbol{q},-k_z,\lambda}^{1\dagger}$. The Hamiltonian (15) involves the same quantities as (9) but shows the coupling of the two-dimensional excitons to the three-dimensional field modes. The electromagnetic field has been quantized inside a microcavity of dimension L' with periodic boundary conditions. The coupling constant is derived from a phenomenological model for the exciton in a quantum well. For simplicity we consider only one polarization for the field and diagonalize the Hamiltonian using the transformation

$$B_{\boldsymbol{q}}^{l} = \sum_{k_z} W_l(k_z,\boldsymbol{q})\, A_{\boldsymbol{q},k_z}^{1} + X_l(\boldsymbol{q})\, A_{\boldsymbol{q}}^{2} + \sum_{k_z} Y_l(k_z,\boldsymbol{q})\, A_{-\boldsymbol{q},-k_z}^{1\dagger} + Z_l(\boldsymbol{q}) A_{-\boldsymbol{q}}^{2\dagger}. \tag{17}$$

The transformation (17) allows constructing the polariton states out of the product of free particle states. We show that squeezing is also present for microcavity polaritons as an intrinsic property. We introduce the quantities

$$d_{1/2} = \pm \frac{\sqrt{\pm 1}}{2^{3/2}} \left[B_{\boldsymbol{q}} + B_{-\boldsymbol{q}} \pm B_{-\boldsymbol{q}}^{\dagger} \pm B_{\boldsymbol{q}}^{\dagger} \right],$$

and we evaluate the variances in the exciton–photon vacuum $|0\rangle$. This leads to the result

$$\langle 0|(\Delta d_1)^2|0\rangle = \frac{1}{4}\Bigg[|Z(\boldsymbol{q})|^2 + |X(\boldsymbol{q})|^2 + \sum_{k_z}\{|W(k_z,\boldsymbol{q})|^2 + |Y(k_z,\boldsymbol{q})|^2\}$$

$$- \frac{1}{2}\mathfrak{Re}\, X(\boldsymbol{q})\, Z(\boldsymbol{q}) - \frac{1}{2}\mathfrak{Re}\sum_{k_z} W(k_z,\boldsymbol{q})\, Y(-k_z,-\boldsymbol{q}) \Bigg]. \tag{19}$$

We have calculated (19) for a GaAs quantum well with a width $L = 60\,\text{Å}$. Cavity and quantum well have the same dielectric constant $\varepsilon = 12$ and an exciton energy

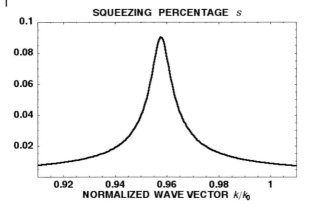

Fig. 11.2 Plot of the squeezing percentage $s \equiv 100\left[1 - 2\sqrt{\langle 0|(\Delta d_1)^2|0\rangle}\right]$ for th lower polariton as a function of the normalized wave vector k/k_0.

of 1.6 eV. We also use the relation $|\mu_{cv}|^2|F(0)|^2/\hbar = fe^2/2m\omega_0$ between the oscillator strength [19] f and the quantity used here; the value of $f = 36 \times 10^{-5}$ Å^{-2} is taken from [20]. The result is presented in Fig. 11.2 in terms of the quantity

$$s \equiv 100\left[1 - 2\sqrt{\langle 0|(\Delta d_1)^2|0\rangle}\right]. \tag{20}$$

One obtains an amount of squeezing of about 0.09%, which is small but larger than in the bulk case as a consequence of larger oscillator strength of the confined exciton. For the upper polariton squeezing is defined in the same way. The effect should be measurable in realistic cavities of finite length and reflectivity close to one.

We conclude this section by noticing that in spite of the smallness of the effect, intrinsic polariton squeezing has conceptual relevance. Intrinsic squeezing is related to the anti resonant terms in the Hamiltonian and therefore polaritons are indeed quantum excitations.

11.4
Squeezing for Interacting Microcavity Polaritons

In the former Section we have shown that bulk as well as microcavity polaritons show a small amount of intrinsic squeezing. This squeezing is a consequence of the fact that the interaction between the radiation field and the exciton modifies the ground state of the system. In contrast to the bulk case the squeezing of microcavity polaritons may be experimentally detected. Unfortunately the amount of intrinsic squeezing is predicted to be very small even for microcavity polaritons. However, it should be possible to increase in a substantial way the amount of squeezing of microcavity polaritons by considering a different excitation regime. In the low excitation regime considered until now, the interactions between excitons may be neglected. This fact justifies the use of the bilinear Hamiltonian (9). In a regime of higher excitation characterized by exciton densities smaller that the exciton saturation density, but larger than the ones considered until now, strong interactions between polaritons exist due to the Coulomb interaction between excitons and to the Pauli principle between electrons and holes. These interactions result in a polariton–polariton scattering, conserving energy and in-plane momentum. Due to the peculiar energy dispersion, this scattering is strongly resonant for selected initial and final momenta. This process turns out to be a parametric process and is responsible for two important effects. The parametric amplification, in which the scattering of resonantly excited polaritons is stimulated by a probe field resonant with the final scattering state, and the parametric photoluminescence, in which the same scattering process occurs spontaneously, driven by vacuum-field fluctuations. These effects have been investigated both experimentally and theoretically over the last years [20, 21]. Since we deal with a parametric effect, we may exploit the analogy between the interacting polariton system and the parametric effects in Quantum Optics mentioned in Section 11.2. As a consequence we expect a non-negligible amount of squeezing to appear due to the polariton–polariton interaction.

We consider a configuration, in which polaritons are excited by a continuous pump at resonance with the exciton polariton dispersion curve (Fig. 11.1). As it is well known [22, 23], for moderate excitation intensities the polaritons may be approximately considered to be bosons. When describing parametric amplification [24] and parametric luminescence [25] the pump has been chosen to excite polaritons at a wave vector k which correspond to a specific angle of incidence of the pump, the so called magic angle [26]. In this configuration we may describe most of the interesting physical characteristics of the nonlinearly interacting polaritons by considering the scattering process involving three modes only: the pump mode at $q = k$, the signal mode at $q = 0$ and the idler mode at $q = 2k$. The Hamiltonian describing this scattering process has been discussed in detail in [20, 24, 27] and reads

$$H = H_{LP} + H_{PP}^{eff} + H_p + H_{probe}, \tag{21}$$

where $H_{LP} = \sum_q \hbar \omega_{LP}(q) P_q^\dagger P_q$, $\omega_{LP}(q)$ is the polariton dispersion, P_q and $P_q^\dagger$ are the annihilation and creation operators for the lower polaritons in the rotating wave approximation. In the notation of (17) we have $P_q = X_q A_q^2 + W_q A_q^1$ and the Hopfield coefficients Y_q and Z_q [20] are zero. In the following we consider the modes $q = 0$ (signal), k (pump), $2k$ (idler). The polariton–polariton interaction [20] takes the form

$$H_{PP}^{eff} = \hbar \omega_{int} P_{2k}^\dagger P_0^\dagger P_k P_k + h.c. \tag{22}$$

The coupling constant $\hbar \omega_{int}$ contains the exciton–exciton coupling constant as well as the phase space filling [20]. Notice that $\hbar \omega_{int}$ is positive and represent a repulsive interaction. Furthermore, we couple the polaritons to a classical external field with the amplitude F_p describing the pump field through the Hamiltonian

$$H_p = F_p(t) P_k^\dagger + \text{h.c.} = \hbar F_p(0) \exp\left[-i\omega_p t\right] P_k^\dagger + \text{h.c.} \tag{23}$$

The Heisenberg–Langevin equations of motion for the polariton operators are straightforwardly obtained. They contain the decay rates γ_0, γ_k, γ_{2k} of the polariton modes and the corresponding Langevin force operators $F_0(t)$, $F_k(t)$, $F_{2k}(t)$ that follow from the quasi-mode approximation [7]. This approximation is introduced in order to account for cavity losses. The Heisenberg–Langevin equations for the polariton modes read

$$i\frac{d}{dt} P_0(t) = \left(\tilde{\omega}_{LP}(0) - i\gamma_0\right) P_0(t) + \omega_{\text{int}} P_k^2(t) P_{2k}^\dagger(t) + F_0(t) + F_{\text{probe}}(t), \tag{24a}$$

$$i\frac{d}{dt} P_{2k}(t) = \left(\tilde{\omega}_{LP}(2k) - i\gamma_{2k}\right) P_{2k}(t) - \omega_{\text{int}} P_k^2(t) P_0^\dagger(t) - F_{2k}(t), \tag{24b}$$

$$i\hbar\frac{d}{dt} P_k(t) = \left(\tilde{\omega}_{LP}(k) - i\gamma_k\right) P_k(t) + 2\omega_{\text{int}} P_k^\dagger(t) P_0(t) P_{2k}(t) + F_k(t) + F_p\, e^{i\omega_p t}, \tag{24c}$$

where $\tilde{\omega}_{LP}(q)$ is the polariton dispersion blue-shifted [24] by the polariton–polariton interaction. These operator equations have been solved analytically only within the approximation of replacing the pump mode operator by its expectation value [25, 28]. In order to get more insight into the dynamics described by (24) and in particular in order to show the onset of coherence in the photoluminescence emission we used [4] a self-consistent mean field approach [29]. Within this scheme we have obtained a set of fifteen non-linear equations involving the expectation values of the polariton modes amplitudes and the following quantities

$$A_{q,q'}(t) = \langle p_q p_{q'}\rangle(t) = A_{q',q}(t) \qquad \text{(anomalous correlations)} \tag{25a}$$

$$N_{q,q'}(t) = \langle p_q^\dagger p_{q'}\rangle = N_{q',q}(t)^* \qquad \text{(normal correlations)} \tag{25b}$$

$$\langle \tilde{p}_q \tilde{p}_{q'}\rangle(t) = \langle p_q p_{q'}\rangle(t) - \langle p_q\rangle(t)\langle p_{q'}\rangle(t) \qquad \text{(correlation of the fluctuations)} \tag{25c}$$

Here, $p_q = P_q \exp[-i\tilde{\omega}_{LP}(q)t]$, and $\tilde{p}_q = \left(P_q - \langle P_q\rangle\right)\exp[i\tilde{\omega}_{LP}(q)t] = p_q - \langle p_q\rangle$. All correlations between the polariton modes have been numerically determined in function of time. For later use, we list here the values of the constants used for the numerical evaluations. The coupling coefficient (Eq. (21) of Ref. [20]) is calculated with a vacuum field Rabi splitting $\hbar\Omega_R = 3.5$ meV, an exciton Bohr radius $\lambda_x = 80$ Å, and a quantization area $A = 100\ \mu m^2$, leading to a coupling constant $\omega_{\text{int}} = 3.54 \times 10^{-5}$ meV. We choose for the broadening $\gamma_{2k} = \gamma_k = \gamma_0 = \gamma = 0.1$ meV. With this choice of parameters we can estimate the threshold polariton density n_k at the pump momentum k following Ref. [25] we obtain $n_k = 3 \times 10^9$ cm^{-2}, less than 2% of the saturation density [25] $n_{sat} = 7/16\pi\lambda_x^2 \cong 2 \times 10^{11}$ cm^{-2}. At this density, the correction to the bosonic commutation rule in the Hartree–Fock approximation [30] is of the order of $n_k/n_{sat} \leq 2\%$, which finally justifies the bosonic approach used here. The stationary solutions of the system of fifteen equations are calculated numerically and are reported in [4]. It is shown that the polariton system starts amplifying at $f_p \equiv F_p\sqrt{\gamma/\omega_{int}} = 1$ but that the signal and idler mode amplitudes remain with the value zero. When $f_p = 2.45$, a coherence threshold is attained, the signal and idler amplitudes become different from zero whereas the fluctuations and occupation numbers rapidly decrease. We expect that quantum features like squeezing will be important in the amplification region because in this regime the quantum fluctuations expressed by the anomalous correlation between signal and idler modes is important. Indeed it may be shown that below the coherence threshold only the quantities

$$A_{0,2k}(t) = \langle p_0 p_{2k}\rangle(t), \quad A_{k,k}(t) = \langle p_k p_k\rangle(t), \tag{26a}$$

$$N_{0,0}(t) = \langle p_0^\dagger p_0\rangle, \quad N_{2k,2k}(t) = \langle p_{2k}^\dagger p_{2k}\rangle, \quad N_{k,k}(t) = \langle p_k^\dagger p_k\rangle \tag{26b}$$

contribute to the evolution of the system. These quantities are accessible in the experiments.

Below the coherence threshold an analytic form of the stationary solution of the system of fifteen equations is found with the following characteristics: all anomalous correlations with the exception of $A_{0,2k}^{stat}$, $A_{k,k}^{stat}$ and all off diagonal normal correlations $N_{k,k'}^{stat}$ as well as the signal and idler amplitudes are identically zero. The explicit form of the remaining correlations may be obtained by solving the stationary equations for the quantities $\langle p_k\rangle, A_{0,2k}^{stat}, A_{k,k}^{stat}$ and $N_{q,q}^{stat}$ for $q = 0, k, 2k$. Moreover, one easily shows that for imaginary F_p, $\langle p_k\rangle$, $A_{k,k}^{stat}$ and $N_{q,q}^{stat}$ for $q = 0, k, 2k$ are real, while $A_{0,2k}^{stat}$ is pure imaginary. We report here the stationary solutions for the anomalous correlations $A_{0,2k}^{stat}$, $A_{k,k}^{stat}$ and $N_{0,0}^{stat} + N_{2k,2k}^{stat}$ for $f_p < f_{thr}$.

$$A_{0,2k}^{stat} = \frac{-i\alpha}{2} \frac{A_{k,k}^{stat}}{1 - \left|\alpha A_{k,k}^{stat}\right|^2}, \tag{27a}$$

$$A_{k,k}^{stat} = -i\alpha \frac{A_{0,2k}^{stat}}{1 - \left|2\alpha A_{0,2k}^{stat}\right|^2} + \frac{(F_{pump}/i\gamma)^2 \left(1 - 2i\alpha A_{0,2k}^{stat}\right)^2}{\left(1 - \left|2\alpha A_{0,2k}^{stat}\right|^2\right)^2}, \tag{27b}$$

$$N_{0,0}^{stat} + N_{2k,2k}^{stat} = i2\alpha A_{k,k}^{stat} A_{0,2k}^{stat}$$

with

$$\alpha = \frac{\omega_{int}}{\gamma}. \tag{27c}$$

Introducing (27a) into (27b) we obtain a relation between $A_{\mathbf{k},\mathbf{k}}^{\text{stat}}$ and F_p for $|\alpha A_{\mathbf{k},\mathbf{k}}^{\text{stat}}| < 1$. Below the coherence threshold, $f_p = 2.45$, squeezing may be expressed analytically using the expression for the correlations outlined above. In order to discuss squeezing system we introduce the operators

$$d_{1/2} = \frac{\sqrt{\pm 1}}{2^{3/2}} \left[e^{i\theta} \left(p_0 + p_{2\mathbf{k}} \right) \pm e^{-i\theta} \left(p_0^\dagger + p_{2\mathbf{k}}^\dagger \right) \right], \qquad (28)$$

and we evaluate the variance $\langle (\Delta d_{1/2})^2 \rangle = \langle (d_{1/2})^2 \rangle - \langle d_{1/2} \rangle^2$. From a straightforward calculation it follows

$$\langle (\Delta d_{1/2})^2 \rangle = \frac{i}{2} A_{0,2\mathbf{k}}^{\text{stat}} \sin(2\theta) \pm \frac{1}{4} \left(N_{0,0}^{\text{stat}} + N_{2\mathbf{k},2\mathbf{k}}^{\text{stat}} + 1 \right). \qquad (29)$$

Introducing (27) into (29) it follows the following expression for polariton squeezing

$$\langle (\Delta d_{1/2})^2 \rangle = \frac{\alpha A_{\mathbf{k},\mathbf{k}}^{\text{stat}} \sin(2\theta) \pm 1}{4 \left[1 - \alpha^2 (A_{\mathbf{k},\mathbf{k}}^{\text{stat}})^2 \right]}. \qquad (30)$$

For $2\theta = -\pi/2$, one has

$$\langle (\Delta d_1)^2 \rangle = \frac{1}{4 \left[1 + \alpha A_{\mathbf{k},\mathbf{k}}^{\text{stat}} \right]}. \qquad (31)$$

In Fig. 11.3 we present the logarithmic plot of $\langle (\Delta d_1)^2 \rangle$ as a function of the external pump field f_p. The percentage of squeezing, i.e. $s = 100 \left[1 - 2\sqrt{\langle (\Delta d_1)^2 \rangle} \right]$, saturates at 30% when f_p approaches 1. This value of the amount of squeezing is maintained until the coherence threshold $f_p = 2.45$ is approached. As soon as the coherence threshold is attained, squeezing rapidly disappears. This fact is related to the strong decrease of the anomalous correlation above threshold, because these correlations are responsible for the squeezing effect, as follows from (28). The feasibility of an experiment in which anomalous correlations and squeezing are measured, is related to the magnitude of the wave vector difference between signal and idler. In fact when the idler polariton is mostly exciton-like, the intensity of its emitted field is very small because its photon component expressed through the Hopfield coefficient $C_{2\mathbf{k}}$ is small. This makes the measurement of the anomalous correlations and therefore of the squeezing very difficult. The wave vector difference between idler and signal and thus the ratio of the intensities of the signal and idler fields may varied by a convenient choice of the position of the pump mode on the polariton dispersion. When this position is very close to the signal mode at $\mathbf{q} = 0$, the three modes are nearly degenerated and the intensities of the emitted fields associated with signal and idler become comparable. Squeezing is more easily observed in this configuration. The three-mode amplifier becomes in his case a

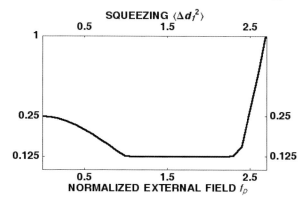

Fig. 11.3 Logarithmic plot of the variance $\langle (\Delta d_1)^2 \rangle$ as a function of the external pump field f_p.

degenerate amplifier, a one-mode system, described by the Hamiltonian (7a). In this degenerate case the Hamiltonian (21) has the simpler form

$$H = \hbar\omega_{LP}(0) P_0^\dagger P_0 + \hbar\omega_{int} P_0^{\dagger 2} P_0^2 + \hbar F_{pump} \exp[i\omega_{pump} t] P_0^\dagger + \text{h.c.} \quad (32)$$

As already mentioned, the degenerate parametric amplifier has been discussed in great detail in quantum optics. For non linear interacting polaritons, the degenerate parametric amplifier has been considered theoretically and experimentally [3, 5, 6, 28]. In particular in [5] the spectrum of squeezing has been measured explicitly in a homodyne configuration. The homodyne detection signal shows a periodic behavior as a function of the local oscillator. In order to calculate the spectrum of squeezing, in analogy with (8), we introduce the operators

$$d_{1/2}(\omega, \theta) = \frac{\sqrt{\pm 1}}{2} [e^{i\theta} p_0(\omega) \pm e^{-i\theta} p_0^\dagger(\omega)], \quad (33)$$

where $p_0(\omega)$ is the Fourier transform of $\tilde{p}_0(t)$ defined above.

The spectrum of squeezing is given in terms of (33) by the normal ordered combination [31, 32]

$$S_{1/2}(\omega, \theta) = \langle : \hat{d}_{1/2}(\omega) d_{1/2}(t=0) : + : d_{1/2}(t=0) \hat{d}_{1/2}(-\omega) : \rangle. \quad (34)$$

In Fig. 11.4 we present the squeezing of the lower polaritons branch, i.e. $1 + S_1(\omega, \theta)$, at $q = 0$, for a fixed frequency $\omega/\gamma = 0.2$ and as a function of the local oscillator phase θ. Values smaller than one indicate the occurrence of squeezing. This theoretical result is in qualitative agreement with recent experiments presented in [5].

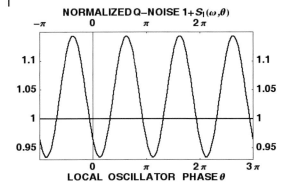

Fig. 11.4 Plot of the squeezing of the lower polaritons branch at $k=0$, i.e. $1+S_1(\omega,\theta)$, for a fixed frequency $\omega/\gamma = 0.2$ and as a function of the local oscillator phase θ. Squeezing occurs for $1+S_1(\omega,\theta)<1$.

Acknowledgements

We acknowledge illuminating discussions with A. Baas, C. Ciuti, B. Deveaud, E. Giacobino, R. Houdré, Z. Hradil, C. Piermarocchi, V. Savona, J.-L. Staehli, and C. Weisbuch.

References

[1] J. J. Hopfield, Phys. Rev. **112**, 1555 (1958).
[2] M. Artoni and J. L. Birman, Phys. Rev. B **44**, 3736 (1991).
[3] P. Schwendimann, C. Ciuti, and A. Quattropani, phys. stat. sol. (c) **1**, 470 (2004).
[4] P. Savona, P. Schwendimann, and A. Quattropani, Phys. Rev. B **71**, 125315 (2005).
[5] A. Baas, J. P. Karr, H. Eleuch, and E. Giacobino, Phys. Rev. A **69**, 023809 (2004).
[6] E. Giacobino, in: Proceedings of the International School of Physics E. Fermi, Course CL: Electron and Photon Confinement in Semiconductor Nanostructures, edited by B. Deveaud, A. Quattropani, and P. Schwendimann (IOS Press, Amsterdam, 2003), p. 167.
[7] L. Mandel and E. Wolf, Optical Coherence and Quantum Optics (Cambridge University Press, Cambridge, 1995).
[8] H. P. Yuen, Phys. Rev. A **13**, 2226 (1976).
[9] S. I. Pekar, Sov. Phys. JETP **6**, 785 (1958).
[10] E. Burnstein and F. D. Martini (Eds.), Polaritons (Pergamon Press, New York, 1974).
[11] E. Burnstein and C. Weisbuch (Eds.), Confined Electrons and Photons (Plenum Press, New York, 1995).
[12] A. Quattropani, L. C. Andreani, and F. Bassani, Nuovo Cimento D **7**, 55 (1986).
[13] P. Schwendimann, A. Quattropani, and Z. Hradil, Nuovo Cimento D **15**, 1421 (1993).
[14] P. Schwendimann and A. Quattropani, Europhys. Lett. **17**, 355 (1992).
[15] P. Schwendimann and A. Quattropani, Europhys. Lett. **18**, 281 (1992).
[16] R. Loudon and P. L. Knight, J. Mod. Opt. **34**, 709 (1987).
[17] Z. Hradil, A. Quattropani, V. Savona, and P. Schwendimann, J. Stat. Phys. **76**, 299 (1994).
[18] C. Weisbuch, M. Nishioka, A. Ishikawa, and Y. Arakawa, Phys. Rev. Lett. **69**, 3314 (1992).

[19] C. Ciuti, P. Schwendimann, and A. Quattropani, phys. stat. sol. (a) **190**, 305 (2002).
[20] C. Ciuti, P. Schwendimann, and A. Quattropani, Semicond. Sci. Technol. **18**, S279 (2003).
[21] Proceedings of the International School of Physics Enrico Fermi, Course CL: Electron and Photon Confinement in Semiconductor Nanostuctures, edited by B. Deveaud, A. Quattropani, and P. Schwendimann (IOS Press, Amsterdam, 2003).
[22] E. Hanamura and H. Haug, Phys. Rep. **33**, 209 (1977).
[23] G. Rochat, C. Ciuti, V. Savona, C. Piermarocchi, A. Quattropani, and P. Schwendimann, Phys. Rev. B **61**, 13856 (2000).
[24] C. Ciuti, P. Schwendimann, B. Deveaud, and A. Quattropani, Phys. Rev. B **62**, R4825 (2000).
[25] C. Ciuti, P. Schwendimann, and A. Quattropani, Phys. Rev. B **63**, 041303 (2001).
[26] P. G. Savvidis, J. J. Baumberg, R. M. Stevenson, M. S. Skolnick, D. M. Whittaker, and J. S. Roberts, Phys. Rev. Lett. **84**, 1547 (2000).
[27] C. Ciuti, P. Schwendimann, B. Deveaud, and A. Quattropani, phys. stat. sol. (b) **221**, 111 (2000).
[28] P. Schwendimann, C. Ciuti, and A. Quattropani, Phys. Rev. B **68**, 165324 (2003).
[29] A. Griffin, Phys. Rev. B **53**, 9341 (1996).
[30] H. Haug and S. W. Koch, Quantum theory of the optical and electronic properties of semiconductors (World Scientific, Singapore, 1996).
[31] M. J. Collett and D. F. Walls, Phys. Rev. A **32**, 2887 (1985).
[32] F. A. Hopf, P. Meystre, P. D. Drummond, and D. F. Walls, Opt. Commun. **31**, 245 (1979).

Chapter 12
High Efficiency Planar MCLEDs

Reto Joray, Ross P. Stanley, and Marc Ilegems

12.1
Introduction

Light-emitting diodes (LEDs) are promising candidates for high efficiency light sources, with modern devices showing internal quantum efficiencies of virtually 100%. However, due to the high refractive index of the commonly used semiconductor materials it is difficult to have a large extraction efficiency; in a standard cubic geometry most of the internally emitted light is trapped inside the device due to total internal reflection at the semiconductor–air interface. Planar non-nitride III–V semiconductor-based LEDs show extraction efficiencies for emission into air on the order of 2%. This problem was understood already in the 1960s [1–3] but was not addressed until the 1990s when the internal quantum efficiency improved sufficiently that LED performance was limited by the extraction efficiency [4].

Several methods have been developed in order to increase the extraction efficiency of a planar LED. They focus either on optimizing the device geometry in order to increase the escape cone or on modifying the internal spontaneous emission by placing the emitter inside an optical cavity with a thickness of the order of its emitting wavelength. The latter approach is called Microcavity LED (MCLED) or Resonant Cavity LED (RCLED). With high brightness LEDs based on the first approach drastic increases in efficiency could be achieved. These methods use either non-parallel surfaces and transparent substrates [5] or a combination of microcavity related effects [6–8] and geometrical effects [9, 10], even if they do not fall into the category of a standard microcavity.

In a MCLED the part of the internal emission that is extracted is increased via interference effects. Contrary to the other approaches this is possible without changing the device geometry and thus without additional costly back-end processing steps. The control of the far-field radiation pattern makes these devices particularly interesting for high brightness applications, which demand highly directional emitters, such as for printing, bar code reading, large area displays and optical communication.

12.2
Microcavities

12.2.1
Fundamentals

A Microcavity LED (MCLED) is a light-emitting diode, for which the light-emitting active region is placed in an optical cavity with a thickness of the order of the wavelength of emission [11, 12]. The optical cavity is in resonance with the internal emission, resulting in a modification of the spontaneous emission process, such that the internal emission is no longer isotropic. This results in an increased directionality and brightness and a higher spectral purity of the LED emission spectrum. If the cavity is properly tuned, the alteration of the intrinsic emission spectrum leads to an increase of the power emitted within the escape cone, leading to a higher extraction efficiency.

The theory of spontaneous emission from microcavities has been described in several excellent review articles, e.g. by Benisty et al. [13–15], Neyts [16], Wasey and Barnes [17], Delbeke et al. [18] and Baets et al. [19]. In the following a brief summary will be given.

Since the microcavities treated in this review article are optimized for maximum extraction efficiency, their quality factors are moderate. Therefore these devices are always operating in the weak coupling regime (the strong coupling regime is treated in detail elsewhere in this issue). As these structures are planar and the optical confinement is only one-dimensional, the total emission and hence the spontaneous lifetimes are only weakly affected [14, 20, 21]. The microcavity effect manifests itself thus mainly in a redistribution of the internal spontaneous emission spectrum. Finally, MCLEDs are often confused with Vertical Cavity Surface Emitting Lasers (VCSEL's) operating below threshold. However, due to the high mirror reflectivities, the extraction efficiency of spontaneous emission in VCSEL's is low.

A planar microcavity is composed of two mirrors having amplitude reflectivities of r_1 and r_2, respectively. A source in the middle of the cavity emits light at a given angle, θ, in both directions towards both mirrors (cf. Fig. 12.1). The light reflected from the back mirror (r_2) interferes with the light going towards the first mirror. The upward-radiated electric field outside the cavity can therefore be written as

$$E_{out}^{up} = E_0 \frac{(1-r_1)(1+r_2 e^{i\phi})}{1-r_1 r_2 e^{2i\phi}} \qquad (1)$$

where ϕ represents the phase shift and is equal to

$$\phi = \frac{2\pi n d \cos\theta}{\lambda} \qquad (2)$$

with n being the refractive index of the medium, d the distance between mirror 2 and the emitter, θ the angle of propagation and λ the wavelength.

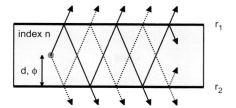

Fig. 12.1 Schematic microcavity of index n, limited by two mirrors with amplitude reflectivities r_1 and r_2, with an emitter inside the cavity emitting two series of waves.

If the emission pattern of the source, E_0, is known, then the emission on the outside is completely defined by the above equation. In the weak coupling regime the spontaneous emission of electron–hole pairs can be adequately represented by electrical dipoles [22]. The light extraction properties of structures of this type can be calculated with standard transfer-matrix techniques. The dipole emission is included simply as additive source terms [15].

In the quest for high external quantum efficiencies, the microcavity can be used to enhance the emission in the normal direction (or more correctly, into the escape cone). Assuming that the finesse of the cavity is reasonable, then the maximum extraction efficiency is limited to $1/m_c$, where m_c is the cavity order and $m_c = 2nL/\lambda$ with L being the cavity length. The increase of efficiency with decreasing cavity order is what makes a microcavity "micro". The smallest cavity is $\lambda/2$ and has an order of 1. In principle, all the light could be extracted from such a cavity. In practice the cavities are never as small as $\lambda/2$ because of material limitations.

A $\lambda/2$ cavity consisting of a high index material shows a cavity mode minimum in the center of the cavity. The maxima are located at the interfaces with the mirrors. The source layer would thus have to be situated at the boundary of the cavity. This is not feasible as an epitaxial interface is full of non-radiative recombination centers. A $\lambda/2$ low index cavity with a cavity mode maximum in the center of the cavity is not conceivable either as the Al-rich low refractive index layers cannot be grown with reasonably low impurity levels up to date. Furthermore, in case of a cavity with one or two metal mirrors, the use of a $\lambda/2$ cavity would imply that the active layer is very close to the metal, which can result in considerable losses due to non-radiative energy transfer from the dipole to the absorptive metal [23]. Moreover a metal mirror requires a heavily doped contact layer with a thickness of several tens of nanometers to ensure good electrical contact. When insulating dielectric mirrors are used, thick (usually $\lambda/4$) and highly conductive lateral current injection layers are needed. Quantum wells are typically sandwiched in graded index separate confinement heterostructures (GRINSCH) with a total thickness of the order of $\lambda/2$. Due to all these reasons λ high index cavities are generally used.

An additional increase of the optical cavity length occurs with the common use of distributed Bragg reflectors (DBR's) rather than ideal mirrors. The optical field penetrates quite deeply into these mirrors, depending on the refractive index contrast between the two mirror materials, $\Delta n = n_{\text{high}} - n_{\text{low}}$. The effective order of the cavity includes the penetration into the two reflectors, $m_c^{\text{eff}} = m_c + \sum_{i=1,2} \Delta m_i^{\text{pen}}$. The penetration depth originates from the phase change of the mirror on reflection which is a function of wavelength and angle, whereas the angular-dependent contribution Δm_θ^{pen} is the dominant one [18, 24]. It can be approximated as

$$\Delta m_\theta^{pen} \approx \frac{n^3}{4\Delta n}\left(\frac{1}{n_{high}^2} + \frac{1}{n_{low}^2}\right) \quad (3)$$

with n being the refractive index of the cavity. In order to maximise efficiency, within technological limitations Δm_θ^{pen} should be minimized.

There has been some confusion over the correct formula for the penetration depth of DBRs. When calculating the Fabry–Perot linewidth, the change in phase with wavelength is important. When calculating the angular extraction efficiency it is the change of phase with angle that plays a role. In the centre of the stopband the phase change is linear in λ or in $\cos\theta$ which can be associated with a length (penetration depth) in the two cases. In the common situation of low index contrast mirrors, the two formulas are the same.

Improvements in terms of external quantum efficiency have been achieved by minimising m_c^{eff} by borrowing a concept from distributed feedback lasers, the phase-shift (PS) cavity, and by maximising the index contrast in the reflector with oxide mirrors [25]. We continue with a review of standard microcavity LEDs and mention the state-of-the-art results achieved using phase-shift cavities and oxide mirrors.

12.2.2
State of the Art Planar Semiconductor MCLEDs

The first electrically pumped Microcavity LED was realized in 1992 in the GaAs–AlGaAs material system by Schubert et al. [11]; it shows an emission peak at 862 nm. The top emitting device consists of an asymmetric DBR/cavity/metal mirror structure, where the DBR is made of AlAs/Al$_{0.14}$Ga$_{0.86}$As periods and the thin transparent metal reflector of silver (Ag) or silver in combination with cadmium tin oxide (Ag/CdSnO$_x$). Since then MCLEDs were realized in different material systems, covering a large range of emission wavelengths. In the following a brief overview will be given on the state of the art of current-injected planar MCLEDs, divided into different wavelength ranges, from the visible to the infrared. For each range typical fields of application, material systems and efficiencies are listed. The maximum external quantum efficiencies as a function of emission wavelength are summarized in Table 12.1 and Fig. 12.2.

12.2.2.1 400–500 nm
InGaN–GaN-based MCLEDs are under development. Even though the light extraction is less of a problem than for the other material systems due to the low index of GaN ($n \approx 2.5$), microcavities are nevertheless an interesting object of study on the way to the realisation of nitride VCSEL's. The AlGaN–GaN material system is commonly used to realize DBR's but is known to suffer from the large lattice mis-

Table 12.1 Overview experimental results planar MCLEDs for different wavelengths, with standard or phase-shift (PS) cavities. Oxide DBR's are labelled AlO$_x$ for aluminum oxide as the low index constituent.

Wavelength (nm)	Material system	Type (↓ or ↑)	Efficiency (%)	Diameter (µm)	Ref.
460	InGaN/GaN/AlInN + Ag/Au	↓	2.6	100	[28]
650	AlGaInP/AlGaAs	↑	9.6/12	300/700	[9, 35]
650	AlGaInP/AlGaAs/AlO$_x$	↑	12.5	200	[36]
650	AlGaInP/AlGaAs + ODR	↑	18/23 (epoxy)	300	[37]
850	GaAs/AlGaAs	↑	14.6	180	[39]
880	GaAs/AlGaAs	↑	16 (epoxy)	80	[40]
970	InGaAs/AlGaAs + Au (PS cavity)	↓	18	150–400	[25]
970	InGaAs/AlGaAs	↑	10	400	[43]
970	InGaAs/AlGaAs (PS cavity)	↑	19	400	[44]
970	InGaAs/AlGaAs/AlO$_x$ (PS cavity)	↑	28	350	[46]
980	InGaAs/GaAs/AlO$_x$ + SiO$_2$/ZnSe	↑	23 (est.)	6	[45]
995	InGaAs/AlGaAs + Au	↓	17/23	85/1.5 mm	[41]
1300	InGaAsP/InP + Au	↓	9	2 mm	[47]
1550	InGaAsP/InP + Au	↓	6.8	85	[48]

match between AlN and GaN. Bragg mirrors with a high aluminum content (>30%) are subject to crack formation or in-plane lattice relaxation. AlGaN/GaN reflectors with a low Al content on the other hand suffer from a low index contrast. Furthermore the lack of a simple epitaxial lift-off technique prevents the realisation of short cavity metal-bound structures so far.

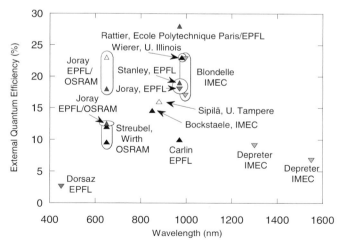

Fig. 12.2 Maximum external quantum efficiency of MCLEDs vs. emission wavelength for top (▲) and bottom emitting (▼) devices; for emission into air (solid symbols) and epoxy (empty symbols).

Recently promising results have been presented on AlInN/GaN DBRs [26, 27]. $Al_{1-x}In_xN$ with 17% indium is lattice-matched to GaN while presenting a reasonable refractive index contrast. However the growth of high quality AlInN was hampered by phase separation issues so far. Preliminary bottom emitting MCLEDs were realized comprising a λ-cavity with five InGaN QWs, an Ag/Au top metallic mirror and a 12 pair $Al_{0.82}In_{0.18}N$/GaN bottom DBR. Devices of this type with an emission area of 100×100 µm² are emitting at 460 nm and show an external quantum efficiency of 2.6% [28]. More details on III-nitride materials can be found elsewhere in this issue.

12.2.2.2 650 nm

Apart from lighting applications, 650 nm emitting MCLEDs are commercially important for plastic optical fiber (POF) based communication, since they show a higher brightness and modulation bandwidth than conventional LEDs [29, 30]. The material system of choice for the cavity is AlGaInP, in combination with AlGaAs DBRs and current spreading and current injection layers. The attainable efficiencies in the red are limited to lower values than for near-IR MCLEDs due to several limiting factors; lower barriers and QW confinement potentials, increased resistivities and a reduced index contrast in the DBRs [25, 29, 31–34]. With red MCLEDs wall-plug efficiencies η_{wp} of 10% and external quantum efficiencies η_{ext} of 9.6% were achieved with (300 µm)² devices [35]. Due to their low forward voltages the external quantum efficiencies are slightly smaller than the wall-plug efficiencies. The structure consists of a λ-cavity with 5 compressively strained GaInP QWs, a 34 period $Al_{0.5}Ga_{0.5}As/Al_{0.95}Ga_{0.5}As$ bottom DBR and a 6 period top DBR. For larger (700 µm)² devices of the same type a maximum wall-plug efficiency of 12% could be shown [9].

2λ-cavity devices with a single GaInP QW, a 3.5 pair oxide bottom DBR and the interface GaAs–air as outcoupling reflector display external quantum efficiencies of 12.5% for 200 µm diameter devices, despite the fact that the DBR is only partially oxidized and the cavity detuning is not optimal [36].

Recently a new device design was presented which corresponds to a combination of a MCLED with a thin-film structure. By adding an omnidirectional reflector (ODR) to a 1λ AlGaInP cavity with 5 GaInP QWs external quantum efficiencies of 23% and 18% could be achieved with and without encapsulation, respectively, despite a non-ideal detuning [37]. The cavity is delimited by a 6 pair AlGaAs/AlAs outcoupling DBR and the ODR. ODRs generally consist of a DBR pair and a dielectric-coated metal mirror and show a high angle-averaged reflectivity [38].

12.2.2.3 850–880 nm

The target applications of 850–880 nm devices are Ethernet data links, remote controls and infrared communication as regulated by the Infrared Data Association (IrDA). This is mainly due to the availability of low-cost Si-based detectors in this wavelength range. The obvious material system for this emission region is GaAs–

AlGaAs. An external quantum efficiency of 14.6% was achieved with an 850 nm top emitting device [39]. It consists of a λ-cavity surrounded by two DBR's, with the bottom mirror being a 20 pair $Al_{0.15}Ga_{0.85}As/AlAs$ DBR and the top one a 1.5 pair $Al_{0.15}Ga_{0.85}As/AlAs$ DBR. The current is confined with an oxide aperture of 180 µm in diameter. With another structure emitting at 880 nm and an emission window of 80 µm an efficiency η_{ext}^{epoxy} of 16% was demonstrated after encapsulation. The λ-cavity was sandwiched between a 20 pair bottom DBR and a 5–7 pair top DBR, both consisting of $Al_{0.2}Ga_{0.8}As/Al_{0.9}Ga_{0.1}As$ [40].

12.2.2.4 970–1000 nm

Even though no large scale application exists so far for 980 nm near infrared devices the availability of high quality InGaAs/GaAs strained QWs makes these devices ideal for the proof of principle of new concepts. The InGaAs active material is used in conjunction with the GaAs–AlGaAs material system. Compared to higher bandgap emitters, this combination has the advantage that the GaAs substrate is transparent in this emission wavelength range. Hence highly efficient bottom emitting devices have been realized. With an asymmetric GaAs/AlAs DBR/λ-cavity/Au mirror structure efficiencies η_{ext} of 17% and 23% were shown for 85 µm and 1.5 mm diameter devices, respectively [41]. The active region consists of three strained InGaAs QWs and the increase in efficiency with device size is attributed to an increased photon recycling. Similar bottom emitting structures comprising a phase-shift cavity with a single InGaAs QW and a 3.5 pair outcoupling DBR show maximum external quantum efficiencies of 18% into air for device diameters ranging from 150 to 400 µm [25].

Top emitting monolithic devices containing three InGaAs QWs and GaAs/AlAs DBRs with 15.5 pairs at the bottom and three pairs at the top led to a maximum external quantum efficiency of 10% [42, 43]. By incorporating a phase-shift cavity with a single InGaAs QW, delimited by a 15.5 pair bottom DBR and the interface GaAs–air at the top, this value could be increased to 19% [44].

With the use of a 6.5 pair high index contrast $GaAs/AlO_x$ bottom DBR and a single period $SiO_2/ZnSe$ dielectric top DBR, a differential quantum efficiency η_{ext}^{diff} of 27% could be demonstrated for top emitting devices [45], corresponding to an absolute external quantum efficiency of approximately 23%. The device contains a λ-cavity, a tunnel contact junction and an aperture of 6 µm in diameter only. Top emitting MCLEDs with a phase-shift cavity, a single InGaAs QW, a 3.5 pair oxide DBR at the bottom and the interface GaAs–air at the top show maximum external quantum efficiencies of 28% for emission into air [46]. This value corresponds to the highest ever reported efficiency for a microcavity LED.

12.2.2.5 1300–1550 nm

Silica optical fibers show attenuation minima around 1300 and 1550 nm and therefore these two wavelengths are the pre-eminent communication windows for telecom (see for example [4]). The principal material system is InGaAsP–InP. In addition to the broader intrinsic emission spectrum of long-wavelength devices, the low refractive index contrasts limit the maximum efficiency of long-wavelength MCLEDs, similar to the case of VCSELs.

Bottom emitting 1300 nm large diameter devices (2 mm) with a peak efficiency η_{ext} of 9% are reported using an asymmetric DBR/λ-cavity/Au mirror structure with three strained $InGa_{0.12}As_{0.56}P$ QWs [47]. The DBR consists of 5.5 pairs of $InGa_{0.23}As_{0.50}P/InP$.

An InP-based MCLED of 85 µm in diameter emitting at 1550 nm with a 6.8% external quantum efficiency is presented in [48]. The device is bottom emitting and consists of a DBR/cavity/Au mirror structure as well. The active region is made up of three $In_{0.84}Ga_{0.16}As_{0.74}P_{0.26}$ QWs and the outcoupling DBR of 12 pairs of $InGa_{0.38}As_{0.82}P/InP$.

12.2.2.6 Mid-IR

Mid infrared emitters are mainly used for gas detection. Several types of MCLEDs based on mercury cadmium telluride (HgCdTe) have been realized with emission wavelengths at 3.3, 3.7, 4.2 and 4.7 µm, adapted for the detection of different gases [49–52]. The structures are grown on cadmium zinc telluride (CaZnTe) substrates and the dielectric DBRs are made of zinc sulfide (ZnS) and yttrium fluoride (YF_3).

12.3
Novel Concepts

The majority of MCLEDs use standard technology inherited from VCSEL technology. However, it is clear from Table 1 that the highest efficiencies have been achieved with two additional extraction techniques, the phase shift cavity and oxide DBRs; or with a combination with a thin-film structure.

12.3.1
Phase-Shift Cavity

Generally MCLEDs comprise a λ cavity, consisting of a high refractive index (low bandgap) material, which shows three antinodes of the optical field within the cavity (cf. Fig. 12.3). For optical purposes a $\lambda/2$ low index cavity with a single antinode would be preferable, due to the reduced cavity order and the absence of guided modes. However this design is not practical for technical reasons. An efficient electrical in-jection demands a low bandgap active layer (high refractive index). In addition, in the InGaAs–AlGaAs material system defects associated with the low index AlAs layers would lower the internal quantum efficiency.

A standard resonant cavity is formed between two reflectors by introducing a phase shift equal to a multiple of π, translating to an optical cavity length equal to an integral times $\lambda/2$. From distributed feedback (DFB) laser theory it is known that this is not the only way to induce Fabry–Perot modes [53]. An alternative appro-

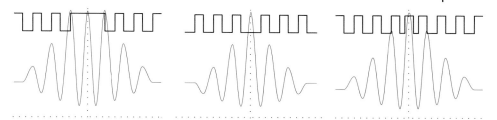

Fig. 12.3 Schematic refractive index profile and optical field of the Fabry–Perot mode for the three different cavity structures; a λ high index cavity (left), a $\lambda/2$ low index cavity (middle) and a phase-shift cavity (right). The cavities are surrounded by $\lambda/4$ layers representing two DBRs. The dotted vertical line denotes the source position.

ach to achieve the same effect is to place two $\lambda/8$ low index phase-shift layers around a $\lambda/4$ high refractive index layer containing the active region, thereby forming a virtual $\lambda/2$ cavity [44]. This structure is called phase-shift (PS) cavity and approaches the optical properties of a $\lambda/2$ low index cavity while keeping the preferential practical aspects of a λ high index cavity. A phase-shift cavity shows a decreased effective cavity length and a reduced coupling to guided modes, allowing significantly higher extraction efficiencies.

These types of phase layers are commonly used in DFB lasers to create a single mode in the center of the DFB stopband [54]. However, for DFB lasers the phase shift layers are placed far apart in order to distribute the optical field throughout the laser structure and reduce the spatial hole burning [53, 55], while in these microcavities the phase shift layers are acting to concentrate the optical field.

In Fig. 12.3 the optical field of the Fabry–Perot mode as a function of position is compared for the three different cavity designs, a standard λ high index cavity, a $\lambda/2$ low index cavity and a phase-shift cavity. The optical properties of these different cavity designs are simulated for the case of the AlGaAs system by assuming $n_{high} = 3.5$ for GaAs and $n_{low} = 2.9$ for AlAs and the cavities being surrounded by two DBRs consisting of $\lambda/4$ layers. The field has a maximum at the center of the cavity in each case and the penetration of the optical field into the DBRs is apparent. The penetration is minimum for the $\lambda/2$ low index cavity and the phase-shift cavity shows a significant improvement compared to the standard λ high index cavity.

Furthermore metal-bound phase-shift cavity devices show reduced metal absorption losses compared to standard structures, even though the source is closer to the metal mirror. This originates again from the stronger overlap of the field with the source.

Near infrared bottom and top emitting MCLEDs including a phase-shift cavity have been realised which has led to a considerable increase in external quantum efficiency (see Table 12.1) [25, 44, 46]. The phase-shift cavity principle cannot be applied to red emitting devices, as in case of the GaInP–AlGaInP material system the confinement potentials are lower [32] and therefore the barriers cannot be kept as thin.

12.3.2
Oxide DBR

The properties of a DBR mainly depend on the refractive index difference between its two constituents, Δn. Unfortunately the range of refractive indices accessible with epitaxially grown materials is very limited. With dielectric DBRs high index contrasts are possible, however only by paying the high price of a more complicated device design and fabrication. In 1990 a method was found which allows to significantly decrease the refractive index of the low index material. Dallesasse et al. discovered that high Al-content AlGaAs layers can be selectively oxidized at elevated temperatures, producing a mechanically stable oxide [56, 57] which shows good insulating properties and a low refractive index of approximately 1.6 [58–61].

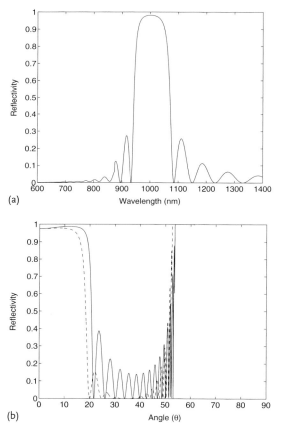

Fig. 12.4 Simulated spectral stopband (a) and angular stopband (b) around 980 nm of a 15.5 pair GaAs/AlAs DBR with $\lambda_{Bragg} = 980$ nm; TE polarization (solid line) and TM polarization (dashed line).

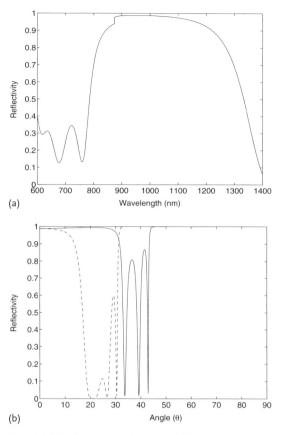

Fig. 12.5 Simulated spectral stopband (a) and angular stopband (b) around 980 nm of a 3.5 pair GaAs/AlO$_x$ DBR with λ_{Bragg} = 980 nm; TE polarization (solid line) and TM polarization (dashed line). Intensity jump at the absorption band edge of GaAs at 870 nm. The reflectivity drop between 30° and 50° for TE polarization is related to losses caused by an evanescent coupling with the substrate.

With this technique high index contrast GaAs/AlO$_x$ DBRs can be formed which show several advantages over conventional semiconductor DBRs. Contrary to dielectric DBRs, devices with so-called oxide DBRs can be fabricated in a single epitaxial growth step. A drastically reduced number of DBR pairs is necessary to achieve a certain reflectivity [60, 62]. Thanks to the high refractive index difference oxide DBRs show a significantly larger spectral and angular stopband, as is obvious from Figs. 12.4 and 12.5. Light emitted at oblique angles larger than the angular stopband edge escapes towards the substrate. These optical modes are called leaky modes. With the increase of the angular stopband width the leaky mode fraction can thus be reduced, in favor of the guided mode fraction. Guided modes occur when the cavity acts as a lateral waveguide. The light is then either extracted laterally or reabsorbed in the active region. Part of the reabsorbed light is extracted by re-emission into the escape window. This contribution is denoted photon recycling and can lead to a significant increase in external quantum efficiency [63].

Figures 12.4 and 12.5 illustrate the difference in spectral and angular stopband width between a 15.5 pair GaAs/AlAs DBR and a 3.5 pair GaAs/AlO$_x$ DBR, both centered at 980 nm. The intensity jump at 870 nm for the simulated curves is related to the absorption band edge of GaAs. Evanescent coupling with the substrate may cause an extension of the low reflectivity region to angles larger than θ_l, the angle for total internal reflection at the first DBR interface.

More importantly, the high refractive index contrast leads to a drastically reduced penetration depth, allowing smaller effective cavity lengths and therefore higher extraction efficiencies [13, 14]. More detailed calculations show that the angular penetration depth is minimal for $n_{low} \approx \sqrt{n_{high}}$ [18].

In Table 12.1 it can be seen that near infrared emitting MCLEDs with a GaAs/AlO$_x$ DBR [45, 46] and red emitting MCLEDs with a AlGaAs/AlO$_x$ DBR [36] show significantly higher external quantum efficiencies compared to the corresponding semiconductor DBR devices. Yet the implementation of an oxide DBR has implications on device design and fabrication. Since oxide DBRs are insulating, they need to be used in combination with a lateral intracavity contact, whereby care must be taken to ensure that the conductivity of the current injection layer is not reduced by the oxidation process. In addition the vertical contraction of the layers during oxidation has to be taken into account [64–66].

This concept of selective lateral oxidation has been successfully applied to other material systems as well, in view of high index contrast DBRs for other wavelength ranges. The selective oxidation of high Al content AlGaInP versus GaInP could be of use for the realisation of visible DBRs [67]. In addition, recently the selective anodic oxidation of AlInN over GaN has been demonstrated [68]. Possible candidates for the realisation of high refractive index contrast DBRs in the 1550 nm telecom wavelength range are the oxides of AlAsSb [69–72], InAlAs [73–77] and AlInAsP [78], all lattice-matched to InP. However, to our knowledge, no MCLEDs with oxide DBRs of this type have been realised so far.

12.4
Conclusions

The thorough study of the physics in optical microcavities in the 1990s led to the realisation of microcavity LEDs. Over the past decade this design has been applied to various material systems and emission wavelength ranges. The performance of these emitters has been continuously improved, leading to maximum external quantum efficiencies up to 30%, more than an order of magnitude higher than for standard planar LEDs. The highest efficiencies have been achieved with the implementation of two novel concepts, phase-shift cavities and oxide DBRs. MCLEDs with a directional emission are ideal candidates for highly focussed emitters with high coupling efficiencies, in particular for low cost, short-haul optical communication systems. Furthermore higher modulation bandwidths can be achieved with MCLEDs. First commercial devices are available now in the red and near infrared from several manufacturers.

In order to surpass the current efficiency limits for MCLEDs a further evolution of the epitaxial growth techniques is necessary. With a better control of the doping profiles and the availability of high index contrast, epitaxially grown DBRs the effective cavity length could be further reduced. Hybrid forms such as combinations of MCLEDs with thin-film structures currently seem to have the highest potential for improvement, especially in the visible wavelength range.

References

[1] W. N. Carr, Infrared Phys. **6**, 1 (1966).
[2] W. N. Carr and G. E. Pittman, Appl. Phys. Lett. **3**, 173 (1963).
[3] A. R. Franklin and R. Newman, J. Appl. Phys. **35**, 1153 (1964).
[4] E. F. Schubert, Light-Emitting Diodes (Cambridge University Press, Cambridge, 2003).
[5] M. R. Krames, M. Ochiai-Holcomb, G. E. Höfler, C. Carter-Coman, E. I. Chen, I.-H. Tan, P. Grillot, N. F. Gardner, H. C. Chui, J.-W. Huang, S. A. Stockman, F. A. Kish, M. G. Craford, T. S. Tan, C. P. Kocot, M. Hueschen, J. Posselt, B. Loh, G. Sasser, and D. Collins, Appl. Phys. Lett. **75**, 2365 (1999).
[6] R. Windisch, C. Rooman, S. Meinlschmidt, P. Kiesel, D. Zipperer, G. H. Dohler, B. Dutta, M. Kuijk, G. Borghs, and P. Heremans, Appl. Phys. Lett. **79**, 2315 (2001).
[7] R. Windisch, C. Rooman, S. De Jonge, C. Karnutsch, K. Streubel, M. Kuijk, B. Dutta, G. Borghs, and P. Heremans, in: Light-Emitting Diodes: Research, Manufacturing, and Applications VI, Proc. SPIE **4641**, 13–18 (2002).
[8] C. Rooman, R. Windisch, M. D'Hondt, B. Dutta, P. Modak, P. Mijlemans, G. Borghs, R. Vounckx, I. Moerman, M. Kuijk, and P. Heremans, Electron. Lett. **37**, 852 (2001).
[9] K. Streubel, N. Linder, R. Wirth, and A. Jaeger, IEEE J. Sel. Top. Quantum Electron. **8**, 321 (2002).
[10] T. Whitaker and R. Stevenson, Compound Semicond. **10**, 18 (2004).
[11] E. F. Schubert, Y.-H. Wang, A. Y. Cho, L.-W. Tu, and G. J. Zydzik, Appl. Phys. Lett. **60**, 921 (1992).
[12] E. F. Schubert, N. E. J. Hunt, M. Micovic, R. J. Malik, D. L. Sivco, A. Y. Cho, and G. J. Zydzik, Science **265**, 943 (1994).
[13] H. Benisty, H. De Neve, and C. Weisbuch, IEEE J. Quantum Electron. **34**, 1612 (1998).
[14] H. Benisty, H. De Neve, and C. Weisbuch, IEEE J. Quantum Electron. **34**, 1632 (1998).
[15] H. Benisty, R. Stanley, and M. Mayer, J. Opt. Soc. Am. A **15**, 1192 (1998).
[16] K. A. Neyts, J. Opt. Soc. Am. A **15**, 962 (1998).
[17] J. A. E. Wasey and W. L. Barnes, J. Mod. Opt. **47**, 725 (2000).
[18] D. Delbeke, R. Bockstaele, P. Bienstman, R. Baets, and H. Benisty, IEEE J. Sel. Top. Quantum Electron. **8**, 189 (2002).
[19] R. G. Baets, D. Delbeke, R. Bockstaele, and P. Bienstman, in: Light-Emitting Diodes: Research, Manufacturing, and Applications VII, Proc. SPIE **4996**, 74–86 (2003).
[20] H. Yokoyama, Science **256**, 66 (1992).
[21] I. Abram, I. Robert, and R. Kuszelewicz, IEEE J. Quantum Electron. **34**, 71 (1998).
[22] W. Lukosz, J. Opt. Soc. Am. **71**, 744 (1981).
[23] Z. Huang, C. C. Lin, and D. G. Deppe, IEEE J. Quantum Electron. **29**, 2940 (1993).
[24] R. J. Ram, D. I. Babic, R. A. York, and J. E. Bowers, IEEE J. Quantum Electron. **31**, 399 (1995).
[25] R. Joray, Ph.D. Thesis, Swiss Federal Institute of Technology, Lausanne (EPFL), 2005. Available online: http://library.epfl.ch/theses/?nr=3170

[26] J.-F. Carlin and M. Ilegems, J. Appl. Phys. **83**, 668 (2003).
[27] E. Feltin, R. Butté, J.-F. Carlin, J. Dorsaz, N. Grandjean, and M. Ilegems, Electron. Lett. **41**, 94 (2005).
[28] J. Dorsaz, J.-F. Carlin, S. Gradecak, and M. Ilegems, J. Appl. Phys. **97**, 084505 (2005).
[29] M. M. Dumitrescu, M. J. Saarinen, M. D. Guina, and M. V. Pessa, IEEE J. Sel. Top. Quantum Electron. **8**, 219 (2002).
[30] K. Streubel and R. Stevens, Electron. Lett. **34**, 1862 (1998).
[31] W. W. Chow, K. D. Choquette, M. H. Crawford, K. L. Lear, and G. R. Hadley, IEEE J. Quantum Electron. **33**, 1810 (1997).
[32] R. P. Schneider Jr., J. A. Lott, M. H. Crawford, and K. D. Choquette, Int. J. High Speed Electron. Syst. Devices **5**, 625 (1994).
[33] P. W. Epperlein, G. L. Bona, and P. Roentgen, Appl. Phys. Lett. **60**, 680 (1992).
[34] S. Adachi, J. Appl. Phys. **54**, 1844 (1983).
[35] R. Wirth, C. Karnutsch, S. Kugler, and K. Streubel, IEEE Photonics Technol. Lett. **13**, 421 (2001).
[36] R. Joray, J. Dorsaz, R. P. Stanley, M. Ilegems, M. Rattier, C. Karnutsch, and K. Streubel, in: Proc. of the 29th Int. Symp. on Compound Semiconductors (ISCS), Lausanne, Switzerland, 7–10 October 2002, Inst. Phys. Conf. Ser. **174**, 363–366 (2003).
[37] R. Joray, M. Ilegems, R. Stanley, W. Schmid, R. Butendeich, R. Wirth, A. Jaeger, and K. Streubel, in: Physics and Applications of Optoelectronic Devices, Proc. SPIE **5594**, 190–198 (2004).
[38] T. Gessmann, E. F. Schubert, J. W. Graff, K. Streubel, and C. Karnutsch, IEEE Electron Device Lett. **24**, 683 (2003).
[39] R. Bockstaele, J. Derluyn, C. Sys, S. Verstuyft, I. Moerman, P. Van Daele, and R. Baets, Electron. Lett. **35**, 1564 (1999).
[40] P. Sipilä, M. Saarinen, V. Vilokkinen, S. Orsila, P. Melanen, P. Savolainen, M. Toivonen, M. Dumitrescu, and M. Pessa, in: Light-Emitting Diodes: Research, Manufacturing, and Applications IV, Proc. SPIE **3938**, 82–89 (2000).
[41] J. Blondelle, H. De Neve, G. Borghs, P. Van Daele, P. Demeester, and R. Baets, in: Colloquium on Semiconductor Optical Microcavity Devices and Photonic Bandgaps, Proc. IEE (1996).
[42] J.-F. Carlin, P. Royo, M. Ilegems, B. Gerard, X. Marcadet, and J. Nagle, J. Cryst. Growth **201/202**, 994 (1999).
[43] J.-F. Carlin, P. Royo, R. P. Stanley, R. Houdré, J. Spicher, U. Oesterle, and M. Ilegems, Semicond. Sci. Technol. **15**, 145 (2000).
[44] R. P. Stanley, P. Royo, U. Oesterle, R. Joray, and M. Ilegems, in: Proc. of the 29th Int. Symp. on Compound Semiconductors (ISCS), Lausanne, Switzerland, 7–10 October 2002, Inst. Phys. Conf. Ser. **174**, 359–362 (2003).
[45] J. J. Wierer, D. A. Kellogg, and N. Holonyak Jr., Appl. Phys. Lett. **74**, 926 (1999).
[46] M. Rattier, H. Benisty, R. P. Stanley, J.-F. Carlin, R. Houdré, U. Oesterle, C. J. M. Smith, C. Weisbuch, and T. F. Krauss, IEEE J. Sel. Top. Quantum Electron. **8**, 238 (2002).
[47] B. Depreter, I. Moerman, R. Baets, P. Van Daele, and P. Demeester, Electron. Lett. **36**, 1303 (2000).
[48] B. Depreter, S. Verstuyft, I. Moerman, R. Baets, and P. Van Daele, in: Proc. of the 11th Int. Conf. on InP and related materials (IPRM), Davos, Switzerland, May 1999 (IEEE, Pennington, NJ, 1999), pp. 227–230.
[49] E. Hadji, J. Bleuse, N. Magnea, and J. L. Pautrat, Appl. Phys. Lett. **67**, 2591 (1995).
[50] E. Hadji, J. Bleuse, N. Magnea, and J. L. Pautrat, Solid-State Electron. **40**, 473 (1996).
[51] E. Hadji, E. Picard, C. Roux, E. Molva, and P. Ferret, Opt. Lett. **25**, 725 (2000).
[52] J. P. Zanatta, F. Noel, P. Ballet, N. Hdadach, A. Million, G. Destefanis, E. Mottin, C. Kopp, E. Picard, and E. Hadji, J. Electron. Mater. **32**, 602 (2003).
[53] G. P. Agrawal, J. E. Geusic, and P. J. Anthony, Appl. Phys. Lett. **53**, 178 (1988).
[54] H. A. Haus and C. V. Shank, IEEE J. Quantum Electron. **12**, 532 (1976).

[55] H. Soda, Y. Kotaki, H. Sudo, H. Ishikawa, S. Yamakoshi, and H. Imai, IEEE J. Quantum Electron. **23**, 804 (1987).
[56] J. M. Dallesasse, P. Gavrilovic, N. Holonyak, R. W. Kaliski, D. W. Nam, E. J. Vesely, and R. D. Burnham, Appl. Phys. Lett. **56**, 2436 (1990).
[57] J. M. Dallesasse, N. Holonyak Jr., A. R. Sugg, T. A. Richard, and N. El-Zein, Appl. Phys. Lett. **57**, 2844 (1990).
[58] A. Bek, A. Aydinli, J. G. Champlain, R. Naone, and N. Dagli, IEEE Photonics Technol. Lett. **11**, 436 (1999).
[59] K. J. Knopp, R. P. Mirin, D. H. Christensen, K. A. Bertness, A. Roshko, and R. A. Synowicki, Appl. Phys. Lett. **73**, 3512 (1998).
[60] M. H. MacDougal, H. Zhao, P. D. Dapkus, M. Ziari, and W. H. Steier, Electron. Lett. **30**, 1147 (1994).
[61] A. R. Sugg, E. I. Chen, N. Holonyak Jr., K. C. Hsieh, J. E. Baker, and N. Finnegan, J. Appl. Phys. **74**, 3880 (1993).
[62] M. H. MacDougal, P. D. Dapkus, V. Pudikov, H. Zhao, and G. M. Yang, IEEE Photonics Technol. Lett. **7**, 229 (1995).
[63] H. De Neve, J. Blondelle, P. Van Daele, P. Demeester, R. Baets, and G. Borghs, Appl. Phys. Lett. **70**, 799 (1997).
[64] P. W. Evans, J. J. Wierer, and N. Holonyak, J. Appl. Phys. **84**, 5436 (1998).
[65] M. H. MacDougal and P. D. Dapkus, IEEE Photonics Technol. Lett. **9**, 884 (1997).
[66] K. D. Choquette, K. M. Geib, H. C. Chui, B. E. Hammons, H. Q. Hou, T. J. Drummond, and R. Hull, Appl. Phys. Lett. **69**, 1385 (1996).
[67] F. A. Kish, S. J. Caracci, N. Holonyak, K. C. Hsieh, J. E. Baker, S. A. Maranowski, A. R. Sugg, J. M. Dallesasse, R. M. Fletcher, C. P. Kuo, T. D. Osentowski, and M. G. Craford, J. Electron. Mater. **21**, 1133 (1992).
[68] J. Dorsaz, H.-J. Bühlmann J.-F. Carlin, N. Grandjean and M. Ilegems, submitted to Appl. Phys. Lett.
[69] M. H. M. Reddy, D. A. Buell, A. S. Huntington, T. Asano, R. Koda, D. Feezell, D. Lofgreen, and L. A. Coldren, Appl. Phys. Lett. **82**, 1329 (2003).
[70] S. K. Mathis, K. H. A. Lau, A. M. Andrews, E. M. Hall, G. Almuneau, E. L. Hu, and J. S. Speck, J. Appl. Phys. **89**, 2458 (2001).
[71] O. Blum, M. J. Hafich, J. F. Klem, K. Baucom, and A. Allerman, Electron. Lett. **33**, 1097 (1997).
[72] O. Blum, K. M. Geib, M. J. Hafich, J. F. Klem, and C. I. H. Ashby, Appl. Phys. Lett. **68**, 3129 (1996).
[73] J. H. Shin, B. S. Yoo, W. S. Han, O. K. Kwon, Y. G. Ju, and J. H. Lee, IEEE Photonics Technol. Lett. **14**, 1031 (2002).
[74] B. Koley, O. King, F. G. Johnson, S. S. Saini, and M. Dagenais, Appl. Phys. Lett. **78**, 64 (2001).
[75] H. Gebretsadik, O. Qasaimeh, H. T. Jiang, P. Bhattacharya, C. Caneau, and R. Bhat, J. Lightwave Technol. **17**, 2595 (1999).
[76] H. Gebretsadik, K. Kamath, W. D. Zhou, P. Bhattacharya, C. Caneau, and R. Bhat, Appl. Phys. Lett. **72**, 135 (1998).
[77] H. Takenouchi, T. Kagawa, Y. Ohiso, T. Tadokoro, and T. Kurokawa, Electron. Lett. **32**, 1671 (1996).
[78] M. Linnik and A. Christou, Mater. Sci. Eng. B, Solid-State Mater. Adv. Technol. **80**, 245 (2001).

Chapter 13
Progresses in III-Nitride Distributed Bragg Reflectors and Microcavities Using AlInN/GaN Materials

Jean-François Carlin, Christoph Zellweger, Julien Dorsaz, Sylvain Nicolay, Gabriel Christmann, Eric Feltin, Raphael Butté, and Nicolas Grandjean

13.1
Introduction

The importance of III-nitride (III–N) materials for optoelectronic, as well as high temperature and high power electronic devices, is nowadays obvious. Their bandgaps are direct, and they cover a very large energy scale: 6.2 eV for AlN, 3.4 eV for GaN and 0.7 eV for InN at room temperature. Together with their excellent mechanical strength and chemical inertness, it makes them essential materials for modern semiconductor devices.

While working at Bell Labs in the seventies, Marc Ilegems made pioneering investigation on GaN optical properties [1, 2]. However, the quality of the material achieved was very poor at this time, and p-type doping was still not available. The real break-through of III–N materials occurred in the nineties, once GaN buffer layer quality was improved by the use of a two-step growth procedure [3] and p-type doping of GaN was under control [4]. III–N blue light emitting diodes (LEDs) [5] and laser diodes (LDs) [6] based on (In)GaN/(Al)GaN heterostructures could then reach industrial maturity, boosting research efforts worldwide on III–N materials.

Marc Ilegems initiated strong research efforts on III-nitrides in our institute, once acquiring and directing a GaN hydride vapour phase epitaxy (HVPE) system in 1998 (an AIXTRON prototype), and a III–N metal-organic vapour phase epitaxy (MOVPE) system in 2002 (a R&D commercial AIXTRON 200/4 RF-S system).

In this article, we present studies on samples grown in the MOVPE reactor, aimed towards the realisation of microcavity (MC) structures. The fabrication of microcavities heavily relies on the availability of distributed Bragg reflectors (DBRs). As discussed below, epitaxial nitride DBRs used by most research groups are made of Al(Ga)N/GaN quaterwave layers, and greatly suffer from intrinsic lattice-mismatch issues. That is why we chose to explore an original route, which consists in using AlInN alloy lattice-matched to GaN as low-index material in AlInN/GaN DBRs. Section 13.2 discusses the growth and characterization of this alloy in

Physics of Semiconductor Microcavities: From Fundamentals to Nanoscale Devices
Edited by Benoit Deveaud
Copyright © 2007 WILEY-VCH Verlag GmbH & Co. KGaA, Weinheim
ISBN: 978-3-527-40561-9

details. Section 13.3 demonstrates a microcavity light emitting diode using a moderate-reflectivity AlInN/GaN DBR. Section 13.4 presents high-reflectivity DBRs and our latest results on high-finesse microcavities are in Section 13.5.

13.2
AlInN Alloy: Growth and Characterization

13.2.1
AlInN: Motivation and Difficulty

Strain management is of prime importance when dealing with epitaxial growth of heterostructures. When an alloy is grown on another one of slightly different lattice constant, a pseudomorphic layer can be formed only provided its thickness remains below the so-called critical thickness, h_c. In that case, the misfit between the two crystals is accommodated by a biaxial strain, so that the in-plane lattice constant remains unchanged; no structural degradation is observed. If the thickness exceeds h_c, the elastic energy accumulated in the layer becomes so large that the crystal relaxes to its unstrained form. It occurs through the generation of dislocations and/or cracks, resulting in a catastrophic structural degradation.

In "classical" III–V systems, III-arsenides and III-phosphides, the Ga–V and Al–V bonds have nearly the same length, for example lattice mismatch between GaAs and AlAs is only 0.16%. The critical thickness is so large that strain issues can be completely forgotten in most of practical cases; GaAs/AlAs or GaP/AlP are "naturally-lattice-matched" material systems, indeed a gift of nature.

But this is absolutely not true for III-nitrides, as illustrated on the bandgap energy versus lattice constant diagram on Fig. 13.1a). Note that InGaN and AlInN curves are here only indicative, as their bowing parameters are still subject to discussion. Nevertheless, the mismatch between AlN and GaN is 2.5%, with experimental h_c that does not exceed a few nanometer [7]. The lattice-matched materials are those along vertical line on the figure, with AlInN for higher bandgap material and AlGaN or InGaN for lower bandgap material. AlInN with indium content around 18% is the one lattice-matched to GaN. So the most evident alloys free of strain issues among III–N materials have the chemical composition $Al_{0.82-x}In_{0.18-x}Ga_{1-x}N$ with $0 < x < 1$, i.e. the dotted line on Fig. 13.1a).

In spite of this, AlInN is by far the least known of III–N ternary alloys, with very few reports available in scientific literature. This is illustrated on Fig. 13.1b) that sketches the amount of publications per year related to AlGaN, InGaN and AlInN. The III–N boom in 1995 is evident on InGaN and AlGaN curves: InGaN (≈300 publications/year) is the material for optically active quantum wells in blue LEDs and LDs; AlGaN (≈400 publications/year) is used as electron blocking layer in LEDs, wave-guiding and cladding layers in LDs and has also applications in microelectronics for high electron mobility transistors (HEMTs). By contrast there were only 8 publications per year related to AlInN during the last decade. The main reason is that the poor quality of the material has prevented its use in devices. In

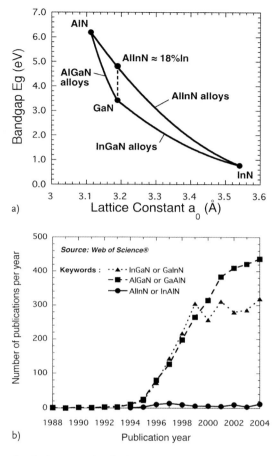

Fig. 13.1 Presentation of III–N ternary alloys. (a) Bandgap versus lattice constant diagram (hexagonal phase). There is a large lattice mismatch between the three binary compounds. The most evident lattice-matched system is GaN/Al$_{0.82}$In$_{0.18}$N, along the dotted line. (b) Number of publications per year related to the different ternary alloys. In spite of this attractive property, there are thirty times fewer reports in the literature related to AlInN, compared to InGaN or AlGaN.

deed, the growth of AlInN is the most difficult. The alloy is not stable thermodynamically due to the large mismatch between the covalent bonds, so that phase separation and composition inhomogeneity are expected [8]. Indium is also hardly incorporated to the AlN matrix due to the large difference of growth temperature between InN ($\approx$600 °C) and AlN ($\approx$1100 °C).

13.2.2
Growth of AlInN and AlInN/GaN DBRs

The AlInN layers and AlInN/GaN DBRs presented in this text were grown in the MOVPE reactor mentioned above, on 2-inch c-plane sapphire substrates. A low-temperature GaN nucleation layer, followed by a few μm-thick GaN buffer layer, always initiated the growth. AlInN was then deposited between 800 °C and 850 °C and at 50 to 75 mbar pressure using N$_2$ carrier gas. Lower growth temperatures led to a degradation of AlInN crystalline quality, while the indium content decreased and could not be maintained near GaN lattice-matching at higher temperatures.

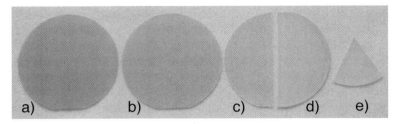

Fig. 13.2 Photographs under diffuse light showing "bare eye" differences in brownish colour of $Al_{0.84}In_{0.16}N$ layers. Samples (a–c) consist of 0.5 μm thick AlInN on GaN buffer, with increasing carrier gas flow from run (a) to run (c) respectively. (d) is a 20 pair AlInN/GaN DBR sample containing altogether 1 μm AlInN, with growth conditions identical to (c). (e) is a piece of GaN buffer layer alone, shown here for comparison.

During the DBR runs, the growth was interrupted at each interface and GaN was deposited at 1050 °C using H_2 and N_2 carrier gases.

We found that the carrier gas flow dynamic has a great influence on the AlInN quality: this is illustrated on Fig. 13.2 where samples a) to c) are AlInN layers with about the same thickness (≈0.5 μm) and indium content (≈16%). Layer a) was grown under low flow of carrier gas and has a strong brownish aspect, indicating light absorption at visible wavelengths. The layers become more transparent when increasing the carrier gas flow (samples b) and c) respectively).

Sample d) is a 20 pair $Al_{0.84}In_{0.16}N$/GaN DBR containing altogether 1 μm AlInN. Despite the fact that AlInN was grown with the same parameters as for sample c), and that the overall thickness is twice more important, the DBR is more transparent than sample c). Indeed, this "bare eye" evaluation cannot distinguish between the DBR and the GaN-on-sapphire reference e), while a slight brownish aspect is still present for sample c). This is an indication that some contamination of the reactor occurs during AlInN deposition. During DBR runs GaN is grown at higher temperature after each deposition of ≈50 nm-thick AlInN quarter-wavelength layers. It probably cleans and/or buries the contaminants at each DBR period, before they can impact on AlInN growth. The presence of contaminants would also explain the observed effect of carrier gas flow on layer quality, as their deposition would be reduced at higher flows. The contaminants are quite probably In or Al species, but their precise nature is not known yet.

13.2.3
Structural Characteristics: X-Ray Evaluations and TEM Images

The high crystalline quality of AlInN achieved is assessed by high resolution X-ray diffraction (HRXRD) (0002) scans. This is shown on Fig. 13.3a), for a 0.5 μm-thick $Al_{0.84}In_{0.16}N$ layer, and for a 20 pair $Al_{0.84}In_{0.16}N$/GaN DBR sample. The HRXRD scans were performed without slit in front of the detector. In this case the diffracted

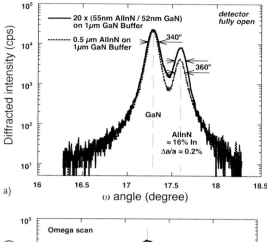

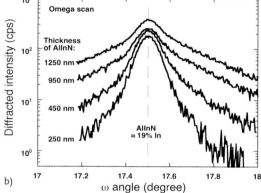

Fig. 13.3 (a) (0002) X-ray diffraction rocking curves of a single 0.5 μm AlInN layer grown on GaN and of a 20 pair AlInN/GaN DBR. (b) Omega scan along (0002) AlInN diffraction line on a 1.25 μm-thick layer etched step by step down to 0.25 μm.

intensity is integrated over a 5° detector angle and the full width at half maximum (FWHM) of the peaks are influenced by both composition fluctuations and c-axis tilt. Both samples exhibit identical crystalline quality, with 360 arcsec FWHM for the AlInN peaks, nearly as narrow as the 340 arcsec FWHM GaN peaks. This is a major improvement compared to the ≈1000 arcsec FWHM linewidth reported in the literature for epitaxial AlInN [9, 10].

We discussed previously that the bare-eye aspect of a 0.5 μm-thick AlInN layer is not as good as that of a DBR sample, indicating a degradation of AlInN optical quality with increasing layer thickness. This degradation is not observed on Fig. 13.3a), but is clearly evidenced on the HRXRD rocking curves shown on Fig. 13.3b). For this experiment a 1.25 μm-thick $Al_{0.81}In_{0.19}N$ layer was grown with optimised growth parameters, the growth rate being 0.2 μm/h. The sample was then cleaved and etched to different depths by reactive ion etching (RIE). The remaining thickness of AlInN was 1250 nm (unetched part), 950 nm, 450 nm and 250 nm, respectively. No variation of the indium content was evidenced by HRXRD omega–two theta scans on these samples, showing a good control of indium content during growth. Figure 13.3b) shows the HRXRD omega scans: these scans are those evidencing the crystalline integrity

 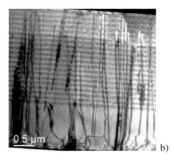

Fig. 13.4 TEM images of a 20 pair AlInN/GaN DBR grown on a 1 μm thick GaN buffer layer. (a) Higher magnification showing sharp and flat interfaces, (b) lower magnification showing the dislocation network originating from the GaN buffer and propagating through the DBR.

and characterize the defects and dislocations affecting c-axis tilt. The diffraction line FWHM is 375 arcsec for the 250 nm-thick sample, but increases up to 580 arcsec for the unetched sample. This comes with the apparition and the development of a very broad signal (≈1000 arcsec FWHM), superimposed to the original thinner line.

It is thus confirmed that the AlInN quality worsens when the layer thickness exceeds several hundred nanometers. Hopefully, the quarterwave layers contained in the DBRs are about 50 nm-thick, which is below the onset of degradation. The overall good structural quality of a DBR sample is also visible on transmission electron microscopy (TEM) images. Figure 13.4a), at higher magnification, reveals flat and sharp interfaces. Figure 13.4b) shows the network of dislocations typical of an MOVPE GaN buffer layer on sapphire substrate. The dislocation density estimated here is 3×10^9 cm^{-2}. The dislocation network propagates in the AlInN/GaN DBR layers without any noticeable increase in density over its 20 periods.

13.2.4
Optical Index Contrast to GaN

The most important parameter of a material system for DBR application is the relative difference in the refractive indices: the larger, the better. So we first evaluated the optical index contrast between AlInN and GaN, $\Delta n/n = (n_{\text{AlInN}} - n_{\text{GaN}})/n_{\text{GaN}}$, by recording the reflectivity of the layers *in situ* during the growth of a few periods of a DBR whose center wavelength matched that of the measurement wavelength [11]. Our experimental set-up consists of a LUXTRON TR-100 using a 950 nm wavelength source under normal incidence, which allows for an absolute reflectivity measurement during the growth runs. Figure 13.5a) shows the evolution of reflectivity during a typical run. The growth of the GaN buffer layer is stopped when its maximum reflectivity is reached around 26%, then AlInN is grown during the negative slope of the reflectivity signal, followed by GaN during the positive slope. Let us note R_i for the reflectivity value after deposition of the ith DBR period. R_i increases with the number of periods starting from the very first period. This alrea

13.2 AlInN Alloy: Growth and Characterization | 267

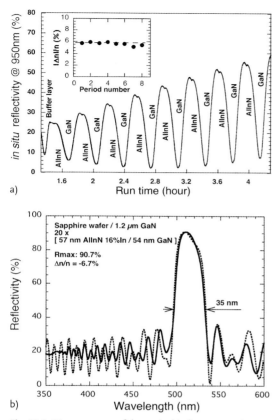

Fig. 13.5 Measurement of the optical index contrast between AlInN and GaN. (a) *In situ* acquisition of reflectivity during the growth of a dedicated DBR matched to the wavelength of measurement, 950 nm. (b) *Ex situ* analysis of the reflectivity spectrum of a moderate-reflectivity DBR (20 pairs) centred at green wavelengths: full line is the experimental spectrum, dotted line the fitted curve.

dy indicates that AlInN has a lower optical index than GaN, otherwise reflections at the AlInN/GaN and GaN/AlInN interfaces would be in antiphase with the GaN/air and sapphire/GaN reflections, leading to a decrease of R_i during the first periods. As reflections at all interfaces are in phase, the well-known formulas for DBRs reflectivity [12] can be used and we can calculate the optical index contrast from the period to period increase in reflectivity using:

$$\frac{|\Delta n|}{n}(i) = 1 - \sqrt{\frac{\left(1+\sqrt{R_i}\right)\left(1-\sqrt{R_{i+1}}\right)}{\left(1-\sqrt{R_i}\right)\left(1+\sqrt{R_{i+1}}\right)}} \ . \tag{1}$$

Of course, this result is only valid if no parasitic effects, such as absorption, appearance of cracks or development of surface roughness, decrease the reflectivity. We can ensure that the index contrast given by Eq. (1) is correct by plotting $|\Delta n|/n$ as a function of the number of periods as shown in the inset of Fig. 13.5a). Any parasitic effect will manifest itself by a decrease of $|\Delta n|/n(i)$. In the case of the run shown in Fig. 13.5a), a marked decrease occurs at the 7th period, and indeed further examination of the sample revealed the presence of cracks. This sample was still quite near lattice-matched, with an estimated compressive strain about 0.4%. On more mismatched AlInN layers, cracks appeared earlier, and in some cases only the first period could be taken into account for the index contrast evaluation.

These *in situ* measurements are very useful as they are fast and they can be used with mismatched alloys, allowing the evaluation of $\Delta n/n$ on a large range of compositions for AlInN. But the values obtained are at a wavelength of 950 nm and at the growth temperature, around 800 °C. Our applications are rather for blue-green wavelengths and room temperature operation, so we also evaluated $\Delta n/n$ at these wavelengths from *ex situ* measurements. We grew for this purpose AlInN/GaN DBRs with a moderate reflectivity. Indeed, the maximum reflectivity of a DBR having a large number of pairs is mainly fixed by the optical losses, and $\Delta n/n$ is difficult to extract from the reflectivity spectra. On the other hand, if the number of pairs is too low the reflectivity is rather influenced by the GaN/sapphire and GaN/air interfaces, which already contribute for a $\approx$30% reflectivity value, and $\Delta n/n$ is also difficult to evaluate. A good compromise is to use DBR stacks having 70% to 90% reflectivity at central wavelength, which corresponds in our case to 10 to 20 pairs. $\Delta n/n$ is then obtained by the best fit of the reflectivity spectra using a standard transfer matrix method, as illustrated on Fig. 13.5b) for a crack-free 20-pair $Al_{0.84}In_{0.16}N/GaN$ DBR. The reflectivity measurements were performed with a Cary 500 reflectometer in double-reflection mode. This DBR already shows pretty good characteristics, the reflectivity reaching over 90% with a 35 nm FWHM stopband.

Figure 13.6a) summarizes the index contrast measured on different samples, and presents the dependence of $\Delta n/n$ as a function of the indium content as estimated from HRXRD (0002) measurements. Open symbols correspond to the *in situ* measurements while other data points correspond to the *ex situ* analysis of DBRs tuned at 455 nm and 515 nm. We note that $\Delta n/n$ values deduced from *in situ* and *ex situ* analyses agree very well. This is not obviously expected because the *in situ* measurements are done at 950 nm and such long wavelength should exhibit lower $|\Delta n|/n$ due to the dispersion relationships of the refractive indices. In fact it appears that the decrease of $|\Delta n|/n$ due to the longer wavelength is compensated by an equivalent increase due to the higher temperature during the measurement. Consequently, the *in situ* evaluations of $\Delta n/n$ correspond in practice to room-temperature measurements at a wavelength of $\approx$480 nm.

The experimental dependence of $\Delta n/n$ with AlInN indium content is well fitted by a linear relationship within the 6% to 21% indium explored range, according to:

$$\frac{\Delta n}{n} (Al_{1-x}In_xN/GaN) = -0.127 + 0.35x. \qquad (2)$$

This relationship should be taken for $\Delta n/n$ at room temperature and at a wavelength of 480 nm, for the reasons explained above. Extrapolation of Eq. (2) to zero

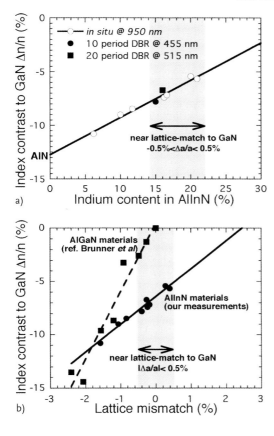

Fig. 13.6 (a) AlInN/GaN refractive index contrast versus AlInN indium content. (b) AlInN/GaN refractive index contrast versus lattice mismatch to GaN. The AlGaN/GaN material system is also shown for comparison.

indium content gives a −12.7% index contrast for AlN/GaN, in reasonable agreement with literature values [13].

The advantage of using near lattice matched AlInN rather than AlGaN as the low-index material is evident from Fig. 13.6b), where the AlInN/GaN index contrast is plotted as a function of lattice mismatch to GaN and compared with that of the AlGaN/GaN material system [13]. A lattice mismatch that lies within ±0.5% is sufficient to avoid relaxation in blue DBR applications. In this case the maximum index contrast achievable with AlGaN/GaN is about 3%, while up to 8% is obtained with AlInN/GaN. The gain is even more dramatic when considering laser diodes application where the lattice mismatch is rather limited to ±0.25%.

13.2.5
Bandgap and Dispersion of Refractive Index

We estimated the bandgap of AlInN layers from reflectance spectra of sapphire/GaN/AlInN samples. Transmission measurements are a better choice for the determination of bandgaps, but the GaN buffer layer, which is necessary to obtain

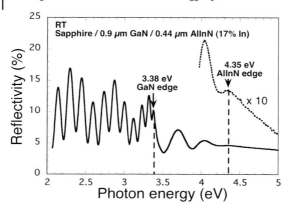

Fig. 13.7 Reflectance spectrum of a sapphire/GaN/ AlInN sample. The bandgaps are estimated by the high-energy edges of the interference fringes, the dispersion of refractive indices by the spacing between the maxima and minima of the fringes. Note the gradual damping of the amplitude of the AlInN fringes, indicative of a large absorption below AlInN gap in this sample.

good quality AlInN, prevents transmission experiments above 3.4 eV. Figure 13.7 shows the reflectivity spectrum of a 440 nm-thick $Al_{0.83}In_{0.17}N$ layer. The experimental band-gap is deduced from the high-energy edge of the interference fringes related to the AlInN layer [10].

Figure 13.8a) presents the experimental dependence of bandgap versus indium content, observed on a set of AlInN samples with composition near lattice-matched to GaN. It is fitted by the linear relationship:

$$E_g(Al_{1-x}In_xN) = 6.2 - 11x. \tag{3}$$

The corresponding bandgap bowing parameter is ≈ -6 eV, which is larger than ≈ -3 eV expected from first-principle calculations [14]. The bandgap of AlInN when lattice-matched to GaN is found to be ≈ 4.3 eV. This is consistent with the highest experimental values reported in literature, which are spread from 2.8 eV to 4.4 eV [9, 10, 15–17].

The reflectivity spectrum of Fig. 13.7 also allows for the evaluation of the dispersion of AlInN refractive index, because the maxima and minima of the thickness fringes occur when the AlInN optical thickness is a multiple of the quarter-wavelength:

$$n = m\lambda_m/4d \quad \text{(maxima and minima are even and odd } m \text{ respectively)}. \tag{4}$$

On Fig. 13.7, the AlInN-related maxima and minima are very clear above GaN bandgap. Below the GaN bandgap, GaN-related fringes are superimposed to the AlInN-related one and the AlInN fringes rather appear as a modulation, but their extrema can still be evaluated.

13.2 AlInN Alloy: Growth and Characterization | 271

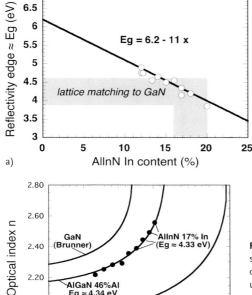

Fig. 13.8 Analysis of the reflectance spectra. (a) Dependence of the bandgap of AlInN with indium content, measured on a set of sapphire/GaN/AlInN samples. (b) Refractive index dispersion on a 17% indium content AlInN layer (dark dots). The lines are the refractive indices reported for AlGaN alloys.

The determination of the refractive index given by (4) also requires a precise measurement of the AlInN thickness d. We already know from the previous study that the refractive index contrast to GaN for this particular $Al_{0.83}In_{0.17}N$ layer is −6.8% at 480 nm, i.e. $n(Al_{0.83}In_{0.17}N) = 2.25$ given $n(GaN) = 2.41$. So we used (4) around 480 nm wavelengths with $n = 2.25$ to calculate d, and introduced this value for d in (4) to calculate the refractive index of the $Al_{0.83}In_{0.17}N$ layer at other wavelengths. The advantage of this determination of d is its consistence with the previous estimations of $\Delta n/n$ without adding an error linked to another thickness measurement method. We are indeed more interested by precise values for $\Delta n/n$ than by absolute values of the refractive index.

The results are reported on Fig. 13.8b), where the refractive index of the $Al_{0.83}In_{0.17}N$ layer is plotted versus photon energy. The full lines shown for comparison are the refractive indices of GaN, AlN and $Al_{0.46}Ga_{0.54}N$ [13]. We find that the refractive index of $Al_{0.83}In_{0.17}N$ follows closely that of $Al_{0.46}Ga_{0.54}N$, which is incidentally the AlGaN alloy having the same bandgap than the one we have determined previously for $Al_{0.83}In_{0.17}N$.

Another point worth noticing on Fig. 13.7, is that the thickness fringes exhibit a very progressive damping before they disappear at $Al_{0.83}In_{0.17}N$ bandgap. This is an indication of a very large absorption tail; huge Urbach tails are indeed widely reported in AlInN literature [9, 10, 15–17]. For example, we roughly estimated from the damping that the residual absorption in this $Al_{0.83}In_{0.17}N$ layer is still a few thousand cm^{-1} at blue wavelengths. Such high residual absorption would limit DBR reflectivity below ≈90%. It is shown further that this is not the case, with DBR reflectivities above 99% and residual absorptions below 100 cm^{-1}. The reason for high residual absorption in this $Al_{0.83}In_{0.17}N$ layer is that its thickness is above the "onset of degradation" discussed previously.

13.2.6
Photoluminescence and Stokes Shift

Photoluminescence (PL) measurements were performed on a set of sapphire/GaN/AlInN samples grown with different parameters. The experiments were done under continuous wave excitation at 244 nm, from a frequency doubled Ar^+ laser.

Figure 13.9a) shows one of the best PL spectra studied. This sample has a 100 nm-thick $Al_{0.84}In_{0.16}N$ layer, thinner than the onset of degradation. It exhibits a clear PL line centred at 340 nm (3.65 eV) with a 33 nm (350 meV) width. A broad yellow band is also observed due to the GaN buffer layer.

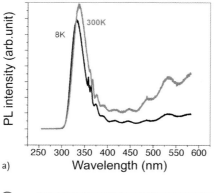

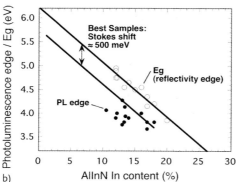

Fig. 13.9 (a) Photoluminescence spectra (300 K and 8 K) of a 100 nm-thick $Al_{0.84}In_{0.16}N$ on a GaN buffer layer. (b) Estimation of Stokes shift: Photoluminescence and reflectance edges versus In content, measured on a large set of samples with different growth parameters and thicknesses.

On samples grown without optimum parameters, or on thicker samples grown beyond the onset of degradation, we observe the appearance of another intense broad signal (not shown here) with maximum of emission around 2.5 eV to 3 eV and a linewidth of 500 meV to 1 eV. This signal seems to be the PL signature of these defects that develop with increasing AlInN thickness, and are responsible for brownish aspect, increased residual absorption and the appearance of the large ≈1000 arcsec diffraction line on HRXRD omega scans. On the thicker samples (or on poorer quality ones) only this broad red-shifted line is present, the PL spectra are then very similar to those usually reported in the literature.

The Stokes shift, the energy difference between the absorption edge and PL emission line, is estimated on Fig. 13.9b). The open dots are the measurements already presented of the reflectivity edge, for evaluation of the absorption edge. The dark dots are the high-energy foot of the AlInN PL line. The reason for this choice is that the AlInN emission line is mixed to the defect-related line on some samples, and the maximum of AlInN line is not always observable. So the Stokes shift is roughly the energy difference between open dots and dark dots. It is ≈500 meV for the best samples, and can reach 1 eV.

Thus, PL properties of our best AlInN samples near lattice-matched to GaN are a 350 meV PL linewidth with a ≈500 meV Stokes shift. For comparison, typical values are around 50 meV for both PL width and Stokes shift in AlGaN or InGaN alloys [18, 19]. AlInN values still reveal a large alloy disorder, but the PL peaks and absorption edges remain above GaN bandgap, which is the crucial point to achieve optical confinement without losses in the lattice-matched AlInN/GaN system.

13.3
Microcavity Light Emitting Diode

To check whether the inclusion of an AlInN/GaN DBR in a light emitting device would affect the crystalline quality and the internal efficiency of active InGaN quantum wells, we demonstrate now the fabrication of a microcavity light emitting diode (MCLED) having a lattice-matched bottom AlInN/GaN DBR [20].

Microcavity light emitting diodes offer interesting possibilities. Indeed, in such devices, the active quantum wells are placed in a Fabry–Perot cavity and light is redirected towards the extraction cone by optical interferences. This leads to a higher output power and better directionality of the light beam. MCLEDs optimized for efficient light extraction should have a low quality factor, to avoid multiple reflections in the cavity and internal loss induced by absorption. This is especially important in this case as the residual absorption increases for doped nitride layers as reported by Ambacher et al. [21]. Our substrate-emitting design is shown on Fig. 13.10a). The top mirror is an Ag/Au metallic mirror with a reflectivity of 97% used also as p-contact and the output mirror is a 12-pair Si-doped $Al_{0.82}In_{0.18}N$/GaN distributed Bragg reflector with an internal reflectivity of 52%. The cavity consists of a Si-doped GaN layer followed by the active region, a Mg-doped AlGaN electron blocking layer and a Mg-doped GaN cap layer. The Si-doped GaN layer was deposi-

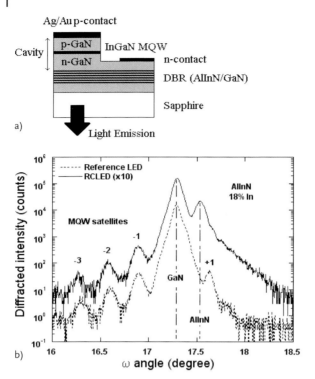

Fig. 13.10 (a) Sketch of the microcavity LED. For clarity, the electron-blocking layer is not shown. (b) (0002) HRXRD rocking curves of the MCLED sample and a reference LED.

ted on top of the DBR at 1050 °C using H_2 carrier gas. It is 40 nm thick with an electron concentration $n = 3 \times 10^{18}$ cm^{-3}. The active region was deposited at 740 °C and consists of five 2.9 nm thick $In_{0.16}Ga_{0.84}N$ multi quantum wells (MQWs) emitting at 454 nm, separated by 12 nm thick Si-doped GaN barriers. Above the quantum wells, a 15 nm thick p-doped $Al_{0.19}Ga_{0.71}N$ barrier was introduced to avoid electron leakage in the cap layer by establishing an electron blocking layer. The Mg-doped GaN cap layer is 100 nm thick and was deposited at 980 °C at 200 mbar pressure. Hole concentration in the Mg-doped regions is close to $p = 5 \times 10^{17}$ cm^{-3}. Processing by standard photolithography techniques included the following steps: Activation of the Mg-doped layers in a rapid thermal annealing system (RTA), RIE of the mesa structures to access the Si-doped nitride layers, deposition of a Ti/Al/Ni/Au n-contact by electron-beam evaporation, and finally deposition of high-reflectivity Ag/Au p-contacts. The contacts are not annealed to preserve high reflectivity.

Figure 13.10b) compares (0002) HRXRD rocking curves of the MCLED sample with that of a reference LED grown without DBR. Satellites from the InGaN MQWs are clearly identified on both samples and no significant degradation of the InGaN MQWs could be detected due to the presence of the DBR.

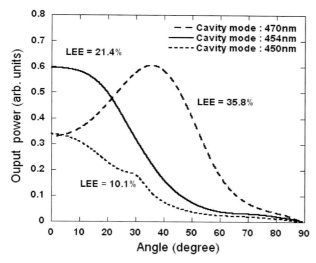

Fig. 13.11 Effects of the detuning on the light output pattern and on the light extraction efficiency. The simulation is performed with a program developed in our group by R. P. Stanley and R. Houdré. It is based on the transfer matrix method to calculate light propagation in the MCLED dielectric layers, and the active QWs are described by a planar distribution of oscillating dipoles.

In a microcavity LED, the main design parameter, which governs both the shape of the radiation pattern and the light extraction efficiency, is the detuning $\lambda_E - \lambda_C$ between the central wavelength of emission λ_E and the optical mode resonant in the normal direction within the cavity $-\lambda_C$ [22]. Each emitted photon is resonant in a given direction that depends on its wavelength, due to optical interferences within the cavity. As described elsewhere [23] a proper design is achieved by adjusting the thickness of the cavity by taking into account the phase-shift induced by the Ag/Au mirror and the phase penetration depth in the Bragg mirror. To optimize the coupling between the source and the cavity, it is essential to place the quantum wells at an antinode position in the cavity and to make sure that the DBR stopband and the cavity mode match.

Figure 13.11 shows the variation of the radiation pattern and light extraction efficiency (LEE) depending on the detuning of the device. The emission properties are simulated using a method based on the plane wave expansion of an electrical dipole inside a multi-layer structure [24]. For the simulations, the dipole layer was set in the first QW from the top of the structure and the source was the photoluminescence spectrum of the quantum wells centred at 454 nm with a 19 nm FWHM. Integration of the power emitted over the full solid angle gives the value of the device light extraction efficiency. For the cavity without detuning, we calculated 21.4% light extraction efficiency assuming no light absorption in the device and no refractive index dispersion.

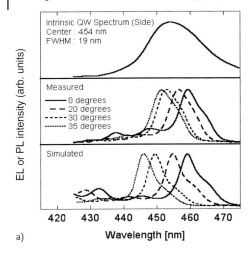

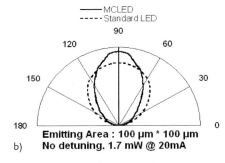

Fig. 13.12 Emission characteristics of the MCLED. (a) Effects of the cavity on the intrinsic spectrum of the quantum wells. (b) Angular pattern of a light emitting diode without detuning: deviation from the Lambertian emission of a standard LED.

Angle-resolved photoluminescence measurements are performed to show the effect of the microcavity on the light emission characteristics of the LED. As shown on Fig. 13.12a), the intrinsic spectrum of the quantum wells, measured from the side of the sample (guided light), is compared to spectra emitted at different angles through the substrate. The FWHM of the modes is reduced from 19 nm for the intrinsic spectrum to about 9 nm for the different angular spectra. Simulations and measurements are in relatively good agreement as expected from a low quality factor microcavity. In addition, we observe shorter resonant wavelengths at higher emission angles, which is a typical microcavity effect and consistent with similar work on AlGaN/GaN based MCLED [25]. The cavity mode, resonant in the normal direction is centered at 460 nm for this particular LED, which was confirmed by additional reflectivity measurements. Figure 13.12b) shows the radiation pattern for a MCLED without detuning. Due to the cavity effect, the emission is redirected in the normal direction and differs from the Lambertian emission pattern of a standard LED.

For this device, without any encapsulation, we obtained an optical output power of 1.7 mW at 20 mA for an emitting area of 100×100 µm^2 corresponding to a 2.6%

external quantum efficiency. To our knowledge, these are the highest values reported for extraction efficiency and emitted power in nitride-based microcavity LEDs. It is still low compared to the simulated LEE value of 21.4%, but it is important to note that in the simulation we assumed that there was no residual light absorption in the nitride layers and that light was emitted by a single quantum well placed at a cavity antinode. Experimentally, a reduced efficiency is expected as the sample is not completely transparent and some of the 5 MQWs of the active region are not optimally positioned within the cavity. As a consequence, the light extraction efficiency would be improved by reducing the number of quantum wells and placing them in an optimal position. An in depth investigation of the structural properties of the active region is also required to further characterize the effect of the AlInN/GaN Bragg mirror on the internal efficiency of the MQWs.

13.4
High Reflectivity DBR and Residual Absorption

One of the main concerns with high reflectivity AlInN/GaN Bragg reflectors is the possible residual light absorption in the UV and blue wavelength ranges, as discussed in Section 13.2. This would lead to a strong decrease in the performance of the Bragg reflector, especially when moving to shorter wavelengths.

To further investigate the optical properties of the Bragg reflectors, we have grown a 40-pair $Al_{0.83}In_{0.17}N$/GaN mirror on a 2-inch c-plane sapphire substrate polished on both sides, allowing thus to carry out reflectivity and transmission measurements. The GaN and AlInN $\lambda/4$ layers are 47 nm thick and 50 nm thick respectively, for a total DBR thickness of about 4 μm. Again, the high crystalline quality of the DBR was confirmed by X-ray diffraction measurements and no cracks could be detected by standard optical microscopy. The reflectivity and transmission curves, measured with a Cary 500 spectrophotometer system, are shown on Fig. 13.13a) as well as the optical loss in the sample given by $100-R-T$. The side lobes are not clearly resolved due to the ~1% growth thickness variation over the few tens of mm^2 probed by our setup. Near the center of the wafer, we measured a 30 nm optical stopband centred at 450 nm with a record 99.4% peak reflectivity, the highest reflectivity reported so far for nitride Bragg reflectors (see Fig. 13.14b).

The transmission is 0.35% at the Bragg central wavelength, which leads to an optical loss of $0.25 \pm 0.05\% \cong \lambda = 450$ nm. The average absorption coefficient α [cm^{-1}] in the Bragg reflector can then be estimated using the following relations, which are valid at the optical stopband center λ_{DBR} only:

$$\alpha = \frac{1-R-T}{L_p} \quad \text{and} \quad L_p = \frac{\lambda_{DBR}}{n_{GaN}^2 - n_{AlInN}^2} \approx \frac{\lambda_{DBR}}{2n_{Av.}^2 \frac{\Delta n}{n}}. \tag{5}$$

Using an average refractive index $n_{Av.} = 2.36$ and an index contrast between GaN and AlInN $\Delta n/n = 7 \pm 0.3\%$, the penetration length in the reflector is $L_p = 577 \pm 75$ nm and the absorption coefficient $\alpha = 43 \pm 14$ cm^{-1} at 450 nm.

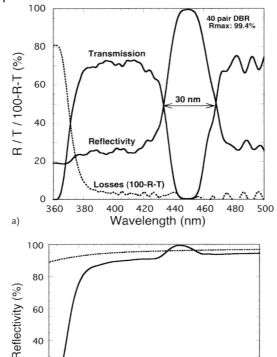

Fig. 13.13 (a) Reflectivity, transmission and optical losses, measured on a 40 pair AlInN/GaN Bragg reflector. (b) Reflectivity of the DBR sample and of a sapphire substrate, both with Ag coatings. The residual absorption in the DBR is obtained from the analysis of these measurements.

Outside the optical stopband, the reflected light at the AlInN/GaN interfaces do not interfere constructively, and the DBR sample is considered as a single absorbing pseudo-alloy layer with 2 interfaces with air. Thus, the absorption in the DBR is estimated in a larger wavelength range from the reflection and transmission measurements using the following relations:

$$1 - R - T = (1 - R_1)\frac{(1-A)(1+AR_2)}{1 - A^2 R_1 R_2} \quad \text{and} \quad A = \exp(-\alpha d), \tag{6}$$

where R_1 is the reflectivity at the air/GaN top interface, $R_2 \approx 10\%$ is the reflectivity of the GaN/sapphire/air bottom interfaces, T is the transmitted intensity, R the reflected one and d the total thickness of the absorbing media, which in our case is the DBR total thickness $d = 4$ μm. This is a direct method to measure the residual absorption in our layer but is precise only if the transmission and the reflectivity are measured at the same spot on the sample. Due to the interferences from the sap-

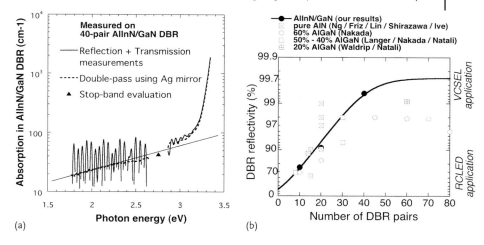

Fig. 13.14 (a) Residual absorption versus photon energy in the 40-pair AlInN/GaN DBR. Measurement methods are detailed in the text. (b) Reflectivity versus number of pairs for III–N DBRs. Dark plots are our experimental data deduced from AlInN/GaN DBRs, the full line is calculated using refractive index contrast and residual absorption values of $\Delta n/n = -7\%$ and $\alpha = 50$ cm^{-1} respectively. Open dots are different designs of Al(Ga)N/GaN DBRs reported in the literature, for comparison.

phire substrate, a small deviation between the transmission and reflectivity measurements leads to a noisy evaluation of the optical losses in the sample. Consequently, in addition to this first method, we used a second one to evaluate the average absorption coefficient in the Bragg reflector with enhanced precision, by depositing a silver mirror on the backside of the DBR sample. Indeed, as shown on Fig. 13.13b), comparing its reflectivity spectra with a reference sample, consisting of a 2-side polished sapphire substrate with a silver mirror on the backside, gives a second estimate of the absorption coefficient. In this case, the absorption coefficient can be calculated using the following relations:

$$R_{sample} = R_1 + \frac{A^2 R_{Ag}(1-R_1)^2}{1 - A^2 R_{Ag} R_1} \quad \text{and} \quad R_{ref} = R_2 + \frac{R_{Ag}(1-R_2)^2}{1 - R_{Ag} R_2} \tag{7}$$

where R_{sample} is the reflectivity of the DBR sample, R_{ref} the reflectivity of the reference sample, R_1 is the reflectivity of the air/GaN interface, $R_2 \approx 8\%$ the reflectivity of the air/sapphire interface and R_{Ag} the reflectivity of the silver coating.

The average absorption coefficient of the Bragg reflector is plotted on Fig. 13.14a) versus the photon energy. Our three evaluation methods are well correlated, and we observe a low residual absorption in the DBR, with values of 28 ± 7 cm^{-1} at 520 nm, 43 ± 14 cm^{-1} at 450 nm and 75 ± 19 cm^{-1} at 400 nm. In addition, a low Urbach tail energy of 26 meV is also deduced from the optical band edge of GaN, a sign of the presence of high quality GaN periodic layers in the DBR. The exponential behaviour and these values of the residual absorption are similar to previous works performed by Ambacher et al. [21], Brunner et al. [13] and Omnès et al. [26] on various GaN and AlGaN samples. In these reports, the different residual absorption levels can be explained in terms of crystalline quality of the layers (growth technique, doping levels). This indicates that high-quality AlInN is comparable to other nitride alloys in terms of residual absorption in this wavelength range.

It is interesting to observe the dissymmetry of the absorption at the DBR stopband edge. The curve lies above the exponential interpolation at higher energies and below it at lower energies. This type of behaviour is the signature of a large difference in absorption in the DBR materials. In this case, it would indicate that the residual light absorption is mainly present in the AlInN layers. This would not be a surprise, given the large Stokes shift the material still exhibits.

Using the refractive index dispersion presented on Fig. 13.8b) and the residual absorption for the AlInN/GaN DBR presented here, we can estimate the maximum reflectivity achievable with AlInN/GaN Bragg reflectors using:

$$R_{\max}(E) = 1 - \alpha(E) \cdot L_p(E). \tag{8}$$

We estimate that a lattice-matched AlInN/GaN Bragg reflector with a sufficient number (≈ 60) of pairs can reach a maximum reflectivity of 99.8% in the green wavelength range, 99.7% in the blue wavelength range, 99.5% at 400 nm and only 98.5% at 380 nm due to the absorption occurring in the GaN layers. Consequently, for UV applications requiring high reflectivity mirrors below 390 nm, the development of lattice-matched AlGaN/AlInN reflectors would be a better option.

Figure 13.14b) compares the lattice-matched (LM) AlInN/GaN DBRs presented here with Al(Ga)N/GaN DBRs from literature, plotting the reflectivity versus number of pairs. AlInN/GaN data are the dark dots, they are fitted by the curve calculated using the refractive index contrast and the residual absorption measured previously, $\Delta n/n = 7\%$ and $\alpha = 50$ cm^{-1}.

The only symbols standing above the LM AlInN/GaN curve, i.e. higher reflectivity for a lower number of pairs, are AlN/GaN DBRs [27–31]. This results from the higher refractive index contrast $\approx 17\%$, but it is also the worst case with respect to strain issues: these DBRs are usually highly cracked [29]. The best AlN/GaN DBR is reported by Ive [30], it was grown on a 6H-SiC substrate without GaN buffer layer and has 99% reflectivity without cracks. It was shown that in this case the AlN layers are under 1.3% tensile strain and GaN layers under an equal compressive strain, so the whole DBR is strain-compensated with an in-plane lattice constant corresponding to that of Al$_{0.5}$Ga$_{0.5}$N. The problem here is that GaN is not lattice-matched with the DBR and the strain issues are reported on further layers if, for exemple, a GaN-based cavity has to be grown on top of the DBR. Note also that high-reflectivity AlN/GaN DBRs were only reported using molecular beam epitaxy growth technique.

The other AlGaN/GaN DBR plots are below the LM AlInN/GaN curve. When AlGaN aluminium content is in the range [40%...60%] [32–35], the reflectivities are comparable with those of LM AlInN/GaN DBRs (so are the refractive index contrasts) as far as the number of pairs does not exceed 20 to 30, i.e. for moderate reflectivity DBRs. But the reflectivity remains below ≈97% with a higher number of pairs, still limited by defects/cracks induced by the accumulated strain. The aluminium content has to be further decreased to 20%, and the pair number increased to 60, in order to reach again reflectivities ≥99%. Even with this lower aluminium content, a "stress-engineering" solution by insertion of AlN interlayers at DBR interfaces is still needed to avoid cracks [36].

To summarize, complex stress management is required to achieve crack-free Al(Ga)N/GaN DBRs. High Al-content ones are difficult to integrate into more complex heterostructures because they impact on the in-plane lattice parameter while low-Al content ones suffer from very low refractive index contrast. LM AlInN DBRs do not only show improvement in terms of maximum achievable reflectivity, but they can be integrated into more complex heterostructures without any special concern as demonstrated by the growth of fully epitaxial high-Q microcavities in the following section.

13.5
Epitaxial Microcavities

The growth of high reflectivity LM AlInN/GaN DBRs being under control, we pursued our effort towards the fabrication of high quality factor ($Q = \Delta\lambda/\lambda$) microcavity structures [37, 38]. We present here two fully epitaxial III–N microcavity samples, centred at 420 nm wavelength:

- An empty 35–3λ/2–30 microcavity. It consists in a standard sapphire/GaN template, followed by a 35-pair lattice-matched AlInN/GaN bottom-DBR, a 3λ/2 GaN cavity, and a 30-pair lattice-matched AlInN/GaN top-DBR. The DBRs and the cavity are centred at 420 nm. The mirrors are designed to have equal internal reflectivities of 99%.
- A full 28–3λ/2–23 microcavity. Nearly the same design than above, but two sets of three 1.5 nm-thick $In_{0.15}Ga_{0.85}N$ MQWs are inserted at the antinodes of the electric field in the 3λ/2 GaN cavity. The bottom and top-DBRs are made of 28 and 23 AlInN/GaN pairs, respectively. We expect DBR internal reflectivity of 96% in this case.

These structures are the thickest ones among those we have presented and growth runs nearly last for 24 hours, but we did not observe any negative impact on their structural characteristics. Figure 13.15a) is a cross-section SEM image on the 28–3λ/2–23 microcavity. The measured thicknesses of the AlInN and GaN λ/4 layers are 45 and 41 nm, respectively, without measurable variations. These values are in good agreement with those deduced from XRD and growth rate calibrations.

The AlInN/GaN interfaces are well defined and seem perfectly flat, even at higher magnifications (Fig. 13.15b)). Once again, as a consequence of the near lattice-matched growth of AlInN λ/4 layers on GaN, the 2 inch wafer is completely crack-free, as checked by optical microscopy.

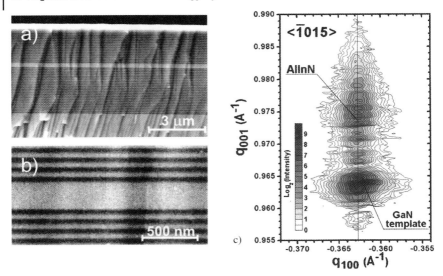

Fig. 13.15 (a) SEM image of the 28–3λ/2–23 microcavity sample. (b) Scaled-up SEM image of the GaN cavity showing the abrupt AlInN/GaN interfaces. (c) XRD reciprocal space map ($\langle \bar{1}015 \rangle$ reflection) of the 35–3λ/2–30 microcavity sample.

The $\langle \bar{1}015 \rangle$ XRD reciprocal space map of the 35–3λ/2–30 microcavity sample is presented in Fig. 13.15c). The peak centered at $q_{001} = 0.964$ Å^{-1} ($c = 5.187$ Å), associated to the GaN template, indicates a slight compressive stress originating from the epitaxial growth on sapphire. The perfectly aligned set of peaks in the q_{001} direction is the signature of pseudomorphic layers in the DBRs. A DBR period of 84.1 nm was deduced from their spacing. The presence of a high number of satellites demonstrates the large coherence of the λ/4 layers in the Bragg mirrors along the growth direction and the good overall quality of the AlInN/GaN interfaces. The envelope of these satellites shows a maximum at $q_{001} = 0.975$ Å^{-1}. This corresponds to a lattice parameter for AlInN of $c = 5.131$ Å. This is 0.35% above the value expected for perfect lattice-matching, showing an excess of indium in AlInN layers, 20% instead of 18%. Nevertheless, the position of the GaN template and of the peaks associated to the DBRs are nearly the same in the q_{100} direction: $a = 3.185$ Å for the GaN buffer layer and $a = 3.183$ Å for the DBR. Thus the relaxation of the in-plane lattice parameter does not exceed 6×10^{-4} in this more than 7 μm-thick structure.

Typical RT reflectivity and transmission spectra measured near normal incidence on the 35–3λ/2–30 microcavity sample are presented in Fig. 13.16a). The stopband is 28 nm wide, resulting from the significant refractive index contrast between $Al_{0.80}In_{0.20}N$ and GaN layers estimated to ≈7.7%. The flat-topped stopband, away from the cavity resonance, and the well-defined oscillations (in particular the short wavelength ones most sensitive to internal absorption effects) are indicative of the high sample quality. Indeed, the residual absorption estimated in this sample is in agreement with the measurements on the 40-pair DBR: 60 cm^{-1} at 420 nm. The

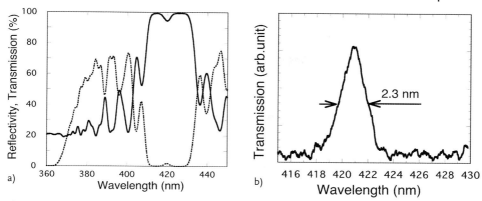

Fig. 13.16 (a) Room temperature reflectivity and transmission spectra of the empty 35–3λ/2–30 microcavity. (b) Scaled up plot of the cavity mode in the transmission spectrum.

cavity mode is well resolved and exhibits a linewidth of 2.3 nm in Fig. 13.16b). Such a value corresponds to a quality factor of the order of 180. This already denotes the potential of such structures for short wavelength optoelectronic devices. Nevertheless, the quality factor Q expected theoretically for these structures is higher than 1000 using DBRs with such reflectivities. This broader linewidth than expected theoretically could result from the integration of the reflectivity and transmission signals over a large area (a few mm^2) and the variation of the cavity thickness across the wafer.

Indeed photoluminescence experiments, for which the excitation spot is much smaller than reflectivity or transmission experiments (few tens μm instead of a few

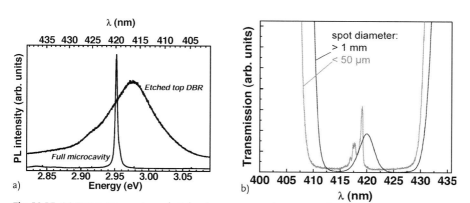

Fig. 13.17 (a) Room temperature photoluminescence spectra measured on the 28–3λ/2–23 microcavity sample (with active MQWs), before and after etching of the top-DBR. (b) Transmission spectra measured on the 35–3λ/2–30 microcavity sample (no active MQW), probed with two different spot sizes. The measurement with the smaller spot (<50 μm) reveals much narrower mode (FWHM ≤ 0.5 nm) and indicates the presence of multiple effective cavity lengths.

mm), reveals a much narrower mode. Figure 13.17a) shows the MQW PL spectrum of the 28–3λ/2–23 microcavity sample measured on an etched area (top DBR removed by RIE) together with the spontaneous emission spectrum measured on the full cavity under low excitation power ($P_{exc} \approx 2$ mW). The MC exhibits here a very narrow spontaneous emission linewidth = 0.52 nm underlying the very high quality factor of this structure, $Q \approx 800$. The latter has to be compared to the broad MQW PL spectrum (full width at half maximum 11.2 nm, i.e. 80 meV) spanning over the entire stopband width (SBW 28 nm). This Q factor corresponds to a cavity with 94.4% peak reflectivity mirrors when considering the effective cavity length. Such a value compares well with the calculated ≈96% reflectivity values for the bottom and top mirrors. Taking into account a decrease in Q resulting from MQW absorption as well as internal optical losses due, e.g. to residual absorption, scattering at the interfaces and cavity width fluctuation. Note that the $Q \approx 800$ factor reported here stands above values measured on hybrid nitride MC structures made of a bottom AlGaN/GaN DBR and a top dielectric DBR ($Q = 500 - 740$) [39, 40], or two dielectric DBRs ($Q = 670$) [41], thus underlining the high potential of AlInN/GaN-based MC structures for vertical cavity laser applications.

In order to clarify the effect of the probe size on the transmission spectra, we performed transmission measurements with another set-up allowing for a spot size ≤50 μm, comparable with that of the laser. Figure 13.17b) compares transmission spectra of the empty 35–3λ/2–30 microcavity sample obtained with the two probe sizes, ≈ few mm and ≤50 μm. There are two main conclusions. First we indeed observe a much narrower transmission linewidth (≤0.5 nm, $Q \geq 800$) with the smaller spot, as expected. Secondly and more surprisingly, several lines (here two, sometimes three) can be observed. These multiple transmission lines may be due to composition inhomogeneities in the AlInN layers (as suggested by the large Stokes shifts) that would induce lateral variation of the effective cavity length and mode localisation. But it may also be linked to more general issues in III–N materials, for example mode localisation by dislocations, or possible birefringence issues. The question is open and currently under investigation. The possible mode localisation observed on this sample would probably not be a great issue for vertical cavity laser applications as the typical size of a VCSEL (5 … 20 μm) will cover only one of these modes on average.

13.6
Conclusion

We have stressed the improvement of AlInN material quality for compositions near lattice-matched to GaN, in particular a residual optical absorption ≤100 cm^{-1} below GaN bandgap in AlInN/GaN stacks. A LM AlInN/GaN DBR was used successfully in a microcavity light emitting diode, record values in terms of DBR reflectivity (99.4%) and microcavity quality factor (≥800) prove the superior performance of this material system compared to the usual Al(Ga)N/GaN system.

These results have application for fundamental studies of light–matter interaction in III-nitrides, in which excitons are expected to have very strong oscillator strength.

The main device-related challenge to address deals with the fabrication of electrically-injected vertical cavity lasers, which are still not demonstrated in III–N semiconductors. Owing to their lattice-matching, a situation close to the well-known GaAs/AlAs material system, the growth of AlInN/GaN-based vertical cavity laser structures on commercially available low dislocation density templates such as GaN quasi-substrates ($N_D \approx 10^6$ cm^{-2}) should offer an unprecedented advantage over the AlGaN/GaN system. Indeed no additional dislocations would be added during the growth of these structures and the use of AlInN/GaN DBRs should lead to an improved lasing threshold and device lifetime.

Acknowledgement

This work was performed in the framework of the Swiss National Center of Competence Quantum Photonics program. Nicolas Grandjean and Raphaël Butté are grateful to Sandoz Family Foundation for financial support. We thank M. Laügt, CNRS-CHREA, for XRD mapping and S. Gradecak for TEM.

References

[1] M. Ilegems, R. A. Logan, and R. Dingle, J. Appl. Phys. **43**, 3797 (1972).
[2] M. Ilegems and R. Dingle, J. Appl. Phys. **44**, 4234 (1973).
[3] H. Amano, N. Sawaki, I. Akasaki, and Y. Toyoda, Appl. Phys. Lett. **48**, 353 (1986).
[4] S. Nakamura, N. Iwasa, M. Senoh, and T. Mukai, Jpn. J. Appl. Phys. **31**, 1258 (1992).
[5] S. Nakamura, T. Mukai, and M. Senoh, Appl. Phys. Lett. **64**, 1687 (1994).
[6] S. Nakamura, M. Senoh, S. Nagahama, N. Iwasa, T. Yamada, T. Matsushita, H. Kiyoku, and Y. Sugimoto, Jpn. J. Appl. Phys. **35**, L74 (1996).
[7] N. Grandjean and J. Massies, Appl. Phys. Lett. **71**, 1816 (1997).
[8] T. Matsuoka, Appl. Phys. Lett. **71**, 105 (1997).
[9] T. Fujimori, H. Imai, A. Wakahara, H. Okada, A. Yoshida, T. Shibata, and M. Tanaka, J. Cryst. Growth **272**, 381 (2004).
[10] T. Onuma, S. Chichibu, Y. Uchinuma, T. Sota, S. Yamaguchi, S. Kamiyama, H. Amano, and I. Akasaki, J. Appl. Phys. **94**, 2449 (2003).
[11] J. F. Carlin and M. Ilegems, Appl. Phys. Lett. **83**, 668 (2003).
[12] H. A. Macleod, in: Thin-film optical filters (A. Hilger Ltd, Bristol, 1985), p. 165.
[13] D. Brunner, H. Angerer, E. Bustarret, F. Freudenberg, R. Hopler, R. Dimitrov, O. Ambacher, and M. Stutzmann, J. Appl. Phys. **82**, 5090 (1997).
[14] Y. K. Kuo and W. W. Lin, Jpn. J. Appl. Phys. **41**, 5557 (2002).
[15] T. Peng, J. Piprek, G. Qiu, J. O. Olowolafe, K. M. Unruh, C. P. Swann, and E. F. Schubert, Appl. Phys. Lett. **71**, 2439 (1997).
[16] S. Yamaguchi, M. Kariya, S. Nitta, H. Kato, T. Takeuchi, C. Wetzel, H. Amano, and I. Akasaki, J. Cryst. Growth **195**, 309 (1998).
[17] M. J. Lukitsch, Y. V. Danylyuk, V. M. Naik, C. Huang, G. W. Auner, L. Rimai, and R. Naik, Appl. Phys. Lett. **79**, 632 (2001).
[18] C. Sasaki, H. Naito, M. Iwata, H. Kudo, Y. Yamada, T. Taguchi, T. Jyouichi, H. Okagawa, K. Tadatomo, and H. Tanaka, J. Appl. Phys. **93**, 1642 (2003).

[19] Y. H. Cho, G. H. Gainer, J. B. Lam, J. J. Song, W. Yang, and W. Jhe, Phys. Rev. B **61**, 7203 (2000).
[20] J. Dorsaz, J. F. Carlin, C. M. Zellweger, S. Gradecak, and M. Ilegems, phys. stat. sol. (a) **201**, 2675 (2004).
[21] O. Ambacher, W. Rieger, P. Ansmann, H. Angerer, T. D. Moustakas, and M. Stutzmann, Solid State Commun. **97**, 365 (1996).
[22] H. Benisty, H. De Neve, and C. Weisbuch, IEEE J. Quantum Electron. **34**, 1612 (1998).
[23] P. Maaskant, M. Akhter, B. Roycroft, E. O'Carroll, and B. Corbett, phys. stat. sol. (a) **192**, 348 (2002).
[24] H. Benisty, R. Stanley, and M. Mayer, J. Opt. Soc. Am. A **15**, 1192 (1998).
[25] B. Roycroft, M. Akhter, P. Maaskant, P. de Mierry, S. Fernandez, F. B. Naranjo, E. Calleja, T. McCormack, and B. Corbett, phys. stat. sol. (a) **192**, 97 (2002).
[26] F. Omnes, N. Marenco, B. Beaumont, P. de Mierry, E. Monroy, F. Calle, and E. Munoz, J. Appl. Phys. **86**, 5286 (1999).
[27] I. J. Fritz and T. J. Drummond, Electron. Lett. **31**, 68 (1995).
[28] T. Shirasawa, N. Mochida, A. Inoue, T. Honda, T. Sakaguchi, F. Koyama, and K. Iga, J. Cryst. Growth **190**, 124 (1998).
[29] H. M. Ng, T. D. Moustakas, and S. N. G. Chu, Appl. Phys. Lett. **76**, 2818 (2000).
[30] T. Ive, O. Brandt, H. Kostial, T. Hesjedal, M. Ramsteiner, and K. H. Ploog, Appl. Phys. Lett. **85**, 1970 (2004).
[31] C. F. Lin, H. H. Yao, J. W. Lu, Y. L. Hsieh, H. C. Kuo, and S. C. Wang, J. Cryst. Growth **261**, 359 (2004).
[32] R. Langer, A. Barski, J. Simon, N. T. Pelekanos, O. Konovalov, R. Andre, and L. S. Dang, Appl. Phys. Lett. **74**, 3610 (1999).
[33] N. Nakada, M. Nakaji, H. Ishikawa, T. Egawa, M. Umeno, and T. Jimbo, Appl. Phys. Lett. **76**, 1804 (2000).
[34] N. Nakada, H. Ishikawa, T. Egawa, T. Jimbo, and M. Umeno, J. Cryst. Growth **237**, 961 (2002).
[35] F. Natali, D. Byrne, A. Dussaigne, N. Grandjean, J. Massies, and B. Damilano, Appl. Phys. Lett. **82**, 499 (2003).
[36] K. E. Waldrip, J. Han, J. J. Figiel, H. Zhou, E. Makarona, and A. V. Nurmikko, Appl. Phys. Lett. **78**, 3205 (2001).
[37] J. F. Carlin, J. Dorsaz, E. Feltin, R. Butté, N. Grandjean, M. Ilegems, and M. Laügt, Appl. Phys. Lett. **86**, 031107 (2005).
[38] E. Feltin, R. Butté, J. F. Carlin, J. Dorsaz, N. Grandjean, and M. Ilegems, Electron. Lett. **41**, 94 (2005).
[39] T. Someya, R. Werner, A. Forchel, M. Catalano, R. Cingolani, and Y. Arakawa, Science **285**, 1905 (1999).
[40] S. Kako, T. Someya, and Y. Arakawa, Appl. Phys. Lett. **80**, 722 (2002).
[41] Y. K. Song, H. Zhou, M. Diagne, A. V. Nurmikko, R. P. Schneider, C. P. Kuo, M. R. Krames, R. S. Kern, C. Carter-Coman, and F. A. Kish, Appl. Phys. Lett. **76**, 1662 (2000).

Chapter 14
Microcavities in Ecole Polytechnique Fédérale de Lausanne, Ecole Polytechnique (France) and Elsewhere: Past, Present and Future

Claude Weisbuch and Henri Benisty

14.1
Introduction

It is a pleasure to give an overview on the past, present and future of microcavities at the symposium in honour of Marc Ilegems. He has long been connected with the topic of the interaction of light with semiconductors, having in particular conducted some of the pioneering optical studies of AlGaAs alloys and GaN while at Bell Labs in the late 1960s and early 1970s. Under his leadership, associated with others, Ecole Polytechnique Fédérale de Lausanne (EPFL) has developed into a major powerhouse in the general fields of optoelectronics, low-dimensional semiconductor heterostructures and quantum photonics. There are far too many achievements in these fields, both at EPFL and elsewhere, to be able to present in a few pages any reasonable coverage. This contribution should rather be understood as personal reminiscences on the topic as well as on our association with Marc Ilegems for a number of years in a most fruitful collaboration.

14.1.1
The Light–Matter Interaction in Semiconductors

The topic of light–matter interaction in semiconductors has long been the subject of a huge research effort because of both its importance in the fundamental understanding of these solids and the many applications that depend on it. On the former aspect, we can cite the photoconductivity property, light emission capability and various electro-optic and nonlinear effects. On the latter aspect, we recall such large-scale applications as light detectors, LEDs and laser diodes, photovoltaic generators, displays, etc.

From the very beginning, it has been recognized that the fundamental optical properties of semiconductors are intimately linked to features originating in the symmetry and dimensionality of the electron system. The first models of absorption edge of semiconductors and insulators relied on free Bloch electrons in the one-electron model [1]. However, soon after, the exciton concept, a bound electron–hole

Physics of Semiconductor Microcavities: From Fundamentals to Nanoscale Devices
Edited by Benoit Deveaud
Copyright © 2007 WILEY-VCH Verlag GmbH & Co. KGaA, Weinheim
ISBN: 978-3-527-40561-9

pair possessing a translational motion, was developed, as it appeared that the translational invariance of the crystals required the photoexcited electronic species to preserve the crystal symmetry [1]. The relative roles of free carriers or excitons in a variety of optical effects has been a subject of many studies, even controversies as witnessed by the recent discussions of exciton luminescence and laser action in II–VI materials at room temperature [2–7]. In brief, one expects excitonic effects at low densities/low temperatures/low impurity content, but both excitonic and free carrier effects are to be considered when studying optical properties of semiconductors, with a relative importance depending on conditions.

To have a full quantum description of the system, one has to consider the complete system of coupled electronic (excitons) and electromagnetic (photons) states. Then, situations are found at low enough temperatures and concentrations where both types of excitations interact coherently giving rise to coupled-mode excitations, the excitonic polaritons pioneered by Pekar [8] and Hopfield [9] in the late 1950s. This occurs when damping of both excitations is slower than the oscillation frequency between the two states, proportional to the splitting of the resulting normal modes and called the Rabi oscillation frequency (more precisely the vacuum Rabi oscillation frequency). The perturbation approach to the light–matter interaction (Fermi's golden rule) collapses and one deals with a strongly coupled system between light and matter instead of the usual weakly coupled system. It should, however, be stressed that full classical pictures of coupled oscillators also give rise to a normal mode splitting similar to that described by quantum mechanical models.

The field has undergone a major revival due to the possibility of fabricating systems with electron and photon states of varying dimensionality, activities in which Lausanne groups have played a very important role.

14.1.2
The Impact of Electronic Motion Quantization

The considerations of electron and photon state symmetries should also apply in the case of lower dimensionality semiconductor structures. One expects modifications in those optical properties based on free electrons due to changes in the densities of states with dimensionality [10]. The changes in exciton parameters (binding energies, Sommerfeld factor, etc.) can also dramatically alter the balance between the two types of excitations. Early theoretical work by Sugano on excitons in two-dimensional (2D) systems predicted an increase of the exciton binding energy, of up to four times, due to the electron–hole pair state compression [11].

These expectations were well demonstrated in the late 1970s and 1980s by the advent of quantum wells (QWs) [10], true 2D systems for electronic properties. The improved light–matter interaction over three-dimensional (3D) 'bulk' semiconductors relies to a first order on the freezing of one degree of freedom in such structures, yielding more spectrally concentrated optical features such as absorption and gain for free carriers, based on the square density of states (DOS). This was shown by the progress as a result of the use of QW active layers in semiconductor lasers [10]. It had been quite a matter of debate as to whether whether excitons were present, under strong excitation conditions, and would thus play a major role in the laser action, for instance if their presence would lead to lower thresholds. It should be stressed that experiments displaying an exciton signature are non-trivial, and that the mere coincidence (if any at all, see [12]!) of exciton absorption or emission at low densities, usually observed from the top of a sample, and an in-plane laser line is certainly no proof of the identity of the two excitations under the two conditions [12]. Theoretical modelling also points to laser action somewhat at the exciton energy, although originating from electron–hole pair recombination, a coincidence due to bandgap renormalization and reabsorption effects. Additionally, analysis of photoluminescence (PL) experiments is often unclear due to the inhomogeneous excitation situation.

A major surprise in QW luminescence is the dominance of free exciton species over bound excitons, in contrast to bulk materials, and also the higher quantum efficiency of QW materials [13], which might explain why excitons were supposed to be connected with the better laser action observed [3]. This exciton importance was only understood in the late 1980s through the measurement of ultrashort radiative lifetimes by Deveaud [14], based on the work by Andreani et al. [15, 16], who rediscovered independently the prediction by Agranovitch and Dubovskii of short radiative lifetimes due to the symmetry breaking between 2D exciton states and 3D photon states which makes luminescence allowed at first order [17]. This shortened radiative lifetime explains in part the much higher radiative efficiency of excitons in QWs than in the bulk at low temperatures. The good luminescence efficiency often observed at room temperature might have some origin in this effect provided that radiative recombination occurs through excitons. It is more likely to be due to improved free carrier recombination because of a mix of causes such as higher carrier concentrations for a given injection intensity and lower active nonradiative defects (the ultimate case being the nitride alloys where In segregation in InGaN QWs seems to localize carriers and forbid them to reach the numerous dislocations). All told, laser action in QW lasers seems to be always due to free carriers, with the gain over 3D active layers being due to the square DOS, the reduction of the quantum states to be inverted to reach threshold (the major factor) and some improvement in radiative efficiency.

Going to 2D QWs with increased exciton binding energy still allowed observable exciton effects at room temperature, such as in absorption, and therefore led to applications based on larger excitonic electro-optic (EO) and nonlinear (NL) effects [18]. The reason for invoking exciton effects for these phenomena, while rejecting them for spontaneous or laser emission, is that EO and NL effects rely on unrelaxed excitations, leading to an exciton linewidth determined by scattering, often smaller than or similar to the exciton binding energy, while light emission originates from thermally relaxed excitations with a broadening of kT, in most circumstances larger than the exciton binding energy. This is why it is legitimate to claim simultaneously that exciton effects are present in EO and NL phenomena while they are absent in light emission.

In view of the successes of QWs, much effort has been devoted since the mid-1980s to the fabrication and evaluation of lower dimensionality systems such as quantum wires (QWRs) and quantum dots (QDs), in order to freeze more degrees of freedom, and therefore reach sharper optical features in the spectrum of such structured solids as determined by the DOS. The interplay between electronic and photonic dimensionalities for QWRs and QDs again introduces specific modifications of the lifetime against bulk material. However, QWRs and even more so QDs, ultimate solid-state systems with atom-like discrete energy levels, suffer from the difficulty of fabricating such tiny structures with small enough size fluctuations. It can also be remarked that the room-temperature QW excitons readily yield sharp features for unrelaxed excitations. Therefore, for phenomena that rely on such unrelaxed excitations (NL or EO coefficients) there will be limited advantages to using QDs over QWs (to be more precise, properties only based on oscillator strength will be similar but properties involving dynamics or relaxation might be very different). It is only in those cases where one relies on single quantum system phenomena, i.e. focusing on single QD properties, such as single photon emitters or Qbits, that the full quantum nature of single QDs is indeed exploited [19].

It should be pointed out that in spite of the drawback of size fluctuations, QD lasers are slowly emerging as useful devices. However, their properties are not so much due to the zero-dimensional (0D) quantization than to better operating parameters, which can be viewed as 'second-order effects': growth on large, high-quality GaAs wafers, high-bandgap discontinuities allowing for good temperature insensitivity, low facet degradation, etc. [20].

14.1.3
The Impact of Photon Mode Quantization

One is not constrained to rely only on 3D photon behaviour when it comes to controlling the light–matter interaction in solids. This was only recognized in the late 1980s, although it was recognized long before in atomic physics. For atoms in cavities it led to the remarkable field of cavity quantum electrodynamics (QED) [21].

For semiconductors interacting with photonic systems of reduced dimensionality, new effects can appear [22], fundamentally due to the fact that two continua of states come into play to yield the usual broad optical features, and that these can be greatly changed by changing the dimensionalities of electron or photonic states. Of course, as for electron motion quantization, it is possible to define photon quantizing systems with two, one or zero dimensions (Fig. 14.1). The 0D case is a limiting system because of the difficulty of manufacturing a true single-mode 3D optical cavity. However, the mode number can be so reduced in micropillar or small photonic crystal (PC) microcavities that only one mode will interact with the available electronic states.

14.2
The Interplay of Photon and Electron Dimensionalities

Having described separately the effect of dimensionality of electron and photon states, we should now concentrate on the importance of the interplay of their dimensionalities through a few important examples.

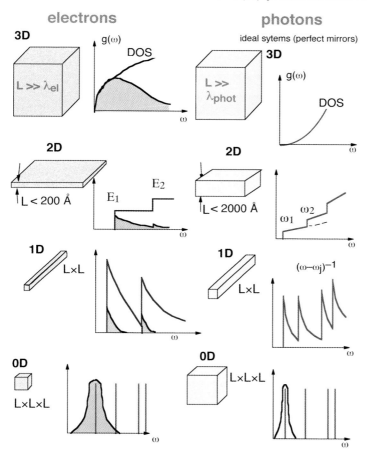

Fig. 14.1 Schematics of systems with varying dimensionalities, through electron motion quantization in quantum wells, wires and boxes (left) or photon mode quantization in microcavities (right).

As mentioned above, light absorption and emission processes for weakly damped 3D excitons and photons rely on coupled-mode excitations, excitonic polaritons. These were first demonstrated in II–VI compounds, and then in GaAs (Fig. 14.2) in the seminal work by Sell et al. [23]. The absorption phenomenon is sharply changed in the polariton picture, as it is due to polariton scattering and no more to the elementary light–matter interaction. Therefore, Beer's law ($T \propto e^{-\alpha L}$) is not obeyed. It was remarked by Hopfield, and expanded on by Toyozawa, that luminescence at finite photon energies could only be an out-of-equilibrium phenomenon, as the ground state of the coupled exciton–photon states is a zero energy photon-like state [24, 25]. Luminescence actually occurs because of the existence of a polariton energy relaxation bottleneck in the strongly mixed energy region (the anti-crossing point), where phonon-assisted energy relaxation is much decreased by the highly

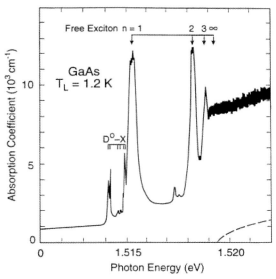

Fig. 14.2 Exciton absorption in pure GaAs. One observes the Rydberg series of free exciton peaks, as well as very sharp donor-bound exciton peaks. Note that the 'reduced' absorption at the $n = 1$ exciton peak, experimentally due to light appearing on the non-excited side of the sample, is in fact due to transmitted polaritons (they are less prone to dissipation than $n = 2$ excitons or polaritons), to multiple elastic polariton scattering and to resonant polariton fluorescence (R. Ulbrich and C. Weisbuch, unpublished, 1976).

decreased DOS of photon-like states while the radiative lifetime (the time of flight to the semiconductor interface) is sharply reduced for these states which travel at the speed of light.

As the dimensionalities of exciton and photon states can be varied independently, new phenomena are bound to happen. The first case is that of QW 2D excitons interacting with a normal 3D photon environment, where there is no symmetry preservation in the photon–exciton interaction. Only the in-plane momentum is conserved in optical transitions. An exciton is then coupled to a continuum of photon states. There is then no possibility of reversible strong coupling. A similar effect occurs for 1D electron QWRs in a 3D photonic environment [26, 27].

Conversely, one expects that whenever excitons and photons have the same dimensionalities a situation of strong coupling can develop like in 3D due to the conservation of crystal translational symmetry in light absorption and emission processes. This was indeed observed for 2D QW excitons and planar microcavities in 1992 [28]. As in 3D, radiative recombination becomes a second-order process, now due to the imperfect reflectivity of mirrors, which allows photons to leak out (hence with a tailored lifetime that has no link with the photon–exciton interaction). Due to the stronger exciton binding energy, such strongly coupled excitations can persist up to room temperature.

The novelty of the system comes from the fact that cavity polaritons (CPs) have a finite ground state, with very remarkable features, as these quasi-particles behave like photons (light mass) or like excitons (interactions with one another, with phonons, with disorder). Hence, one could expect remarkable collective quantum effects (such as Bose–Einstein condensation, BEC) as the mean free path can be much larger than the average particle distance at concentrations for which coherent interactions should occur. Unfortunately, like their 3D counterparts, CPs experience a relaxation bottleneck due to the diminishing DOS when relaxing towards photon-like (i.e. low-mass) states [29]. This has greatly hampered so far the applications of the strong coupling in the real world (i.e. room-temperature and electrically injected devices), and also direct observations of fundamental phenomena such as spontaneous BEC. However, recent advances point to the existence of a quantum fluid obtained through relaxation induced by NL effects [30, 31].

Even for unbound, free carriers, one encounters the importance of the interplay between electron and photon dimensionalities. It is not usually based on strong coupling as the oscillator strength of a single electron–hole pair, even bound to an exciton, leads to a rather small Rabi splitting [32] (in 2D microcavities, one has a collective coupling of the exciton dipole strength per unit surface to the vacuum electric field of the resonant cavity mode, also expressed per unit surface). It originates in the modification of Fermi's golden rule, the effect being important only for 0D photonic systems (Fig. 14.3). Indeed, it can be shown that in weak coupling conditions the free carrier spontaneous emission time is little changed when going from 3D to 2D or even 1D photon systems [33, 34]. However in 0D photonic systems, i.e. in a 3D microcavity, only a single photon mode can interact with the electronic system, giving rise eventually to a modification of the emission rate proportional to the quality factor of the cavity, the Purcell effect [21, 35]. This is, however, only true as far as the electronic transition is spectrally narrower than the cavity mode, otherwise the rate is determined by the 'material's quality factor', its inverse relative emission linewidth $\omega/\Delta\omega$. A typical relative linewidth for free carriers in the solid state is $kT/(1\text{ eV})$, a few percent, due to thermal broadening, unless all the translational degrees of freedom are frozen, as in QDs. It is therefore no surprise that the Purcell effect can be large enough to be unambiguously observed in systems where both electron and photon states are 0D, i.e. for QDs placed in 3D microcavities, or micropillars in the case of the breakthrough experiment by Gérard et al. [36] (see also [19]).

14.3
Looking Backwards: a Short History of Microcavities in Solids

The first solid-state emitting systems having high-quality cavities were of course vertical-cavity surface-emitting lasers (VCSELs). At first they were not microcavities as they usually operated at rather high cavity order to increase the cavity quality factor.

The use of microcavity effects to increase LED efficiency was very well developed at Bell Labs by Schubert and Hunt [37], where they demonstrated, and patented, that the best structure was an asymmetric cavity with a metal mirror and a DBR structure. Their designs seem, however, to have culminated at around 10% efficiency. The work was vigorously pursued in the context of a European project involving, in particular, the University of Gent and EPFL. In a series of increasingly refined experiments, in particular emphasizing the importance of using detuned cavities, De Neve and Blondelle [38] reached an efficiency of 20%. In more optimized work by a team from Ecole Polytechnique and EPFL, a record of 28% in air was reached, this time using a cavity consisting of the GaAs interface as a top emitting mirror, and a GaAs/AlO$_x$ mirror on the bottom [39].

Modifications of the spontaneous emission time, the Purcell effect, were also researched in the late 1980s when the possibility arose of making planar microcavities, often similar to early VCSELs with non-epitaxial mirrors (i.e. oxide mirrors). However, for planar cavities, the effect was always quite small [35]. Actually, while it was believed that an increase of three times could be seen in the emission rate from simple modelling of metallic cavities, it was shown by Bjork that for dielectric cavities the maximum effect would be a factor of 1.5 [40].

For exciton effects in microcavities, Yamamoto et al. showed in the early 1990s hat emission could be highly directional, but they could not observe the strong coupling [41].

Let me describe the sequence of events that led to the discovery of CPs. I was at the University of Tokyo for four months in the summer of 1991, hosted by Professor Arakawa, and I was looking for some experiments to do. After some wonderings about QWs in high magnetic fields and QD excitation spectroscopy, two experiments that did not appear practical in the short time I was there, I decided to look at QWs in microcavities. Arakawa had some samples from a previous experiment [42], and I was thinking that their properties could not be as simple as previously reported.

I then saw some strange effects when moving the sample around the resonance (mainly due to the imbalance between front and back mirror, but at that time I did not see this point). I therefore asked Arakawa for better samples (with only one type of wells, as the old ones were two-well samples) and M. Nishioka grew three samples with variable number of wells, all of them excellent, a remarkable feat as he had not grown such complex, demanding samples for two years.

I then tried to observe the increase in luminescence intensity that should occur when excitons and microcavities are resonant. Such a search was made possible by the fact that the growth speed was non-uniform over the wafer, a very fortunate occurrence: moving the observation point by point over the wafer, I could observe different situations of relative energies between the microcavity photon mode and the QW exciton. I observed that there was no clear-cut resonance (this point was later elucidated in a thoughtful paper from EPFL [43]). I decided then that luminescence could not tell me the resonance, as luminescence is a complex phenomenon, but that reflectivity would. I thus mounted a reflectivity measurement, but then I would never observe a resonant peak, but a doublet whenever I would be 'around' resonance (as displayed by some resonance in PL intensity).

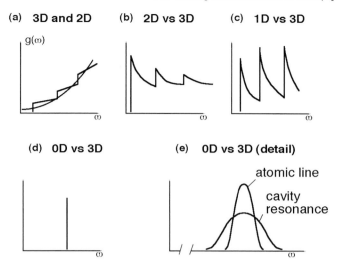

Fig. 14.3 Schematics of the variation of the radiative rate compared to infinite medium value as modified by cavity QED effects when in 2D, 1D and 0D photonic structures.

After many observations of the double peak behaviour of the resonance crossing, and the absence of any single peak no matter how hard I looked, I then decided that this was a true physical effect, and had to decide which.

Having worked some 15 years before on excitonic polaritons in the bulk [44–47] and having in particular determined their dispersion curve directly by resonant Brillouin scattering [44], I was familiar with such strong coupling excitations. In particular, I knew their dynamics, in particular that the scattering times are much longer than for free carriers, due to their neutral type. Therefore, I was not afraid to state from a 'back-of-envelope' calculation that indeed the coupling strength of cavity photons with excitons was stronger than any scattering mechanism, at variance with 'common knowledge' which predicted that the strong coupling just previously observed in atomic physics would never occur in solid-state physics due to excitation damping (Fig. 14.4).

A simple modelling by A. Ishikawa using linear dispersion theory allowed the determination of an exciton oscillator strength of the right order of magnitude. The results from this simple model have since then been verified through both semi-classical and full quantum theory, by the Lausanne theory group in particular [48].

I had difficulties convincing some colleagues that indeed one was observing strong coupling...

As for publication, one referee of *Physical Review Letters* (PRL) observed that everything was fine with the experiment and the model, but as such phenomena had already been observed in atomic physics, it did not warrant publication in PRL. The second referee said exactly the same thing on experiment, model and atomic physics, but reached the opposite conclusion in that it was so surprising that it would

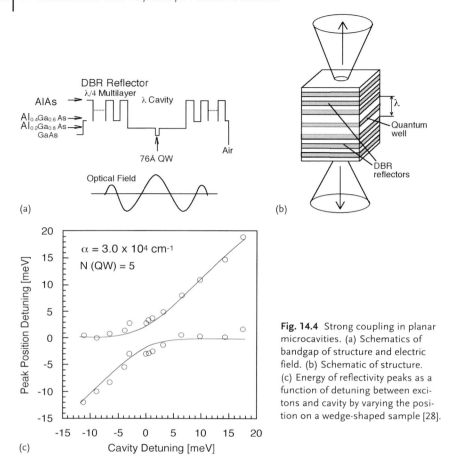

Fig. 14.4 Strong coupling in planar microcavities. (a) Schematics of bandgap of structure and electric field. (b) Schematic of structure. (c) Energy of reflectivity peaks as a function of detuning between excitons and cavity by varying the position on a wedge-shaped sample [28].

occur in the solid state that it certainly was PRL material. After another two months, the third referee that I had requested just came up with the remark: 'That's what we like in physics: always good for surprises' and it was published [28].

The name cavity polaritons was given at the Erice summer school [22] during some heated sessions (involving in particular the two Elis, Eli Burstein and Eli Yablonovitch) on the nature of these excitations.

14.4
The Birth of the Microcavity Effort in Lausanne

After the discovery of CPs in Tokyo, a European project named SMILES (Semiconductor Microcavity Light Emitters) aiming at exploiting their outstanding properties was launched with John Hegarty (Trinity College, Dublin) as a coordinator, and associating Roel Baets (University of Gent), Marc Ilegems, Romuald Houdré, Ross

Stanley and U. Oesterle (EPFL), Markus Pessa (Tampere), and Henri Benisty and Claude Weisbuch (Ecole Polytechique, Palaiseau). The idea was that the lifetime being a thousand times (1 ps) shorter for CPs than for usual excitations, they would lead to ultrafast emitters with high internal quantum efficiency.

EPFL was well equipped for such studies: a former effort by the group of Franz Reinhart had very early developed electrically injected VCSELs [49], and had also studied QD systems. The group of Marc Ilegems had a great deal of experience in high-quality MBE growth as demonstrated by the growth of microcavities with a quality factor of 5000, limited by residual GaAs absorption and interface scattering [50]. Finally, the theory group headed by Antonio Quattropani had a long experience in exciton phenomena.

With this background, it is no surprise that microcavity studies developed quickly in Lausanne. Experimentally, in the group of Ilegems, high-quality microcavity samples allowed the observation of CPs up to room temperature [51]. The CP dispersion curve was determined from the analysis of the angular dependence of emission [52]. In the heated discussions on the existence of motional narrowing to explain the narrow lines observed in spite of the large QW structural disorder, both EPFL experimental and theory groups were instrumental in showing that the narrow lines observed were rather due to the disorder averaging over the very large CP wavefunction (essentially the photon mode volume), leading to lines with homogeneous linewidth. A simple way to reach this conclusion is to observe that CPs are a special system for the observation of Rabi splitting, as one is dealing with an inhomogeneous system.

Houdré et al. established in a semi-classical model that when the inhomogeneous linewidth is smaller than the Rabi splitting, one mainly observes a standard doublet with the homogeneous linewidth [53]. From there, the importance of inhomogeneous broadening being downplayed, the observed asymmetric homogeneous phonon broadening could be well explained through a classical linear dispersion model or through full quantum calculation [54, 55]. As regards devices, although resonant luminescence was shown experimentally [56] and theoretically [57] to decay very fast, the nonresonant decay was shown to decay much slower. This was theoretically shown [29] to be due to the existence of a relaxation bottleneck for polaritons, arising from the diminishing DOS when the dispersion curve becomes photon-like in the resonant region, as for bulk excitonic polaritons [58]. As this would be the excitation situation encountered in electrical injection for devices, hopes of a new family of LED faded, although there might be some possibilities based on hybrid organic–inorganic microcavities by associating the good electrical injection properties of semiconductor materials with the ultrafast energy relaxation properties of organic ones [59].

The activity in EPFL developed from then on in two main directions. In the fundamental one, many results were obtained on the linear and NL behaviour of CPs, experimentally or theoretically, in Ilegems', Deveaud's and Quattropani's groups (see the various contributions to this volume). In Ilegems' group, besides the systematic study of polariton linewidth, particularly noteworthy are: (i) the observation a resonant Rayleigh scattering over a ring; (ii) the coherent backscattering from

CPs which act as a disordered 2D optical system; and (iii) the observation of single beam NL emission phenomena, with specific far-field emission patterns, most certainly connected with the inhomogeneous excitation density (see Romuald Houdré's contribution in this volume).

As regards applications, a 'strategic retreat' was operated towards weak coupling microcavities, within the European SMILES project mentioned above, to be followed by a second project, SMILED (semiconductor microcavity LEDs). EPFL made major contributions in red microcavity LEDs, in modelling and optimization, etc., culminating in setting the world record in planar microcavity with an overall efficiency of 28% [39]. Still today, the theses by Paul Royo and Daniel Ochoa are mines of information for researchers in the field [60]. More recently, work has developed in activities on photonic crystals geared towards integrated photonics applications.

14.5
Why We Like Microcavities!

At this point it is worth wondering why there is so much interest in microcavities. They are basically systems with controlled light–matter interaction, and they are at the same time: (i) a new playground to test new ideas in physics, quite often leading to unexpected observations; (ii) challenging physical systems, in particular when trying to obtain fully quantized 0D systems for both electrons and photons; and (iii) a powerful concept when it comes to large-scale applications.

A first theme that we can identify for the use of microcavities occurred in the first 'era' in the field (late 1980s and early 1990s). It dealt with the control of the directionality of spontaneous emission in the 'weak' coupling regime between semiconductor excitations and planar cavities. Such an effect had its roots in both atomic physics (the landmark paper by Kastler on the directionality of spontaneous emission of atoms in a Fabry–Perot cavity in 1962) and modifications of the emission of atoms and molecules in dielectrics and interfaces (a field pioneered by Drexhage and Lukosz, still active, see e.g. Lagendijk). The best early effort in the control of directionality is represented by the activity of Hunt and Schubert at Bell Labs demonstrating microcavity LEDs with dielectric DBR and metallic mirrors as the best device candidates [37].

A second theme of microcavity research in the solid state was the search for a modification of the lifetime through the Purcell effect, but the use of planar cavities and/or of broad emission lines prevented the observation of a large effect [25]. This has been recently revived by the use of single narrow QD lines interacting with single photon modes in pillar microcavities and PhC structures [61–63].

A third theme, more ambitious, aims at changing the light–matter interaction from weak to strong. As mentioned above, this situation was observed for bulk semiconductors, but the situation is of limited interest due to the zero energy of the exciton polariton state in 3D. This is to be contrasted with the situation occurring for CPs, due to the 'strong' coupling regime between 2D excitons and 2D photons [28], where the CPs have a nonzero ground state energy. As discussed above, this has proved to be a very fertile ground for new physics, with many unexpected results of major interest.

A fourth fundamental theme is arising from the recent advances in obtaining more control of the light modes through systems with lower photonic dimensionalities. One aims at reproducing in the solid state many of the beautiful quantum physics experiments of cavity QED, and eventually based on the specificities of solid-state systems. The compactness of solid systems might also prove very useful, for instance if one wishes to implement compact quantum computers. A number of structures and geometries have been put to use: some are rather large and show confinement based on total internal reflection, such as silica microspheres, photonic wires and toroidal cavities [64]; others are more microscopic and based on multiple Bragg reflections such as micropillars and photonic crystal-based microcavities. The competition between the two types of systems for QED effects is very open, the first solutions being able to compensate for their large volume because of the very high quality factors Q reached (a common factor of merit is the ratio Q/V, where V is the photon mode volume).

It might prove useful to wonder why it appears that ones gets more leverage when quantizing photons as compared to electron states, or at least why it seems a more useable approach to new phenomena and applications (I know that I will raise some controversy here).

To compare in a basic manner the two approaches of confinement, let us consider (i) a single-electron quantum box, interacting with a continuum of blackbody photon modes or with a standard 1 mW light beam, and (ii) an optical microcavity, of approximate size $(\lambda/2n)^3$, interacting with an optically active medium located within the microcavity. In the first case, the electron–photon interaction of the single-electron states are unmodified when compared to bulk material, at least to a first order: the lifetime and oscillator strength are unchanged. However, the optical beam does not interact enough with the electronic system: it can only generate $\sim 10^9$ (spontaneous rate) to 10^{10} or 10^{11} (stimulated rate) transitions per second, not enough to control efficiently or generate a sizeable optical beam. Moreover, the confinement factor, due to the overlap of the optical beam with the quantum box, is small, fundamentally due to the difference in wavelengths between the electron and photon. In the second case, the lifetime of the electronic excitations is almost unchanged, as well as the coupling to the optical field: the resonance effect of the cavity increases the resonant electric field as much as it decreases the fields of all other modes. The cavity acts as a concentrator of optical fields into a single optical mode. However, in this case, the active material volume, $\sim(\lambda/2n)^3$, is such that it can contain enough quantum states ($>10^6$) in bulk, or multi-QWs, QWRs, QDs (if they are all resonant), so that they can control or generate a sizeable optical beam (10^{15}–10^{18} transitions per second). In addition, photon spontaneous emission occurs selectively in the cavity mode, which is not achieved for QDs coupled to a continuum of photon modes.

It can be said that the situations of electron or photon confinements are not symmetrical: whereas both bring sharper optical features, the photon confinement scheme adds mode selectivity and a single microcavity handles enough power to achieve a useable device, whereas the electron confinement scheme requires a large number of quantum boxes to achieve sizeable effects. This difference can be traced to the Boson nature of photons, which allows many photons in a single optical mode, whereas the Fermion nature of the electron allows only one electron in a given quantum state (a single electron mode) in a quantum box.

As discussed above, using single QDs in single-mode 3D microcavities should bring the best of both worlds, in particular bringing the coupling efficiency of a single QD to the mode close to 100% (when approaching the strong coupling condition, without requiring it in full) and enhancing emitted intensity by the Purcell effect.

14.6
The Future: What Are We Looking For?
(What We Really See is Sometimes a Matter of Controversy)

This short overview of what has kept us busy in the past is a good indicator of what might lie ahead.

In those systems that allow it, collective effects might be the way to go: it is clear that CPs are a system of choice to explore Boson interactions under quantum conditions. This is mainly the area of 2D systems, in particular CPs. All the parameters are adequate, because of the coexistence of large interparticle interactions (Coulomb forces between the exciton constituents) and large mean free paths (very high particle photon-like velocities, i.e. small DOS) of the hybrid excitations. Some effects have already been seen, often appearing with spectacular experimental features, but we are certainly still at the beginning of exploiting a very unique system.

Single quantum effects are also a topic of prime choice for future investigations. Single photon emission is a flourishing field, and more is to come from novel structures. The challenge here is to obtain single or correlated pair emission in practical devices, i.e. at least compact, hopefully electrically injected and operating at high temperature, possibly room tempera-ture. Obtaining strong coupling in single QD structures is an obviously highly desirable objective. While there have been some recent demonstrations of such effects, more effort is also needed here to obtain practical structures where one can concentrate on quantum state manipulation.

Applications will also drive the field. There are major applications in view, be it in energy-efficient devices (solid-state lighting), telecommunications (quantum secure communications, high functionality integrated optics circuits, etc.). The field of biology should also be a fertile playground.

It should be stressed, however, that, as usual for novel fields, the more interesting results and applications will come from concepts that we do not foresee today!

References

[1] R. S. Knox, Theory of Excitons, Vol. Suppl. 5 (Academic Press, New York, 1963).
[2] I. Galbraith and S. W. Koch, J. Cryst. Growth **159**, 667 (1996).
[3] J. Ding, H. Jeon, T. Ishihara, M. Hagerott, A. V. Nurmikko, H. Luo, N. Samarth, and J. Furdyna, Phys. Rev. Lett. **69**, 1707 (1992).
[4] A. Girndt, S. W. Koch, and W. W. Chow, Appl. Phys. A **66**, 1 (1998).
[5] M. F. J. Pereira and K. Henneberger, phys. stat. sol. (b) **202**, 751 (1997).
[6] J. Szczytko et al. phys. stat. sol. (c) **1**, 493 (2004).
[7] D. Hägele et al., Physica B **272**, 328 (1999).
[8] S. I. Pekar, Sov. Phys.-JETP **6**, 785 (1958).
[9] J. J. Hopfield, Phys. Rev. **112**, 1555 (1958).
[10] C. Weisbuch and B. Vinter, Quantum Semiconductor Structures: Fundamentals and Applications (Academic Press, Boston, 1991).
[11] M. Shinada and S. Sugano, J. Phys. Soc. Jpn. **21**, 1936 (1966).
[12] S. Tarucha, Y. Horikoshi, and H. Okamoto, Jpn. J. Appl. Phys. **22**, L482 (1983).
[13] C. Weisbuch, R. C. Miller, R. Dingle, A. C. Gossard, and W. Wiegmann, Solid State Commun. **37**, 219 (1981).
[14] B. Deveaud, F. Clérot, N. Roy, K. Satzke, B. Sermage, and D. S. Katzer, Phys. Rev. Lett. **67**, 2355 (1991).
[15] L. C. Andreani, F. Tassone, and F. Bassani, Solid State Commun. **77**, 641 (1991).
[16] L. C. Andreani, in: Confined Electrons and Phonons: New Physics and Applications, edited by E. Burstein and C. Weisbuch (Plenum, New York, 1995), p. 57.
[17] V. M. Agranovitch and O. A. Dubovskii, JETP Lett. **3**, 223 (1966); reprinted in: Confined Electrons and Photons: New Physics and Applications, edited by E. Burstein and C. Weisbuch (Plenum, New York, 1995), p. 795.
[18] D. S. Chemla, D. A. B. Miller, P. W. Smith, A. C. Gossard, and W. Wiegmann, IEEE J. Quantum Electron. **20**, 265 (1984).
[19] P. Michler, ed., Single Quantum Dots: Fundamentals, Applications and New Concepts (Springer, Berlin, 2003).
[20] D. Bimberg, M. Grundmann, and N. N. Ledentsov, Quantum Dot Heterostructures (Wiley, Chichester, 1999).
[21] S. Haroche and D. Kleppner, Phys. Today **42**, 24 (1989).
[22] E. Burstein and C. Weisbuch, Confined Electrons and Photons: New Physics and Applications (Plenum, New York, 1995).
[23] D. D. Sell, S. E. Stokowski, R. Dingle, and J. V. DiLorenzo, Phys. Rev. B **10**, 4568 (1973).
[24] Y. Toyozawa, Prog. Theor. Phys. Suppl. **12**, 111 (1959).
[25] J. J. Hopfield, Proc. Int. Conf. Physics of Semiconductors (Kyoto, 1966); J. Phys. Soc. Jpn. **21** (Suppl.), 77 (1966).
[26] H. Akiyama et al., Phys. Rev. Lett. **72**, 2123 (1994).
[27] D. Gershoni and M. Katz, Phys. Rev. B **50**, 8930 (1994).
[28] C. Weisbuch, M. Nishioka, A. Ishikawa, and Y. Arakawa, Phys. Rev. Lett. **69**, 3314 (1992).
[29] F. Tassone et al., Phys. Rev. B **53**, 7642 (1996).
[30] P. G. Savvidis and P. G. Lagoudakis, Semicond. Sci. Technol. **18**, S311 (2003).
[31] C. Ciuti and I. Carusotto, phys. stat. sol. (b) **242**(11) (2005) (this issue).
[32] L. C. Andreani, G. Panzarini, and J. M. Gérard, Phys. Rev. B **60**, 13276 (1999).
[33] S. Brorson and P. M. W. Skoovgard, in: Optical Processes in Microcavities, edited by R. K. Chang and A. J. Campillo (World Scientific, Singapore, 1996), p. 77.
[34] S. T. Ho, L. Wang, and S. Park, in: Confined Photons, edited by H. Benisty et al., Lecture Notes in Physics Vol. 531 (Springer, Berlin, 1998), p. 243.
[35] H. Yokoyama, Y. Nambu, and T. Kawakami, in: Confined Electrons and Phonons: New Physics and Applications, edited by E. Burstein and C. Weisbuch (Plenum, New York, 1995), p. 427.

[36] J. M. Gérard et al., Phys. Rev. Lett. **81**, 1110 (1998).
[37] N. E. J. Hunt et al., Electron. Lett. **28**, 2169 (1992).
[38] H. De Neve et al., Appl. Phys. Lett. **70**, 799 (1997).
[39] M. Rattier et al., IEEE J. Sel. Top. Quantum Electron. **8**, 238 (2002).
[40] G. Bjork, IEEE J. Quantum Electron. **30**, 2314 (1994).
[41] G. Bjork et al., Phys. Rev. **44**, 969 (1991).
[42] T. Yamauchi, Y. Arakawa, and M. Nishioka, Appl. Phys. Lett. **58**, 2339 (1991).
[43] R. P. Stanley et al., Phys. Rev. B **53**, 10995 (1996).
[44] C. Weisbuch and R. G. Ülbrich, Phys. Rev. Lett. **39**, 654 (1977).
[45] R. Ulbrich and C. Weisbuch, Phys. Rev. Lett. **38**, 865 (1977).
[46] C. Weisbuch and R. Ulbrich, J. Lumin. **18–19**, 27 (1979).
[47] C. Weisbuch and R. Ulbrich, in: Light Scattering in Semiconductors, Topics in Current Physics, Vol. 51 (Springer, Berlin/Heidelberg/New York, 1982).
[48] V. Savona et al., Solid State Commun. **93**, 733 (1995).
[49] C. Wütrich et al., Electron. Lett. **26**, 1601 (1990).
[50] R. P. Stanley et al., Appl. Phys. Lett. **65**, 1883 (1994).
[51] R. Houdré et al., Phys. Rev. B **49**, 16761 (1994).
[52] R. Houdré et al., Phys. Rev. Lett. **73**, 2043 (1994).
[53] R. Houdré, R. P. Stanley, and M. Ilegems, Phys. Rev. A **53**, 2711 (1996).
[54] V. Savona et al., Phys. Rev. Lett. **78**, 4470 (1997).
[55] C. Ell et al., Phys. Rev. Lett. **80**, 4795 (1998).
[56] T. Norris et al., Phys. Rev. B **50**, 14663 (1994).
[57] V. Savona and C. Weisbuch, Phys. Rev. B **54**, 10835 (1996).
[58] U. Heim and P. Wiesner, Phys. Rev. Lett. **30**, 1205 (1973).
[59] V. M. Agranovich, H. Benisty, and C. W. Weisbuch, Solid State Commun. **102**, 631 (1997).
[60] P. Royo, Thesis, Ecole Polytechnique Fédérale de Lausanne, 2001.
D. Ochoa, Thesis, Ecole Polytechnique Fédérale de Lausanne, 2001
(available at: http://membres.lycos.fr/danielochoa/accueil.html).
[61] J. P. Reithmaier et al., Nature **432**, 197 (2004).
[62] T. Yoshie et al., Nature **432**, 200 (2004).
[63] A. Badalato et al., Science **308**, 1158 (2005).
[64] K. J. Vahala, Nature **424**, 839 (2003).

Subject Index

a
absorption 291
acoustic phonon 62, 70
angle-resolved photoluminescence 50
angle-resolved reflectance 94
angular momentum 187
anomalous correlation 12, 221, 238
anti-crossing 46, 57

b
backscattered direction 79
bandgap renormalisation 56
BEC of excitons 216
Beer's law 291
Berezinskii–Kosterlitz–Thouless 17, 215
bi-exciton 80, 101
Bir–Aronov–Pikus 188
bistability 129
bleaching 222
Bloch electron 287
Bloch vector 89
blue shift 109, 132, 238
Bogoliubov 16, 127, 137, 214
Boltzmann 58
Born approximation 70
Born–Markov approximation 200
Bose commutation rule 124
Bose fluctuation field 221
Bose operator 231
Bose particles 212
Bose statistics 212
Bose–Einstein condensation 13, 63, 80, 123, 211, 212, 293
Boser 63, 80
Boson 108, 300

bosonic stimulation 189
bottleneck 58, 217, 222
branch sticking 144
Brillouin zone 89, 96

c
cavity mode 125, 299
cavity polariton 45, 50, 293, 296
cavity pulling 58
cavity QED 299
cavity quantum electrodynamics (CQED) 4, 45
cavity resonance 35
cavity-polariton laser 56
chemical potential 127
Cherenkov regime 146
circular polarization degree 203
coherent backscattering 79
coherent control 161
coherent state 229
collective excitation 130, 215
composite boson 79
Coulomb interaction 220, 237
Coulomb scattering 7

d
damping rate 46, 159
dark state 187
de Broglie wavelength 79
decoherence 15
density matrix 193
detuning 71, 108
Dick state 68
dipole matrix element 46
dislocation density 266
disorder 65, 151

Subject Index

dispersion curve 55
distributed Bragg reflector (DBR) 2, 31, 247, 261, 277, 294
distributed feedback (DFB) 252
Dyakonov–Perel' 188

e

elastic RRS ring 139
electron–hole pair 55, 153, 218, 221
Elliott–Yaffet 188
energy and momentum conservation 106, 112
entanglement 14, 123
envelope function 195
excitation induced dephasing 164
exciton 187
exciton binding energy 289
exciton Bohr radius 5
exciton linewidth 93
exciton–exciton interaction 126, 189
excitonic polariton 288
exciton–polariton 1, 87, 230
exciton–polariton laser 217
external quantum efficiency 247
extraction efficiency 245, 275

f

Fabry–Perot 31, 47, 96, 273, 298
Fabry–Perot mode 35, 49, 63, 252, 253
Fermi's golden rule 89, 293
final-state stimulation 218
finesse 41, 247
four-wave-mixing 112, 151

g

gain 108
graded index separate confinement heterostructure (GRINSCH) 247
Gross–Pitaevskii 16, 215, 221
guided mode 255

h

Hartree–Fock approximation 239
Hartree–Fock–Bogoliubov 214, 220

Heisenberg uncertainty principle 214, 228
Heisenberg–Langevin 238
high electron mobility transistor (HEMT) 262
high resolution X-ray diffraction (HRXRD) 264
higher-order mixing 179
Hohenberg–Mermin–Wagner theorem 215
Hopfield coefficient 240
Hopfield Hamiltonian 232
hot luminescence 76
hysteresis 132

i

idler 10, 108, 158, 181, 193
$In_{0.04}Ga_{0.96}As$ quantum well 154
inhomogeneous broadening 7, 151
in-plane wavevector 51
interference 183
interference pattern 141
intrinsic radiative lifetime 55

j

Jaynes-Cummings 2, 3, 98
Josephson oscillation 16

k

Kerr instability 129
Kosterlitz–Thouless 219
Kramers–Krönig 64, 73

l

Larmor precession 189
laser action 289
laser operating without population inversion 217
lattice constant 262
lattice mismatch 248
lattice-matching 263
leaky mode 5, 255
light emitting device (LED) 273, 293
light–matter interaction 287, 298
linewidth 7, 72

Liouville equation   200
Liouville–von Neumann equation   194
longitudinal–transverse splitting   188, 195
Lorentz oscillator   49, 63, 94
luminescence   143, 294

m

Mach-Zender interferometer   182
macro-occupation   212
magic angle   11, 105, 158, 192
many-body   123, 215
mean-field equation   124
metal-organic vapour phase epitaxy   261
microcavity LED (MCLED)   245, 275
microcavity polariton   1, 4, 217
micro-disk   87
micro-pillar   20, 87, 290, 298
micro-roughness   42
molecular beam epitaxy (MBE)   31, 280
Mollow   101
momentum conservation   51, 234
motional narrowing   8, 297
multiple scattering   111

n

nanocavity   98
nanoscale cavity   96
III-nitride (III–N) material   261
nonclassical states   21
nonlinear   105, 126
nonresonant pumping   56
normal correlation   238
number-conserving approach   221

o

Off-Diagonal Long Range Order (ODLRO)   213
optical orientation   187
order parameter   16
oscillator strength   46, 56, 93, 98, 116

p

pair polariton scattering   105
parametric amplification   10, 105, 237
parametric amplifier   192, 229
parametric gain   158
parametric instability   129, 144
parametric photoluminescence   221, 237
parametric scattering   156
parametric stimulation   158
particle-particle collision   214
Pauli principle   237
Pauli-matrix   194
Penrose–Onsager criterion   18, 216
periodic boundary condition   235
phase-shift layer   253
phonon scattering   62
phonon-polariton   230
photoluminescence   39, 189, 272, 289
photoluminescence linewidth   69
photonic crystal   87, 290
photonic eigenmode   91
pillar microcavity   95
polariton   105
polariton amplification   123
polariton blueshift   167
polariton collective excitation   135
polariton liquid   114
polariton parametric oscillation   13
polariton parametric oscillator   207
polariton quantum boxes   20
polariton scattering   174
polariton vacuum   232
polariton–polariton   154, 219
polariton–polariton scattering   140, 156, 180, 193, 206
polarization   115
pseudo alloy   49
pseudomorphic layer   282
pseudospin   188
pump–probe   106, 152, 156, 161, 174, 208
Purcell effect   293, 298

q

Q-factor   96
quantum collective   16
quantum complementarity   15, 180
quantum correlation   13, 219
quantum dot   98, 289
quantum field   213
quantum fluctuation   239
quantum fluid   16, 123, 293
quantum information   15
quantum interference   18
Quantum Optics   228
quantum well   33, 87, 106, 288
quantum well exciton   47
quantum wire   289
quasi-elastic ring   77
quasimode   109

r

Rabi   2
Rabi frequency   46, 125
Rabi oscillation   168
Rabi splitting   293, 297
radiation pattern   275
radiative life-time   289
radiative linewidth   89
Rashba effective magnetic field   196
Rayleigh scattering   9, 76
reflectivity   39, 47, 266, 294
reflectivity spectrum   69
refractive index   271
relaxation bottleneck   6, 291
renormalization   174
resonant Brillouin scattering   295
Resonant Cavity LED (RCLED)   245
resonant excitation   62
resonant Rayleigh scattering   138, 151, 297
RHEED oscillation   34

s

scattering   105, 159
screening   57
second-order susceptibility   229
second-order time coherence   19
self-stimulated regime   184
semiconductor microcavity   49
shaped   174
signal   10, 108, 158, 193
signal and idler mode   222, 239
signal-idler correlation   160
signal-idler pair   180
signal mode   181
single photon emission   300
Snell–Descartes law   51
Sommerfeld factor   288
sound velocity   142
speckle   77, 78, 140
spectral disorder   68
spin-forbidden   187
spin-optronic   209
spin projection   187
spontaneous emission   246, 284, 298
spontaneous parametric scattering   159
spontaneous symmetry breaking   16
squeezing   14, 123, 227
Stark effect   153
stimulated scattering   190, 191
Stokes shift   273
stopband   35, 253
strain   37
strain management   262
streak camera   117
strong coupling   45, 93, 295
strong excitation   153
superfluid helium   134
superfluid regime   142
superfluidity   16, 214
superlattice   34

t

thermalization dynamics   58
thermodynamic limit   213
transfer matrix   40, 49
translational invariance   288
translational symmetry   292
transmission   47, 269
transmission electron microscope (TEM)   36
two-level system   101

u
Urbach tail 279

v
vacuum state 228
vacuum-field fluctuation 222
vacuum-field Rabi splitting 4, 45, 99, 218
vertical-cavity surface-emitting laser (VCSEL) 2, 20, 48, 246, 293, 297
visibility 183

w
Wannier exciton 234
wave-mixing 222
weak coupling 48, 56, 81, 246, 298
which-way 15, 180

x
X-ray diffraction 34, 38

z
zero-amplitude 221

Related Titles

R. Cingolani (Ed.)

Proceedings of the 8th Conference on Optics of Excitons in Confined Systems (OECS-8) Lecce, Italy, 15–17 September 2003

physica status solidi (c) – conferences and critical reviews, Vol. 1, No. 3

2004
ISBN-13: 978-3-527-40505-3
ISBN-10: 3-527-40505-4

R. Waser (Ed.)

Nanoelectronics and Information Technology

Advanced Electronic Materials and Novel Devices

2003
ISBN-13: 978-3-527-40363-9
ISBN-10: 3-527-40363-9

Y. Yamamoto, A. Imamoglu

Mesoscopic Quantum Optics

1999
ISBN-13: 978-0-471-14874-6
ISBN-10: 0-471-14874-1